S0-ADZ-081

Starvation in Bacteria

Starvation in Bacteria

Edited by

Staffan Kjelleberg
University of Göteborg
Göteborg, Sweden
and University of New South Wales
Sydney, Australia

PLENUM PRESS • NEW YORK AND LONDON

Library of Congress Cataloging-in-Publication Data

Starvation in bacteria / edited by Staffan Kjelleberg.
    p.   cm.
    Includes bibliographical references and index.
    ISBN 0-306-44430-5
    1. Microbial metabolism. 2. Starvation.   I. Kjelleberg, Staffan.
QR88.S7 1993
589.9'0133--dc20   93-21918
                                                              CIP

ISBN 0-306-44430-5

© 1993 Plenum Press, New York
A Division of Plenum Publishing Corporation
233 Spring Street, New York, N.Y. 10013

Preface

Concerted efforts to study starvation and survival of nondifferentiating vegetative heterotrophic bacteria have been made with various degrees of intensity, in different bacteria and contexts, over more than the last 30 years. As with bacterial growth in natural ecosystem conditions, these research efforts have been intermittent, with rather long periods of limited or no production in between. While several important and well-received reviews and proceedings on the topic of this monograph have been published during the last three to four decades, the last few years have seen a marked increase in reviews on starvation survival in non-spore-forming bacteria. This increase reflects a realization that the biology of bacteria in natural conditions is generally not that of logarithmic growth and that we have very limited information on the physiology of the energy- and nutrient-limited phases of the life cycle of the bacterial cell.

The growing interest in nongrowing bacteria also stems from the more recent advances on the molecular basis of the starvation-induced nongrowing bacterial cell. The identification of starvation-specific gene and protein responders in *Escherichia coli* as well as other bacterial species has provided molecular handles for our attempts to decipher the "differentiation-like" responses and programs that nondifferentiating bacteria exhibit on nutrient-limited growth arrest.

Several laboratories have contributed greatly to the progress made in life-after-log research. Of these it is pertinent to mention the strong pioneering work of Richard Morita and his collaborators, which predominantly studied marine *Vibrio* species, and Abdul Matin and co-workers, who in a series of important publications introduced physiological and molecular aspects of starvation survival and the starvation-induced program in *E. coli*. Recently, the field has benefited greatly from the elegant and innovative series of studies reported by the research laboratories of Roberto Kolter and Regine Hengge-Aronis. The detailed analysis of global control systems (for the regulation above the operon level) by, primarily, Fred Neidhardt and colleagues has been of prime importance for our understanding of the starvation-induced defense systems exhibited by prokaryotes.

v

This monograph provides an up-to-date presentation of the means by which traditionally nondifferentiating bacteria adapt to starvation conditions. The genetic program and physiological features of adaptation to starvation by different bacteria are explored. This book also addresses prevailing ecosystem conditions that lead to intermittent growth or long-term starvation in bacteria.

It is suggested that an improved understanding of starvation survival and nongrowth biology is an essential goal in microbiology, with far-reaching implications in bacterial physiology and ecology, as well as in applied bacteriology and biotechnology. Public health microbiology and environmental biotechnology are areas that greatly benefit from the advances recently made in research programs that deal with starvation in bacteria. This monograph serves as an overview and introduction also for those interested in further exploring such applications.

I wish to thank the authors for providing their excellent contributions, Plenum Senior Editor Mary Phillips Born and other Plenum staff members for constructive support throughout the various stages of preparing this monograph, and Kevin Marshall for proposing that it is timely to publish a book on starvation in bacteria.

Staffan Kjelleberg

Göteborg and Sydney

Contributors

Marta Almirón, Department of Microbiology and Molecular Genetics, Harvard Medical School, Boston, Massachusetts 02115

R. T. Bell, Institute of Limnology, Uppsala University, S-751 22 Uppsala, Sweden

Thomas Egli, Swiss Federal Institute for Water Resources and Water Pollution Control (EAWAG), CH-8600 Dübendorf, Switzerland

Klas Flärdh, Department of General and Marine Microbiology, University of Göteborg, S-413 19 Göteborg, Sweden

John W. Foster, Department of Microbiology and Immunology, College of Medicine, University of South Alabama, Mobile, Alabama 36688

Regine Hengge-Aronis, Department of Biology, University of Konstanz, 7750 Konstanz, Germany

Louise Holmquist, Department of General and Marine Microbiology, University of Göteborg, S-413 19 Göteborg, Sweden

Åsa Jouper-Jaan, Department of General and Marine Microbiology, University of Göteborg, S-413 19 Göteborg, Sweden

Staffan Kjelleberg, Department of General and Marine Microbiology, University of Göteborg, S-413 19 Göteborg, Sweden. *Present address*: School of Microbiology and Immunology, University of New South Wales, Kensington, New South Wales 2033, Australia

Roberto Kolter, Department of Microbiology and Molecular Genetics, Harvard Medical School, Boston, Massachusetts 02115

C. G. Kurland, Department of Molecular Biology, Uppsala University, S-751 24 Uppsala, Sweden

C. Anthony Mason, Swiss Federal Institute for Water Resources and Water Pollution Control (EAWAG), CH-8600 Dübendorf, Switzerland

Riitta Mikkola, Department of Molecular Biology, Uppsala University, S-751 24 Uppsala, Sweden

D. J. W. Moriarty, Department of Marine Microbiology, University of Gothenburg, Gothenburg, Sweden. *Present address*: Department of Chemical Engineering, University of Queensland, St. Lucia, Queensland 4067, Australia

Richard Y. Morita, Department of Microbiology, College of Science and College of Oceanography, Oregon State University, Corvallis, Oregon 97330-3804

Thomas Nyström, Department of Microbiology and Immunology, University of Michigan Medical School, Ann Arbor, Michigan 48109-0620. *Present address*: Department of General and Marine Microbiology, University of Göteborg, S-413 19 Göteborg, Sweden

James D. Oliver, Department of Biology, University of North Carolina at Charlotte, Charlotte, North Carolina 28223

Jörgen Östling, Department of General and Marine Microbiology, University of Göteborg, S-413 19 Göteborg, Sweden

Deborah A. Siegele, Department of Microbiology and Molecular Genetics, Harvard Medical School, Boston, Massachusetts 02115. *Present address*: Department of Biology, Texas A&M University, College Station, Texas 77843

Michael P. Spector, Department of Biomedical Sciences, College of Allied Health, and Department of Microbiology and Immunology, College of Medicine, University of South Alabama, Mobile, Alabama 36688

Björn Svenblad, Department of General and Marine Microbiology, University of Göteborg, S-413 19 Göteborg, Sweden

J. D. van Elsas, Institute for Soil Fertility Research, 6700AA Wageningen, The Netherlands

L. S. van Overbeek, Institute for Soil Fertility Research, 6700AA Wageningen, The Netherlands

Contents

Chapter 1
Bioavailability of Energy and the Starvation State
Richard Y. Morita

Chapter 2
Bacterial Growth and Starvation in Aquatic Environments
D. J. W. Moriarty and R. T. Bell

Chapter 3
Bacterial Responses to Soil Stimuli
J. D. van Elsas and L. S. van Overbeek

Chapter 4
Dynamics of Microbial Growth in the Decelerating and Stationary Phase of Batch Culture
C. Anthony Mason and Thomas Egli

Chapter 5
Starvation and Recovery of Vibrio
Jörgen Östling, Louise Holmquist, Klas Flärdh, Björn Svenblad, Åsa Jouper-Jaan, and Staffan Kjelleberg

Chapter 6
Global Systems Approach to the Physiology of the Starved Cell
Thomas Nyström

Chapter 9
Starvation-Stress Response (SSR) of Salmonella typhimurium:
Gene Expression and Survival during Nutrient Starvation
Michael P. Spector and John W. Foster

Chapter 10
The Impact of Nutritional State on the Microevolution of Ribosomes
C. G. Kurland and Riitta Mikkola

Chapter 11
Formation of Viable but Nonculturable Cells
James D. Oliver

1

Bioavailability of Energy and the Starvation State

Richard Y. Morita

Feast or famine (Koch, 1971), fast or famine (Poindexter, 1981), or fast and famine with rare feast

1. INTRODUCTION

Starvation-survival has been defined as the physiological state resulting from an insufficient amount of nutrients, especially energy, for the growth (increase in size) and multiplication of microorganisms (Morita, 1982). The normal state for most of the bacteria in the ocean is the starvation mode and the microbes, themselves, make most ecosystems oligotrophic (Morita, 1987). The latter situation results from the fact that there are all physiological types of bacteria in any environment as well as abilities to rapidly utilize various substrates. Yet, survival of the species, especially in relation to the lack of energy, is a subject rarely addressed by microbiologists.

All ecosystems are energy-driven. Thus, one must take into consideration the amount of energy available, the quality of the energy and its bioavailability, the turnover rate, and replenishment of the energy. Recognizing that the energy is limited, the question that must be asked is: What is the metabolic state of bacteria in the oligotrophic environment? The rationale and explanation for the above statements will be developed in this chapter.

For more information concerning starvation-survival in bacteria, Kjelleberg *et al.* (1987), Roszak and Colwell (1987), and Morita (1982, 1984, 1988) as well as subsequent chapters in this monograph should be consulted.

Richard Y. Morita • Department of Microbiology, College of Science and College of Oceanography, Oregon State University, Corvallis, Oregon 97330-3804.
Starvation in Bacteria, edited by Staffan Kjelleberg. Plenum Press, New York, 1993.

1

2. SURVIVAL OF BACTERIA IN NATURAL AND ARTIFICIAL ECOSYSTEMS

I first became interested in the subject of survival of bacteria in natural systems when I cultivated bacteria from organic carbon-poor red clay from the open ocean beyond the continental shelf during the Mid-Pacific Expedition of 1950. These bacteria were isolated from the deepest strata of cores taken with a Kullenberg piston-type corer, which represented material laid down more than a million years ago (Morita and ZoBell, 1955). Subsequent reading of Lipman (1931) and others (Reiser and Trask, 1960; Bosco, 1960; Garbosky and Giambiagi, 1966; Bollen, 1977) made me realize that survival of the species in the absence of bioavailable energy was a subject that needed investigating. With energy being limited in ecosystems, starvation-survival provides a mechanism by which many species of bacteria remain in the environment until the proper environmental conditions become available for them to grow and multiply. It only takes one cell of any species per environment for survival of the species (Morita, 1980). As long as one cell of a species survives, that single cell can regenerate the whole population when the environmental conditions become suitable.

The longest laboratory experiment dealing with the survival of bacteria without the presence of an energy source is that of Iacobellis and DeVey (1986). These investigators stored 14 different isolates of *Pseudomonas syringae* subsp. *syringae* in distilled water for 24 years. The cell counts initially were mostly 10^8 per ml and usually dropped two or three logs after 24 years. These starved cells retained their antigenic properties during this period while subculturing in media did not permit these plant pathogens to retain their antigenic structure. I have been informed that this is a common method of preserving plant pathogens.

3. ORGANIC MATTER IN OLIGOTROPHIC SYSTEMS

Organic carbon present in different environments is generally used as an index of the amount of energy available for heterotrophic bacteria and other organisms. What is measured chemically, however, is not what is bioavailable to the organisms, mainly because there are mechanisms that render this organic matter unavailable to organisms.

The amount of organic carbon in the soil and aquatic ecosystems is approximately 2.5×10^{15} kg and 3.0×10^{15} kg, respectively. Although this may appear to be quite large, when diluted for the amount of soil and water, the figures become very small in terms of the amount of organic carbon per kilogram of soil or per liter of water. Nevertheless, it is a "blessing in disguise" that most of it is not utilized since it would greatly increase the amount of carbon dioxide in the atmosphere and help bring about the greenhouse effect.

3.1. Soil

Each terrestrial environment has its own level of organic matter. Generally a good loam soil contains approximately 25 mg organic carbon/g. This organic matter is very important in the workability of the soil as well as soil water retention. Recognizing that there are limited substrates for the growth of bacterial in soil, soil microbiologists have thought in terms of survival of the vegetative bacteria in soil (Gray, 1976; Nedwell and Gray, 1987; Gray and Williams, 1971; Williams, 1985). Lockwood (1977) also recognized the fact that the amount of energy in soil for fungi was grossly inadequate. According to Gray and Williams (1971), the dormant state is believed to be characteristic for the major part of the total biomass in soil for most of the year. There are many contributions (e.g., Tate, 1987) dealing with the organic carbon content of soil; hence, the organic carbon in soil will not be discussed in this chapter to any great degree. Additional information on the organic carbon content in soil is given by van Elsas and van Overbeek (this volume).

3.2. Aquatic Systems

In aquatic systems the organic carbon is divided into two fractions: dissolved organic carbon (DOC) and particulate organic carbon (POC). For lakes and rivers, the figure varies widely but a good approximation is 10 mg C/liter. Because photosynthetic processes take place at the surface (euphotic zone), both DOC and POC are much higher. Williams (1971) gives the following values for the average amount of POC and DOC in the ocean. From the surface to 300 m, the DOC and POC are 1.0 (range 0.3–2.0) and 0.1 (range 0.003–0.3) mg C/liter, respectively. From 300 to 3000 m, the average DOC and POC are 0.5 (range 0.2–0.8) and 0.1 (range 0.005–0.03) mg C/liter. The surface DOC is about $10\times$ greater than that of POC and in deep water, about $50\times$. The transition depth of 300 m varies from 100 to 400 m. Above the transition depth, the concentration of DOC and POC may vary widely, depending on the amount of photosynthetic activity taking place. Only about 0.4% of the carbon fixed by photosynthesis enters into the deep-sea DOC (Williams, 1971) and about 5% of the DOM below the thermocline is utilizable by bacteria (Ammerman *et al.*, 1984). The apparent age of this organic matter is 3400 years (Williams *et al.*, 1969). The oxygen utilization rate of all organisms in the Pacific (Craig, 1971) and Atlantic (Arons and Stommel, 1967) deep waters is

0.004 and 0.003 ml/liter per year, respectively. In addition, the residence time of deep water masses can be longer than 1000 years. Thus, when one considers that approximately 7.6% of the oceans is 0–200 m, the deep sea has very little energy and what energy is present is utilized extremely slowly. Even detritus (dead organic matter) may be recalcitrant to microbial decay (Tenore *et al.*, 1982).

4. BIOAVAILABILITY OF ORGANIC MATTER IN OLIGOTROPHIC ENVIRONMENTS FOR HETEROTROPHIC BACTERIA

Although there are many organic compounds found in nature that are biodegradable, their presence does not necessarily signify that they are bioavailable to the microorganisms. In nature, all organisms (large numbers) are competing for a limited amount of energy; hence, microorganisms will not generally accumulate reserve energy sources.

4.1. Energy for Growth and/or Reproduction

What is the threshold amount of energy necessary for bacteria to grow (increase in size) and multiply? This question has been addressed in terms of increase in the number of cells (a measure of growth in microbiology). An increase in size (biomass) has generally not been addressed by microbiologists. In terms of growth, ZoBell and Grant (1942) documented evidence that 0.1 mg glucose or peptone per liter was necessary for growth of marine bacteria, *Staphylococcus citreus*, *Proteus vulgaris*, *Escherichia coli*, and *Lactobacillus lactis*. For *Aeromonas hydrophilia* and *Pseudomonas aeruginosa*, the values were 10 and 25 μg carbon/liter, respectively (Van der Kooij *et al.*, 1980, 1982). Thus, it can be seen that each species has its own threshold energy level for growth and reproduction.

If the threshold level of organic matter is insufficient for growth or multiplication, metabolic activity without growth does take place. For instance, *Vibrio* sp. Ant-300 starved for 72 days has the ability to utilize 10^{-12} M glutamate, but none of it is retained by the cells after 32 h of incubation (Morita, 1984).

4.2. Bioavailability of Substrates and Generation Times

Chemical analysis of environmental samples indicates that there are very labile organic compounds. However, chemical detection does not have any bearing on their availability to organisms. For instance, Alexander (1965) points out that many biodegradable compounds (e.g., cellulose, pectins, amino acids, chitin) have been recovered from fossils and sediments. Thus, they are not bioavailable. Several mechanisms that render compounds recalcitrant were summarized by Alexander (1973). These include lack of sufficient energy or carbon for growth, lack of essential nutrient, exceeding of microbial tolerance to environmental factors, toxicity of substrate or products of its metabolism, inhibition or inactivation of extracellular enzymes, complexing of substrate with resistant organic or polyaromatic compounds, and inaccessibility of site on substrate acted on enzymatically.

What a heterotrophic microorganism can do in nature depends mainly on its specific carbon substrate(s), and the availability of this substrate which depends on environmental factors. For example, denitrifying bacteria can use a wide variety of organic acids, carbohydrates, and other organic compounds as carbon and energy sources under anaerobic conditions, but they are restricted to a few carbon sources under denitrifying conditions (Beauchamp *et al.*, 1989). *Pseudomonas stutzeri* metabolizes cysteine, isoleucine, leucine, and valine aerobically but the same amino acids are not used as carbon and energy sources under anaerobic conditions (Bryan, 1981). Other environmental factors such as pH and temperature must also be taken into consideration. We recognize that, in nature, there are bacteria that have simple growth requirements, and others that have complex requirements (mixture of amino acids, growth factors, and/or undefined requirements). Thus, at any given time, there exists an equilibrium between bacteria with simple requirements and those with more complex ones, and this balance may be modified by many factors (West and Lochhead, 1950).

Much of the organic matter in natural environments is mainly in the form of humic substances, tannins, lignins, etc., all being recalcitrant compounds. Many labile compounds are complexed to lignins, phenolics, etc., making them unavailable to microorganisms. Lignin–protein, silica–protein, and clay–protein are known to be resistant to microbial degradation (Estermann *et al.*, 1959). There are high concentrations of acidic amino acids, low concentrations of basic and neutral amino acids in fulvic acid, and the distributions of amino acids in the humic and humin are quite similar (Sowden *et al.*, 1976). When humic acid that has been extracted from soil is hydrolyzed, guanine, adenine, cytosine, thymine, and traces of uracil can be found (Anderson, 1957, 1958, 1961). Purines and pyrimidines are also found in humic acid, fluvic acid, and humin (Cortez *et al.*, 1976). Ninety-six percent of the amino acid content of Williamson River water originating from Klamath Lake is associated with the aquatic humus (Lytle and Perdue, 1981). Griffin and Roth (1979) argue that humic materials provide energy for microbial growth. Since humic has a residence time of over 100 years, Barber and Lynch (1977) believe this to be unlikely. Tannins and related phenolic compounds were found to inhibit a wide range of enzymes: polygalacturonase, cellulase, urease, amylase, peroxidase, catalase, proteinase (pepsin, trypsin, and chymotrypsin), dehydrogenases, decarboxylases, invertase, phosphatases, β-glucosidase, aldolase, oxidases, polyphenoloxidase, and lipases (Benoit and Starkey, 1968).

Many labile compounds can also be adsorbed to clay minerals (Stotzky, 1986). Labile substances are known to form complexes in soil and natural waters. Chemical analysis may liberate the labile compounds complexed or adsorbed to various substances in soil; hence, what is measured may not be bioavailable to the microorganisms present. Thus, readily utilizable carbon is apparently severely limited in soil (Clark, 1965; Gray and Williams, 1971; Lockwood, 1977). There are many mechanisms for making labile compounds unavailable to microorganisms. For more information concerning humus, Aiken *et al.* (1985) should be consulted.

Although labile compounds can be measured in ecosystems, the question is how much of it is bioavailable to the microorganisms. Not all of the acetate found in sediments chemically can be utilized by the microorganisms (Ansback and Blackburn, 1980; Balba and Nedwell, 1980; Christensen and Blackburn, 1982; Fenchel and Blackburn, 1979; Gibson *et al.*, 1989; Parkes *et al.*, 1984; Sansome, 1986). Up to 84% of the acetate in sediment may not be bioavailable (Christensen and Blackburn, 1982). In terms of alanine in sediments, 97% if unavailable to microorganisms (Christensen and Blackburn, 1980). In terms of amino acids, the amount measured chemically is not all bioavailable for microorganisms in lake water (Burnison and Morita, 1974) and in marine waters (Dawson and Gocke, 1978; Christensen and Blackburn, 1980). This discrepancy also exists for glucose (Gocke *et al.*, 1981). Interestingly, no significant change in the DOC taken from the deep sea could be noted by Barber (1986) when the sample was incubated over 2 months in the presence of viable bacteria. However, acid hydrolysis of the POC is a good substrate for microbial growth (Gordon, 1970).

Because of the limited amount of energy present in ecosystems, the generation time of the biomass of microbial cells is long. Yet, it should be recognized that several species are active while others are not. The question that must be asked is how active is "active," since this term is used to describe activity of bacteria in the natural environment, in spite of uncertainties with methods used. The longest generation times for microbes in clay and broadbalk soils are 3.024 and 15.168 hr (Jenkinson and Ladd, 1981). For most other types of soil, the generation times range from 26 to 93 h (Nedwell and Gray, 1987). Other values for soils can be found in Williams (1985). For coastal muds, Fallon *et al.* (1983) report long generation times from 118 to 3049 h. Other values listed by Nedwell and Gray (1987) for sediments range from 3.8 to 1008 h. Likewise, Lockwood and Filkonow (1981) tabulated generation times that range from less than 1 to 36 generations per year. The range of microbial biomass turnover ranged from 0.4 (912.5 days; Jenkinson and Ladd, 1981) to

5.5 times per year (McGill *et al.*, 1981), respectively. For the deep sea, Carlucci and Williams (1978) calculated a generation time of 210 days.

4.3. Syntrophy and Bacterial Activity

Within any given natural environment, not all of the various physiological types of bacteria are active. The chemical and physical environment may not be conducive for their well-being. More specifically, the main culprit is the lack of energy. This is to be expected since microbes have the ability to consume tremendous amounts of energy when conditions are favorable for their growth and multiplication. The main source of organic matter in the environment is green plant photosynthesis. One of the major components of green plant photosynthesis is cellulose. Not all organisms can utilize cellulose and must rely on those organisms that have the ability to produce cellulase. As a result, the end product of the cellulose digestors are then used by other organisms, a process known as syntrophy. Hence, growth in the environment is sporadic.

In support of this statement, Lundgren and Söderström (1983), using the method of hydrolysis of fluorescein diacetate (FDA), found only 34% of the bacteria in the A1 horizon of forest soil to be active, 54% in the A2 horizon, and 52% in the B horizon. MacDonald (1980), employing the electron transport system (ETS) activity, found 15% of the bacteria active in a barley field, 25% active in a manured soil, 11% active in a turfed soil, 23% active in compost, 23% active in vegetable soil and 31% active in field soil. In the marine environment, Meyer-Reil (1978), employing tritiated glucose with autoradiography, found 2.3 to 56.2% (average 31.3%) of the bacteria active in the water overlying sand. Using the INT [reduction of 2-(*p*-iodophenyl)-3-(*p*-nitrophenyl)-5-phenyl tetrazolium chloride] method, Zimmermann *et al.* (1978) reported only 6 to 12% of the bacteria taken from samples from the coastal areas of the Baltic Sea and 5 to 36% from freshwater lakes and ponds actively respired. On the other hand, Novitsky (1987), using two different techniques, found that 90% of the bacteria in the sediment–water interface were not actively growing. As for fungi, Söderström (1979), employing the FDA method, found only 2.4% active in the A1 horizon of a pine forest, 4.3% active in the A2 horizon, and 2.6% active in the B horizon. It has yet to be shown that there are any continually active biomass in soil (McLaren, 1971).

As mentioned, "active" has been used by many investigators to describe heterotrophic bacteria in oligotrophic environments. It is a relative term. Are they as active as in a culture media? Are they as active in mineral salt solutions? Or, does their activity lie somewhere in between? Just how active can organisms taken from an environmental sample be when the organic matter is extremely low and most of the energy source is refractory to microbial activity?

4.4. Oligotrophic Bacteria

Although we recognize that most ecosystems on earth are oligotrophic, various investigators have employed different terms to describe bacteria in these ecosystems. Such terms as dormant cells, oligotrophic bacteria, oligocarbophilic bacteria, copiotrophic bacteria, resting cells, resuscitable cells, senescent cells, moribund cells, quiescent cells, viable but nonculturable and starved cells have been employed in the literature. Although "oligotrophic bacteria" is a commonly used term in the literature, there is no definitive proof that they exist. Some of the oligotrophic bacteria described in the literature by Akagi *et al.* (1977) have been found to grow on rich media (Martin and MacLeod, 1984), while Morgan and Dow (1986) state that the term should be used to describe the environment. The concept of oligotrophy *sensu stricto* should be applied with caution (Williams, 1985). After reviewing the literature, Gottschal (1992) has come to the same conclusion. I support this notion for the following reason. Nearly all experimental procedures for the determination of oligotrophic bacteria in nature are based on the utilization (as well as the concentration) of peptone. There is no reason why peptone should be selected other than convenience. In addition, one medium is used under only one set of conditions. Thus, testing the organism against various types of media (combination of amino acids, trace elements, vitamins, etc.) under different physical conditions (Eh, pH, temperature, osmotic pressure, etc.) is generally not conducted, mainly because it is a Herculean task. Transport of specific substrates may be temperature dependent, pH dependent, etc.

For more information concerning oligotrophy, see Moriarty and Bell (this volume).

5. THE STARVATION PROCESS

In the natural environment, there are degrees of starvation. Immediately after an organism ceases its uptake of nutrients, the starvation process is initiated. Thus, we have short-, medium-, and long-term starvation-survival. In short-term starvation-survival, the cells probably do not expire. However, this is not the case when medium- and long-term starvation occur. In long-term starvation-survival, we are dealing with the survival of the species in the face of the lack of nutrients, especially energy.

When bacteria taken from the log growth phase are placed in a starvation menstruum, the cell numbers increase (Novitsky and Morita, 1976) (Fig. 1). After the first week of starvation of Ant-300 (a marine vibrio), the direct count by AODC (acridine orange direct count) remains level but the plate count decreases. The optical density decreases with time. Thus, we are not observing

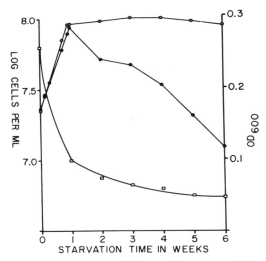

Figure 1. Starvation survival of Ant-300. Cells harvested from exponential growth were starved in a buffered-salt mixture. Total direct counts, ○; plate counts, ●; optical density, □. After Novitsky and Morita (1977).

growth (increase in the number of cells) since the biomass, as evidenced by the optical density measurements, decreases. Nevertheless, the AODC counts indicate the lysis does not take place. There are many reports that cells do not lyse during starvation, the first ones being Gronlund and Campbell (1963) and Harrison and Lawrence (1963). It should also be noted that cryptic growth does not occur during the starvation-survival process (Novitsky and Morita, 1977; Kurath and Morita, 1983).

Not all cells remain viable during the starvation process. In a *Pseudomonas* sp., Kurath and Morita (1983) followed the viable counts (plate counts), the respiring cell counts by the INT method of Zimmermann *et al.* (1978), and the AODC counts (Fig. 2). The difference between the viable and INT counts suggests the existence, within a starving population, of a subpopulation of nonviable but actively respiring cells which is 10-fold more numerous than the viable cells. It took approximately 15 to 18 days for the starvation culture to stabilize as evidenced by the AODC, INT counts, viable cell counts, ATP level per culture, glucose uptake per milliliter of culture, endogenous respiration, and glutamate uptake rate per milliliter.

Employing ^{14}C-labeled Ant-300 cells, it was found that after 2 days of starvation, the endogenous respiration decreased over 80% (Novitsky and Morita, 1977). This endogenous respiration was reduced to 0.007% per hour and remained constant for 28 days. Unfortunately, this experiment was termi-

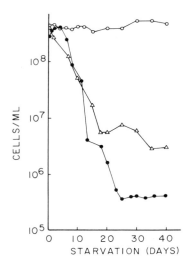

Figure 2. Viable count, direct count, and respiring cell count after various periods of starvation. All counts were carried out on the same starvation culture. Viability ($\bullet$) was determined by spread plates, direct counts ($\circ$) by epifluorescence microscopy, and respiring cells counts ($\triangle$) by the INT method. After Kurath and Morita (1983).

nated after 28 days since this endogenous value is much too high when we know Ant-300 can remain viable for over 2.5 years. It is believed that this endogenous respiration eventually shuts down for long-term survival. Most of the research performed on starvation-survival has been done on marine vibrios and *Escherichia coli* (see other chapters in this volume). Nevertheless, it should be pointed out that most studies dealing with starvation-survival are usually for less than 48 h but some are as short as 90 min. This chapter differs from the subsequent contributions in this volume by presenting results from significantly longer-term starvation experiments.

5.1. Patterns of Starvation

Sixteen freshly isolated bacteria from the open ocean were subjected to the starvation-survival procedure (Amy and Morita, 1983a). Three patterns of starvation-survival were noted (curves A, C, and D of Fig. 3).

Curve B is represented by a marine chemolithotrophic ammonium-oxidizing bacterium (*Nitrosomonas cryotolerans*; Jones and Morita, 1985). *N. cryotolerans* does not undergo fragmentation to produce more cells nor does cell size reduction occur during long periods of starvation (lack of ammonium). Unfortunately, most of the studies on starvation-survival have been carried out on cells that display a viability pattern represented by curve C and a few studies on organisms represented by curve D.

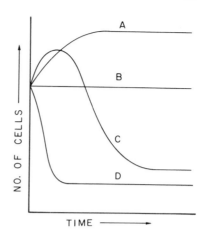

Figure 3. Starvation-survival patterns of micro-organisms. After Morita (1985).

Depending on the age of the culture employed in the starvation process, increases from 100 to 800% of the initial number of cells were noted by Novitsky and Morita (1977) (curve C of Fig. 3). Regardless of the initial increases in cell numbers, 100 to 200% of the initial number of cells were still viable after 6 weeks of starvation. The smaller number of cells in the initial inoculum into a starvation menstruum produces the largest number of cells during the starvation process (Novitsky and Morita, 1978). This huge increase in cell number is considered to be a strategy for the survival of the species—a situation analogous to certain plants producing a large number of seeds or in certain animals where large number of eggs are produced.

There appear to be three stages in the starvation-survival process and the time element will vary depending on the species in question (see Östling *et al.*, this volume). Kurath and Morita (1983) demonstrated by a *Pseudomonas* sp. that it took approximately 18 to 25 days for the organisms to stabilize in terms of ATP content, endogenous respiration, uptake rates, and the percent respiration for exogenous glucose and glutamate. In long-term starvation-survival studies, it should be noted that there appear to be three stages that the psychrophilic marine *Vibrio* sp. Ant-300 goes through (Moyer and Morita, 1989a). In the first stage there is a large fluctuation in the viable counts with a moderate decrease by 14 days. This is followed by a 99.7% decrease in viable cells between 14 and 70 days with a third stage indicating a stabilization of the viable count so that approximately 0.3% are still viable by the method employed. The decrease in viable counts is accompanied by a cell volume decrease, depending on the dilution rate at which the Ant-300 was grown in a chemostat. The viable organisms in stage 3 will probably survive extremely long periods of starvation.

The time element between the three stages will probably depend on the species in question.

5.2. Cell Size of Bacteria in Oligotrophic Environments and in Starvation Microcosms

Although there are early reports of small cells occurring in soil (Casida, 1971; Bae *et al.*, 1972), in seawater (Oppenheimer, 1952; Anderson and Heffernan, 1965; Zimmermann and Meyer-Reil, 1974), and in freshwater (Martin, 1963), the definitive evidence of their occurrence caused by starvation was noted by Novitsky and Morita (1976). The respiring cells in the freshwater ponds were 1.6 to 2.4 μm whereas bacteria in the Baltic Sea were mainly 0.4 μm (Zimmermann *et al.*, 1978). The larger size of the bacteria from freshwater ponds may relate to the higher amount of organic matter present because of phytoplankton and organic matter from freshwater runoff. The ultramicrobacteria in the marine environment appear not only as cocci, but also as vibrios, bacilli, horseshoe, and sigmoid forms (P. W. Johnson and J. M. Sieburth, 1978, *Abstr. Annu. Meet. Am. Soc. Microbiol.* N95, p. 178; Amy and Morita, 1983a). The size reduction in 16 different isolates varied from 40 to 79% upon starvation (Amy and Morita, 1983a). Ultramicrocells from the *Pseudomonas*, *Vibrio*, *Aeromonas*, and *Alcaligenes* isolated from an estuarine environment were found to pass through a 0.2-μm polycarbonate membrane filter (Mac-Donell and Hood, 1982). Deep sea ultramicrobacteria were noted by Tabor *et al.* (1981).

When a microcultural study of bacterial size changes and microcolony and ultramicrocolony formation was undertaken employing dilute Lib-X medium (rich medium), Torella and Morita (1981) demonstrated that small cells became larger and formed a colony quickly as would be expected on growth on nutrient medium, whereas others were ultramicrobacteria which were characterized by slow growth. They defined ultramicrobacteria as being 0.3 μm or smaller. However, I believe the term should be expanded to indicate the size of cells below 1.0 μm or cells that are smaller than those found in rich medium. Other terms, such as dwarf cells, minicells, etc., have been employed but each has its drawbacks (Morita, 1982). It is now known that most bacteria in the soil and aquatic environment are between 0.4 and 0.8 μm. Within limits, the longer the starvation period, the smaller the cells become during the time period of the experiments. For long-term starvation, the cells will probably reach their minimum size. For Ant-300, initially the cells are vibrio, but with starvation they become spherical with an enlarged periplasmic space containing stainable material (Novitsky and Morita, 1976). The viability of the ultramicrobacteria in soil has been correlated with size, the larger the size the better viability (Bakken and Olsen, 1987). Within limits, the longer the starvation period the smaller are

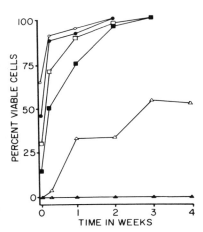

Figure 4. Effect of time of starvation on the size distribution of Ant-300 cells. Portions of the starving culture were passed through filters of various pore sizes. Plate counts of the filtrates are expressed as a percentage of the unfiltered counts. Filter pore sizes: ○, 3.0 μm; ●, 1.0 μm, □, 0.8 μm; ■, 0.6 μm; △, 0.4 μm; ▲, 0.2 μm. After Novitsky and Morita (1976).

the cells (Fig. 4) (Novitsky and Morita, 1976). This reduction in size of the cells with starvation has also been observed by other investigators.

Even when cells are grown in a chemostat at various dilution rates, the cell size can be changed. The cell volumes of Ant-300 cells, when grown in a chemostat at various dilution rates, are 1.16 μm³ (D = 0.170), 0.585 μm³ (D = 0.057), and 0.478 μm³ (D = 0.015), whereas batch-cultured cells are 5.94 μm³ (Moyer and Morita, 1989a). When these cells are starved to stage 3, the cell volumes are 0.189 μm³ (D = 0.170), 0.181 μm³ (D = 0.057), 0.046 μm³ (D = 0.015), and 0.275 μm³ (batch-grown cells). The results only indicate that we should be investigating the effects of eutrophic conditions on starved cells since most of the cells in the environment are ultramicrocells.

5.3. Cell Membrane Changes

Drastic changes take place in the lipid content of *Vibrio cholera* upon starvation (Hood *et al.*, 1986). The total lipid content of nonstarved cells was 3679.18 nmol/10¹⁰ cells, whereas for cells starved for 7 and 30 days the values were 7.75 and 2.43 nmol/10¹⁰ cells, respectively. The phospholipid content of nonstarved, 7- and 30-day-starved cells were 3675, 6.95, and 2.05 nmol/10¹⁰ cells, respectively. These data definitely indicate a change in the membrane when cells are starved since phospholipids are located in the membrane. During the period of starvation, the *cis*-monoenoic fatty acids are preferentially utilized (Guckert *et al.*, 1986). The ability to either synthesize *trans*-monoenoic acids or modify the more volatile *cis*-monoenoic acids to their cyclopropyl derivative may be a survival mechanism which helps maintain a functional membrane during starvation-induced lipid utilization. A *trans/cis* fatty acid ratio greater than 0.1 could be used as a starvation or stress lipid index. This decrease in total lipid may be the major energy source for the increase in cell numbers during the starvation process.

Nonstarved Ant-300 cells are readily disrupted by the usual means of preparing membranes for analysis but 30-day-starved Ant-300 cells are extremely difficult to rupture (M. J. Slominsky, personal communication). The persistence of viable starved bacteria in soil may be the result of the cell surface becoming resistant to degradation, thereby protecting the intracellular material from degradation, analogous to the spore coat of spore-forming bacteria.

Poindexter (1981) suggested that bacteria adapted to oligotrophic conditions possess uptake systems for a greater variety of substrates than bacteria growing on rich media. The field research by Harowitz *et al.* (1983) and Upton and Nedwell (1989) in oligotrophic environments has shown that bacteria isolated and kept on low-nutrient media do have the ability to utilize a greater number of substrates than bacteria isolated and maintained on rich media. This may be a reflection of the change in the bacterial membrane.

5.4. ATP and Adenylate Energy Charge

The dormant soil biomass maintains both AEC (adenylate energy charge) and ATP levels characteristic of exponentially growing organisms *in vitro* (Brookes *et al.*, 1987).

Log-phase cells of a *Pseudomonas* sp. contain 6.5×10^{-10} µg of ATP per respiring cell, which is within the range for marine bacteria (Fig. 5). The amount of ATP per milliliter of culture and per respirating cell are also shown in Fig. 5. The ATP of the culture dropped by 80% during the first 5 days of starvation. Since the majority of cells lost the ability to reproduce and respire (Fig. 2), it is obvious that not all cells in the culture maintained equal amounts of ATP. The total count would be an erroneous denominator for calculating the change in ATP per cells and the viable cell count would also be improper. The actively respiring subpopulation would be able to generate and maintain pools of ATP. The ATP per respiring cell appears to be the most logical value to employ. The ATP per respiring cell for the first 8 days dropped, followed by a gradual increase back to the original level of 6.5×10^{-10} µg of ATP per respiring cell by 40 days of starvation.

5.5. Protein, DNA, and RNA

When Ant-300 cells were starved, the cellular DNA decreased rapidly during the first 14 days, but after this initial fast drop, there was a slight decrease (Novitsky and Morita, 1977). A 46% drop in cellular DNA was noted

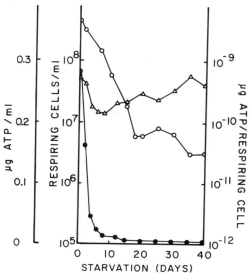

Figure 5. ATP content after various periods of starvation. △, ATP per milliliter of culture; ●, ATP per respiring cell; ○, respiring cell count. After Kurath and Morita (1983).

after 6 weeks of starvation. Only 5% of the DNA could be detected in the supernatant fluid.

Amy and Morita (1983b) demonstrated that new proteins were synthesized, other cellular proteins disappeared, and some cellular proteins remained when cells were starved. These newly synthesized proteins were later termed "starvation proteins." For further information concerning these starvation proteins, see other chapters in this volume.

In stage 1 in the starvation-survival process of Ant-300, the DNA, RNA, and protein fluctuate, showing two to three peaks during the first 14 days (Moyer and Morita, 1989b). On a viable cell basis, the DNA, RNA, and protein increase during stages 1 and 2 but level out in stage 3, whereas on a total cell count basis wide fluctuations occur in the protein and RNA content with the DNA decreasing through the three stages. There is one order of magnitude less DNA per starved cell than unstarved cell and these amounts depend on the dilution rate in growing the organism in a chemostat.

Levels of protein, RNA, and DNA during 10 weeks of starvation remained essentially unchanged in *N. cryotolerans* (Johnstone and Jones, 1988b). An active electron transport system remained during the 10-week period. The energy charge at the onset of starvation was 0.68 and decreased to a level of 0.50 after 5 weeks of starvation and then stabilized; the ATP remained nearly constant for 2 weeks, then decreased until 4 weeks at which time it stabilized at 0.85 fmol/ml for 10^5 cells/ml. The organism is capable of conserving energy by lowering its energy charge and repressing biosynthesis, but remains in a state of readiness which permits it to respond readily to nutrient addition.

5.6. Chemotaxis

Flagellation is retained by the 16 freshly isolated bacteria from the open ocean after a starvation period of 8 months (Amy and Morita, 1983a) as well as in Ant-300 cells (Torrella and Morita, 1981). Chemotaxis in Ant-300 cells does not take place unless the cells have been starved for 24–48 h (Geesey and Morita, 1979). The concept of energized chemotaxis, where one substrate in sufficient supply allows the organism to display better chemotaxis toward a substrate in short supply, is presented by Torrella and Morita (1982). Since chemotaxis is primarily a mechanism for the cell to seek energy, Morgan and Dow (1986) point out that chemotaxis under high nutrient conditions is superfluous and wasteful and it is only when limitation occurs that nutrient gradients, and hence chemotaxis, become ecologically significant. Chemotaxis has been investigated in detail by Kjelleberg's group (see Östling *et al.*, this volume).

5.7. Resistance of Starved Cells to Various Environmental Factors

By far the greatest environment factor that organisms face in the natural environment is the lack of energy. Bacteria isolated on low-nutrient media from the environment are physiologically more tolerant than those isolated on rich media (Harowitz *et al.*, 1983). The lack of energy (carbon starvation) brings about a general resistance in *E. coli* (Matin, 1991). For more information concerning the general resistance of starved cells, Nyström *et al.* (this volume) should be consulted. The starved cells also resist hydrostatic pressure (Novitsky and Morita, 1977). On the other hand, long-term-starved *N. cryotolerans* are less resistant to light, one of the inhibitors of growth of this species (Johnstone and Jones, 1988a).

5.8. Recovery from Starvation

After starving Ant-300 cells for 7 weeks, the cells responded to the addition of glucose immediately and multiplication of the cells occurred (Novitsky and Morita, 1978). The 16 isolates retained the ability to utilize glutamate after 8 months (Amy and Morita, 1983a). Rapid response to the addition of nutrients by starved Ant-300 cells was noted by Amy *et al.* (1983) producing the usual growth curve. On a per viable cell basis, protein, DNA, and RNA increased to the maximum values just before cell division and then returned to close to the initial starved-cell values. The length of the lag phase during recovery was directly proportional to the length of the prior starvation period. Rapid response to nutrient shift-up experiments was noted by Kjelleberg *et al.* (1982). It appears that the mechanisms necessary for the cell to obtain energy remain intact during starvation-survival.

N. cryotolerans, after 5 weeks of starvation, responds immediately to the addition of ammonium, producing nitrite at a constant rate (Johnstone and Jones, 1988c). The cells use stored ATP and reducing equivalents [possibly $NAD(P)H + H^+$] to immediately begin biosynthesis. The electron transport system steadily increased after the addition of ammonium up to 24 h without an increase in cell numbers.

6. CONCLUSIONS

In our attempt to study starvation-survival phenomena, do we perform the appropriate experiments? We grow gluttonous bacteria using laboratory media and subject them to starvation conditions. From an ecological viewpoint, we should rather be studying the response of bacteria grown under oligotrophic conditions to eutrophic conditions. In the laboratory, cells should be grown in a chemostat under the lowest dilution rate possible, and then studied as to their response to eutrophic conditions. The use of the chemostat represents a method that more closely resembles the natural physiological state of microorganisms. Such studies should be more relevant to the ecological situation. This also includes the molecular biology approach.

The ecological distribution of a species depends on its survival mechanisms which include its developmental and morphological characteristics and its biochemical and physiological mechanisms. As stated previously, as long as one cell of a species survives, that single cell can regenerate the whole population when the environmental conditions become appropriate.

It is recognized that in long-term starvation-survival studies, less than 1% of a culture survives. An important question to address is (Morita, 1982): what dictates which few cells will survive and whether there is communication between cells in the starvation-survival process or is there a random or genetic selection as to which cells survive, or is it the survival of the fittest (see Siegele *et al.*, this volume)?

Since the starved state of bacteria is their normal physiological state in the environment, the starved state permits the organism to withstand greater adverse conditions than when it is in the log phase of growth. Long-term starvation-survival is a mechanism by which the species has control of time—a situation analogous to the bacterial spore. All metabolic systems cease to function. I have used the term "metabolic arrest" to describe this physiological state of non-spore-forming bacteria when exposed to long periods of nutrient (energy) deprivation. The term "shutdown" is employed by Dow *et al.* (1983).

ACKNOWLEDGMENT. This paper was originally published as technical paper No. 9864, Oregon Agricultural Experimental Station.

REFERENCES

Aiken, G. R., McKnight, D. M., Wershaw, R. L., and MacCarthy, P., 1985, *Humic Substances in Soil, Sediment and Water*, Wiley, New York.

Akagi, Y., Taga, N., and Simidu, U., 1977, Isolation and distribution of oligotrophic marine bacteria, *Can. J. Microbiol.* **23**:981–987.

Alexander, M., 1965, Biodegradation: Problems of molecular recalcitrance and microbial fallibility, *Adv. Appl. Microbiol.* **7**:35–76.

Alexander, M., 1973, Non biodegradable and other recalcitrant molecules, *Biotechnol. Bioeng.* **15**:611–647.

Ammerman, J. W., Fuhrman, J. A., Hagström, A., and Azam, F., 1984, Bacterioplankton growth in seawater: 1. Growth kinetics and cellular characteristics in seawater cultures, *Mar. Ecol. Prog. Ser.* **18**:31–39.

Amy, P. S., and Morita, R. Y., 1983a, Starvation-survival patterns of sixteen freshly isolated open-ocean bacteria, *Appl. Environ. Microbiol.* **45**:1109–1115.

Amy, P. S., and Morita, R. Y., 1983b, Protein patterns of growing and starved cells of a marine *Vibrio* sp., *Appl. Environ. Microbiol.* **45**:1685–1690.

Amy, P. S., Pauling, C., and Morita, R. Y., 1983b, Recovery from nutrient starvation by a marine *Vibrio* sp., *Appl. Environ. Microbiol.* **45**:1685–1690.

Anderson, G., 1957, Nucleic acid derivatives in soil, *Nature* **260**:597.

Anderson, G., 1958, Identification of derivative of desoxyribonucleic acid in humic acid, *Soil Sci.* **86**:169–174.

Anderson, G., 1961, Estimation of purines and pyrimidines in soil humic acid, *Soil Sci.* **91**: 156–161.

Anderson, J. I. W., and Heffernan, W. P., 1965, Isolation and characterization of filterable marine bacteria, *J. Bacteriol.* **90**:1713–1718.

Ansback, J., and Blackburn, T. H., 1980, A method for the analysis of acetate turnover in a coastal marine sediment, *Microb. Ecol.* **5**:253–264.

Arons, A. B., and Stommel, H., 1967, On the abyssal circulation of the world oceans, *Deep-Sea Res.* **14**:441–457.

Bae, H. C., Cota-Robles, E. H., and Casida, L. E., Jr., 1972, Microflora of soil as viewed by transmission electron microscopy, *Appl. Microbiol.* **23**:637–648.

Bakken, L. R., and Olsen, R. A., 1987, The relationship between cell size and viability of soil bacteria, *Microb. Ecol.* **13**:103–114.

Balba, M. T., and Nedwell, D. B., 1982, Microbial metabolism of acetate, propionate, and butyrate in anoxic sediment from Colne Point saltmarsh, Essex, U.K., *J. Gen. Microbiol.* **128**:1415–1422.

Barber, D. A., and Lynch, J. M., 1977, Microbial growth in the rhizosphere, *Soil Biol. Biochem.* **9**:305–308.

Barber, R. T., 1986, Dissolved organic carbon from deep sea waters resists microbial oxidation, *Nature* 220:274–275.

Beauchamp, E. G., Trevors, J. T., and Paul, P. W., 1989, Carbon sources for bacterial denitrification, *Adv. Soil Sci.* 10:113–142.

Benoit, R. E., and Starkey, R. L., 1968, Enzyme inactivation as a factor in the inhibition of the decomposition of organic matter by tannins, *Soil Sci.* 105:203–208.

Bollen, W. B., 1977, Sulfur oxidation and respiration in 54-year soil samples, *Soil Biol. Biochem.* 9:405–410.

Bosco, G., 1960, Studio della sensibilita, in vitro algi antibiotics de parte di microorganismi isolate in epoca preantibiotica. *Nouvi. Ann. Ingiene Microbiol.* 11:227–240.

Brookes, P. C., Newcombe, A. D., and Jenkinson, D. S., 1987, Adenylate energy charge measurements in soil, *Soil Biol. Biochem.* 19:211–217.

Bryan, B. A., 1981, Physiology and biochemistry of denitrification, in: *Denitrification, Nitrification and Atmospheric Nitrous Oxide* (C. C. Delwiche, ed.), Wiley, New York, pp. 67–84.

Burnison, B. K., and Morita, R. Y., 1974, Heterotrophic potential of amino acid uptake in a naturally eutrophic lake, *Appl. Microbiol.* 27:488–495.

Carlucci, A. F., and Williams, P. N., 1978, Simulated in situ growth of pelagic marine bacteria, *Naturwissenschaften* 65:541–542.

Casida, L. E., Jr., 1971, Microorganisms in unamended soil as observed by various forms of microscopy and staining, *Appl. Microbiol.* 21:1040–1045.

Christensen, D., and Blackburn, T. H., 1980, Turnover of tracer (^{14}C, ^{3}H labelled) alanine in inshore marine sediments, *Mar. Biol.* 58:97–103.

Christensen, D., and Blackburn, T. H., 1982, Turnover of ^{14}C-labelled acetate in marine sediments, *Mar. Biol.* 71:113–119.

Clark, E. F., 1965, The concept of competition in microbiology, in: *Ecology of Soil-Borne Plant Pathogens* (K. F. Baker and W. C. Snyder, eds.), University of California Press, Berkeley, pp. 339–347.

Cortez, F. J., Griffith, S. M., and Schnitzer, M., 1976, The distribution of nitrogen in some highly organic tropical volcanic soils, *Soil Biol. Biochem.* 8:55–60.

Craig, H., 1971, The deep sea metabolism: Oxygen consumption in abyssal ocean water, *J. Geophys. Res.* 76:5078–5086.

Dawson R., and Gocke, K., 1978, Heterotrophic activity in comparison to the free amino acid concentration in Baltic Sea samples, *Oceanol. Acta* 1:45–54.

Dow, C. S., Whittenbury, R., and Carr, N. G., 1983, The 'shut-down' or 'growth precursor' cell— An adaptation for survival in a potentially hostile environment, *Symp. Soc. Gen. Microbiol.* 34:187–247.

Estermann, E. F., Peterson, G. H., and McLaren, A. D., 1959, Digestion of clay–protein, lignin–protein, and silica–protein complexes by enzymes and bacteria, *Soil Sci. Soc. Am. Proc.* 23:31–36.

Fallon, R. D., Newell, S. Y., and Hopkinson, C. S., 1983, Bacterial production in marine sediments: Will cell-specific measures agree with whole system metabolism? *Mar. Ecol. Prog. Ser.* 11:117–119.

Fenchel, T. M., and Blackburn, T. H., 1979, *Bacteria and Mineral Cycling*, Academic Press, New York.

Garbosky, A. J., and Giambiagi, N., 1966, The survival of nitrifying bacteria in soil, *Plant Soil* 17:271–278.

Geesey, G. G., and Morita, R. Y., 1979, Capture of arginine at low substrate concentrations by a marine psychrophilic bacterium, *Appl. Environ. Microbiol.* 38:1092–1097.

Gibson, G. R., Parkes, R. J., and Herbert, R. A., 1989, Biological availability and turnover rate of acetate in marine and estuarine sediments in relation to dissimilatory sulphate reduction, *FEMS Microb. Ecol.* **62:**303–306.

Gocke, K., Dawson, R., and Liebezeit, G., 1981, Availability of dissolved free glucose to heterotrophic microorganisms, *Mar. Biol.* **62:**209–216.

Gordon, R., 1970, Some studies on the distribution and composition of particulate organic carbon in the North Atlantic Ocean, *Deep Sea Res.* **17:**233–243.

Gottschal, J. C., 1992, Substrate capturing and growth in various ecosystems, *J. Appl. Bacteriol. Symp. Suppl.* **72:**93–102.

Gray. T. R. G., 1976, Survival of vegetative microbes in soil, *Symp. Soc. Gen. Microbiol.* **26:** 327–364.

Gray, T. R. G., and Williams, S. T., 1971, Microbial productivity in soil, *Symp. Soc. Gen. Microbiol.* **21:**255–286.

Griffin, G. J., and Roth, D. A., 1979, Nutritional aspects of soil mycostasis, in: *Soil-Borne Plant Pathogens* (B. Schippers and W. Gams, eds.), Academic Press, New York, pp. 79–96.

Gronlund, A. F., and Campbell, J. J. R., 1963, Nitrogenous substrates of endogenous respiration in *Pseudomonas aeruginosa*, *J. Bacteriol* **86:**58–66.

Guckert, J. B., Hood, M. A., and White, D. C., 1986, Phospholipid ester-linked fatty acid profile changes during nutrient deprivation of *Vibrio* cholerae: Increase in cis–trans ratio and proportions of cyclopropyl fatty acids, *Appl. Environ. Microbiol.* **52:**794–801.

Harowitz, A., Krichevsky, M. J., and Atlas, R. M., 1983, Characteristics and diversity of subartic marine oligotrophic, stenoheterotrophic, and euryheterotrophic bacterial populations, *Can. J. Microbiol.* **29:**527–535.

Harrison, A. P., and Lawrence, F. R., 1963, Phenotypic, genotypic, and chemical changes in starving populations of *Aerobacter aerogenes*, *J. Bacteriol.* **85:**742–750.

Hood, M. A., Guckert, J. B., White, D. C., and Deck, F., 1986, Effect of nutrient deprivation on lipid, carbohydrates, DNA, RNA, and protein levels in *Vibrio cholerae*, *Appl. Environ. Microbiol.* **52:**788–793.

Jenkinson, D. S., and Ladd, J. N., 1981, Microbial biomass in soil, measurement and turnover, *Soil Biochem.* **5:**415–471.

Johnstone, B. H., and Jones, R. D., 1988a, Effects of light and CO on the survival of a marine ammonium-oxidizing bacterium during energy source deprivation, *Appl. Environ. Microbiol.* **54:**2890–2893.

Johnstone, B. H., and Jones, R. D., 1988b, Physiological effects of long-term energy-source deprivation on the survival of a marine chemolithotrophic ammonium-oxidizing bacterium, *Mar. Ecol. Prog. Ser.* **49:**295–303.

Johnstone, B. H., and Jones, R. D., 1988c, Recovery of a marine chemolithotrophic ammonium-oxidizing bacterium from long-term energy-source deprivation, *Can J. Microbiol.* **34:**1347–1350.

Jones, R. D., and Morita, R. Y., 1985, Survival of a marine ammonium oxidizer under energy source deprivation, *Mar. Ecol. Prog. Ser.* **26:**175–179.

Kjelleberg, S., Humphrey, B. A., and Marshall, K. C., 1982, Effect of interfaces on small, starved marine bacteria, *Appl. Environ. Microbiol.* **43:**1166–1172.

Kjelleberg, S., Hermansson, M., Marden, P., and Jones, G. W., 1987, The transient phase between growth and nongrowth of heterotrophic bacteria, with special emphasis on the marine environment, *Annu. Rev. Microbiol.* **41:**25–50.

Koch, A. L., 1971, The adaptive responses of *Escherichia coli* to feast and famine existence, *Adv. Microb. Physiol.* **6:**147–217.

Kurath, G., and Morita, R. Y., 1983, Starvation-survival physiological studies of a marine *Pseudomonas* sp., *Appl. Environ. Microbiol.* **45:**1206–1211.

Lipman, C. G., 1931, Living microorganisms in ancient rocks, *J. Bacteriol.* **22:**183–196.

Lockwood, J. L., 1977, Fungistasis in soils, *Biol. Rev.* **52**:1–43.

Lockwood, J. L., and Filonow, A. B., 1981, Responses of fungi to nutrient-limiting conditions and to inhibitory substances in natural habitats, *Adv. Microb. Ecol.* **5**:1–61.

Lundgren, B., and Söderström, B., 1983, Bacterial numbers in a pine forest soil in relation to environmental factors, *Soil Biol. Biochem.* **16**:625–630.

Lytle, C. R., and Perdue, E. M., 1981, Free proteinaceous and humic-bound amino acids in river water containing high concentrations of aquatic humus, *Environ. Sci. Technol.* **15**:224–228.

MacDonald, R. M., 1980, Cytochemical demonstration of catabolism in soil microorganisms, *Soil Biol. Biochem.* **16**:283–284.

MacDonell, M. T., and Hood, M. A., 1982, Isolation and characterization of ultramicrobacteria from a Gulf Coast estuary, *Appl. Environ. Microbiol.* **43**:566–571.

McGill, W. B., Hunt, H. W., Woodmansee, R. G., and Reuss, J., 1981, Phoenix—A model of dynamics of carbon and nitrogen in grassland soils, *Ecol. Bull.* **33**:49–115.

McLaren, A. D., 1973, A need for counting microorganisms in soil mineral cycles, *Environ. Lett.* **5**:143–154.

Martin, A., Jr., 1963, A filterable *Vibrio* from fresh water, *Proc. Pa. Acad. Sci.* **36**:174–178.

Martin, P., and MacLeod, R. A., 1984, Observations on the distinction between oligotrophic and eutrophic marine bacteria, *Appl. Environ. Microbiol.* **47**:1017–1022.

Matin, A., 1991, The molecular basis of carbon-starved-induced general resistance in *Escherichia coli*, *Mol. Microbiol.* **5**:3–10.

Meyer-Reil, L.-A., 1978, Autoradiography and epifluorescence microscopy combined for the determination of number and spectrum of actively metabolizing bacteria in natural waters, *Appl. Environ. Microbiol.* **39**:797–802.

Morgan, P., and Dow, C. S., 1986, Bacterial adaptation for growth in low nutrient environments, in: *Microbes in Extreme Environments* (R. A. Herbert and G. A. Codd, eds.), Academic Press, New York, pp. 187–214.

Morita, R. Y., 1980, Low temperature, energy, survival and time in microbial ecology, in: *Microbiology—1980* (D. Schlesdsinger, ed.), American Society for Microbiology, Washington, D.C., pp. 323–324.

Morita, R. Y., 1982, Starvation-survival of heterotrophs in the marine environment, *Adv. Microb. Ecol.* **6**:117–198.

Morita, R. Y., 1984, Substrate capture by marine heterotrophic bacteria, in: *Heterotrophic Activity in the Sea* (J. E. Hobbie and P. J. L. Williams, eds.), Plenum Press, New York, pp. 83–100.

Morita, R. Y., 1985, Starvation and miniturisation of heterotrophs, with special reference on the maintenance of the starved viable state, in: *Bacteria in Natural Environments: The Effect of Nutrient Conditions* (M. Fletcher and G. Floodgate, eds), Academic Press, New York, pp. 111–130.

Morita, R. Y., 1987, Starvation-survival: The normal mode of most bacteria in the ocean, in: *Current Perspectives in Microbial Ecology* (F. Megusar and M. Gantar, eds.), Slovene Soc. Microbiol., Ljubljana, Yugoslavia, pp. 243–248.

Morita, R. Y., 1988, Bioavailability of energy and its relationship to growth and starvation survival in nature, *Can. J. Microbiol.* **34**:446–441.

Morita, R. Y., and ZoBell, C. E., 1955, Occurrence of bacteria in pelagic sediments collected during the Mid-Pacific Expedition, *Deep Sea Res.* **3**:66–73.

Moyer, C. L., and Morita, R. Y., 1989a, Effect of growth rate and starvation-survival on the viability and stability of a psychrophilic marine bacterium, *Appl. Environ. Microbiol.* **55**:1122–1127.

Moyer, C. L., and Morita, R. Y., 1989b, Effect of growth rate and starvation-survival on cellular DNA, RNA, and protein of a psychrophilic marine bacterium, *Appl. Environ. Microbiol.* **55**:2710–2716.

Nedwell, D. B., and Gray, T. R. G., 1987, Soils and sediments as matrices for microbial growth, *Symp. Soc. Gen. Microbiol.* **41**:21–54.

Novitsky, J. A., 1987, Microbial growth rates and biomass production in a marine sediment: Evidence for a very active mostly nongrowing community, *Appl. Environ. Microbiol.* **53**:2368–2372.

Novitsky, J. A., and Morita, R. Y., 1976, Morphological characterization of small cells resulting from nutrient starvation in a psychrophilic marine vibrio, *Appl. Environ. Microbiol.* **32**: 619–622.

Novitsky, J. A., and Morita, R. Y., 1977, Survival of a psychrophilic marine vibrio under long-term nutrient starvation, *Appl. Environ. Microbiol.* **33**:635–641.

Novitsky, J. A., and Morita, R. Y., 1978, Starvation induced barotolerance as a survival mechanism of a psychrophilic marine vibrio in the waters of the Antarctic Convergence, *Mar. Biol.* **49**: 7–10.

Oppenheimer, C. H., 1952, The membrane filter in the marine environment, *J. Bacteriol.* **64**: 783–786.

Parkes, R. J., Taylor, J., and Jorck-Ramberg, D., 1984, Demonstration using *Desulfovibrio* sp. of two pools of acetate with different biological availabilities in marine pore water, *Mar. Biol.* **83**:271–276.

Poindexter, J. S., 1981, Oligotrophy: Fast and famine existence, *Adv. Microb. Ecol.* **5**:63–90.

Reeve, C. A., Amy, P., and Matin, A., 1984, Role of protein synthesis in the survival of carbon-starved *Escherichia coli*, *J. Bacteriol.* **160**:1041–1046.

Reiser, R., and Tasch, P., 1960, Investigation of the viability of osmophile bacteria of great geological age, *Trans. Kans. Acad. Sci.* **63**:31–34.

Roszak, D. B., and Colwell, R. R., 1987, Survival strategies of bacteria in the natural environment, *Microbiol. Rev.* **51**:365–379.

Sansome, F. J., 1988, Depth distribution of short chain organic acid turnover in Cape Lookout Bight sediments, *Geochim. Cosmochim. Acta* **50**:99–105.

Söderström, B. A., 1979, Seasonal fluctuations of active fungal biomass in horizons of a podsolised pine-forest soil, *Soil Biol. Biochem.* **11**:149–154.

Sowden, F. J., Griffith, S. M., and Schnitzer, M., 1976, The distribution of nitrogen in some highly organic tropical volcanic soils, *Soil Biol. Biochem.* **8**:55–60.

Stotzky, G., 1986, Influence of soil mineral colloids on metabolic processes, adhesion, and ecology of microbes and viruses, *SSSA Spec. Publ.* No. 17, pp. 305–428.

Tabor, P. S., Ohwada, K., and Colwell, R. R., 1981, Filterable marine bacteria found in the deep sea: Distribution, taxonomy and response to starvation, *Microb. Ecol.* **7**:67–83.

Tenore, K. R., Cammen, L., Findlay, S. E. G., and Phillips, N., 1982, Perspective of research on detritus: Do factors controlling availability of detritus to macroconsumers depend on its source? *J. Mar. Res.* **40**:473–490.

Torrella, F., and Morita, R. Y., 1981, Microcultural study of bacteria size changes and microcolony and ultramicrocolony formation by heterotrophic bacteria in seawater, *Appl. Environ. Microbiol.* **41**:518–527.

Torrella, F., and Morita, R. Y., 1982, Starvation induced morphological changes, motility, and chemotaxis patterns in a psychrophilic marine vibrio, *Deuxieme Colloque de Microbiologie marine, Publ. de Centre Nat. pour l'Exploitation des Oceans* **13**:45–60.

Upton, A. C., and Nedwell, D. B., 1989, Nutritional flexibility of oligotrophic and copiotrophic antarctic bacteria with respect to organic substrates, *FEMS Microbiol. Ecol.* **62**:1–6.

Van der Kooij, D. A., Visser, A., and Hijnen, W. A. M., 1980, Growth of *Aeromonas hydrophilia* at low concentrations of substrates added to tap water, *Appl. Environ. Microbiol.* **39**:1198–1204.

Van der Kooij, D. A., Oranje, J. P., and Hijnen, W. A. M., 1982, growth of *Pseudomonas* in tap water in relation to utilization of substrates at concentrations of a few micrograms per liter, *Appl. Environ. Microbiol.* **44:**1086–1095.

West, P. M., and Lochhead, A. G., 1950, The nutritional requirements of soil bacteria—A basis for determining the bacterial soil equilibrium of soils, *Soil Sci.* **50:**409–420.

Williams, P. M., 1971, The distribution and cycling of organic matter in the ocean, in: *Organic Compounds in the Aquatic Environment* (S. Faust and J. Hunter, eds.), Dekker, New York, pp. 45–60.

Williams, P. M., Oeschger, H., and Kinney, P., 1969, Natural radiocarbon activity in the northeast Pacific Ocean, *Nature* **224:**256–258.

Williams, S. T., 1985, Oligotrophy in soil: Fact or fiction? in: *Bacteria in the Natural Environments: The Effect of Nutrient Conditions* (M. Fletcher and G. Floodgate, eds.), Academic Press, New York, pp. 81–110.

Zimmerman, R., and Meyer-Reil, L.-A., 1974, A new method for fluorescence staining of bacterial populations on membrane filters, *Kiel. Meeresforsch.* **30:**24–27.

Zimmermann, R., Iturriaga, R., and Becker-Birck, J., 1978, Simultaneous determination of the total number of aquatic bacteria and the number thereof involved in respiration, *Appl. Environ. Microbiol.* **36:**926–935.

ZoBell, C. E., and Grant, C. W., 1942, Bacterial activity in dilute nutrient solutions, *Science* **96:**189.

2

Bacterial Growth and Starvation in Aquatic Environments

D. J. W. Moriarty and R. T. Bell

1. INTRODUCTION

The oceans cover 70% of the Earth's surface and marine bacteria number around 10^7 to 10^9 per liter, making them significant components of the biosphere. For most aquatic bacteria, starvation or low nutrient concentrations are important factors affecting their growth and survival. Food, or energy, limitation is certainly a major factor influencing the growth of marine bacteria, as Morita points out in this book. We will discuss some examples of nutrient limitation and the variability of bacterial growth rates in response to changing nutrient concentrations in both marine and freshwater ecosystems. Other factors besides nutrients that affect bacterial growth rates, size, and population density will also be discussed.

The structure of the habitat in which aquatic bacteria live has an important influence on the supply of nutrients to them. There are considerable differences between coastal and open ocean environments, or littoral and central lake water columns. The water column is generally structured with, in some cases, a greater production of phytoplankton above the thermocline, in other places greater productivity below it. In sediments, there are gradients in the concentra-

D. J. W. Moriarty • Department of Marine Microbiology, University of Gothenburg, Gothenburg, Sweden. *Present address*: Department of Chemical Engineering, University of Queensland, St. Lucia, Queensland 4067, Australia.   *R. T. Bell* • Institute of Limnology, Uppsala University, S-751 22 Uppsala, Sweden.

Starvation in Bacteria, edited by Staffan Kjelleberg. Plenum Press, New York, 1993.

tion of oxygen and organic and inorganic nutrients; these are especially complex in the rhizosphere of higher plants such as sea grasses and mangroves.

Although the similarities between the bacterial ecology of marine and freshwater habitats are often emphasized (Hobbie, 1988), some major differences are also evident. There are general differences between inland waters and the oceans in (1) the magnitude of seasonal temperature fluctuations, (2) sources of dissolved organic material, (3) the relative importance of nitrogen (N) and phosphorus (P) in limiting primary productivity, and (4) types of grazers capable of ingesting picoplankton ($\leq$ 2 μm). Various protozoans (microflagellates, ciliates) are common in both systems, but the cladoceran zooplankton (e.g., various species of *Daphnia*, the water flea), abundant in most lakes, may at times dominate the consumption of bacteria, thus shortening food chains.

The concentration of organic matter fluctuates considerably and the structure of the environment or the habitat for the bacteria on a microscale is also quite variable. For example, bacteria invading a diatom cell after it lyses and bacteria in the rhizosphere of sea grasses have higher concentrations of organic matter than those free in the water or sediment. Bacteria attached to or close to particles, particularly particles derived from phytoplankton, will obviously experience greater, but presumably transient, concentrations of organic matter than those free in the water column and away from particles. When grazing animals are present, the release of organic matter from phytoplankton would be greater than in their absence. Thus, interactions of grazers with phytoplankton and bacteria are important in determining whether bacterial production is closely linked to primary production in the water column.

Two of the most important factors controlling the growth of heterotrophic bacteria in the sea and lakes are the concentration of organic nutrients and temperature. The effect of substrate concentration is modified by the biochemical composition of the organic matter. The quality of organic matter as a food source varies depending on the particular species of bacteria utilizing it, but not many studies have been carried out on the species composition of bacteria that are active in the decomposition of organic matter at any particular time or place in the sea (one recent study is that of Painting *et al.*, 1989). At present, we can observe the response of heterotrophic bacterial communities to starvation or to variation in the concentration and quality of nutrients in natural environments by examining changes in numbers, size or biomass, growth rates, and productivity. In the future, advances in molecular biology, together with established procedures, hold the promise of identifying which species are the most active at any given time or place with greater assurance of reliability than at present. The strategies bacteria have adopted to survive during starvation have been discussed by Kjelleberg *et al.* (1993) and are outlined in several chapters in this volume.

2. EFFECTS OF STARVATION ON NATURAL POPULATIONS

2.1. Specific Growth Rates and Productivity

Where and when bacteria are starved in natural environments is not easy to determine, but an indication of starvation can be obtained by measurements of productivity and specific growth rates. If these rates are low, then it is likely that many or all of the heterotrophic bacteria are starved. Bacterial productivity is the rate of increase in biomass, and is directly related to the availability of nutrients. It can be determined in natural samples from the rate of tritiated thymidine incorporation into DNA (Moriarty, 1986; Bell, 1993). The specific growth rate is the number of generations occurring per unit time (the cell production rate divided by the number of cells present). Changes in specific growth rate occur in response to fluctuations in nutrient supply to the bacteria, but the specific growth rate itself is not necessarily directly related to the trophic status of the environment, i.e., whether it is eutrophic or oligotrophic. It is possible for the specific growth rate to be as fast in an oligotrophic environment as it is in a eutrophic, if the numbers of bacteria are lower in the oligotrophic environment than in the eutrophic.

The productivity of bacteria is a better indicator of trophic status than abundance, because it is more directly related to the turnover of organic carbon than either specific growth rate or changes in numbers. Bacterial production in the water column is dependent on the concentration and availability of dissolved organic compounds and thus is related to the rate of primary production by phytoplankton or benthic plants (e.g., see Moriarty *et al.*, 1990). In temperate environments it is difficult to separate the effects of temperature from substrate concentration. In coastal environments and small lakes, where there may be external sources of organic nutrients in addition to the phytoplankton, the analysis of factors controlling bacterial growth is more complicated.

Quantitative models have been developed to explain the control of heterotrophic bacterial biomass by organic substrate concentration and grazing in planktonic systems (Billen *et al.*, 1988; Wright and Coffin, 1984; Wright, 1988). Natural bacterial populations are comprised of many species, characterized by different nutrient requirements, different potential maximum specific growth rates, different sizes, and so forth, and thus the concepts used by macroecologists are not always appropriate. In particular, it may be too simplistic to view regulation of bacterial abundance only in terms of balances between predation on the one hand and resource supply or organic carbon concentration on the other. When aquatic microbial ecologists discuss heterotrophic bacteria, they generally consider them as a single group, in the manner of a vertebrate ecologist studying population dynamics of, for example, an antelope species on the African savannah, but in fact they are dealing with a large range of species, some growing at different rates, some not growing. The species composition of bacterial communities can rapidly change, and unlike animals, bacteria can vary their growth rates over large ranges in response to resource availability. As organic carbon compounds increase in concentration, bacteria grow faster, and species composition would change in response to changes in medium composition, selective grazing, and so forth. Until we can follow the dynamics of particular species of bacteria in natural populations, the applicability of concepts borrowed from macroecology is limited.

2.2. Factors Controlling Bacterial Growth

The principal source of organic nutrients for bacteria in the mixed layer of the open ocean is the phytoplankton. The abundance of planktonic bacteria in aquatic environments is generally related to algal biomass, although the correlation in freshwaters is not strong (Bird and Kalff, 1984; Currie, 1990). Bacterial biomass exceeds algal biomass in very oligotrophic systems (Cho and Azam, 1990; Bell and Tranvik, 1993), but the ratio of bacterial to algal biomass decreases with increasing nutrient concentrations (Fig. 1). Bacterial productivity, on the other hand, is more closely related to algal biomass (e.g., Hobbie and Cole, 1984; Bell, unpublished). Across this trophic spectrum of aquatic systems, with varying ratios of bacteria to algae, there are obviously different interactions between algae and bacteria as well as differences in the fate of microbial production.

The bacteria of inland and coastal waters are to a much greater degree exposed to allochthonous dissolved organic matter, i.e., that originating from the surrounding terrestrial environment. These compounds have undergone some degradation and transformation during transport to the pelagic environment and thus are less labile than the autochthonous substrates originating from algal production. Although traditionally considered recalcitrant, these substances may support bacterial growth (Tranvik and Höfle, 1987; Wetzel, 1992). Indeed, the biomass of planktonic bacteria is positively correlated with the humic content in a number of Scandinavian lakes (Hessen, 1985; Tranvik, 1988) and bacterial production may exceed primary production in humic lakes (Tranvik, 1989; Sundh and Bell, 1992). Thus, the food chains in lakes can, to a large extent, be based on a combination of auto- and allochthonous production.

As Ducklow (1984) points out, the time and space scales of measurements may not be appropriate for demonstrating relationships, because abundance is the end result of processes that could have occurred at other places and earlier in time. Billen (1990) observed a delay in the response of bacterial abundance to phytoplankton growth, both during the spring bloom in temperate waters and

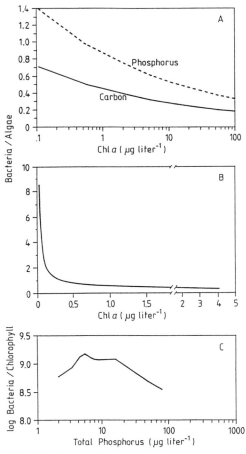

Figure 1. Various relationships between bacteria and chlorophyll (algal biomass) in aquatic environments. (A) Biomass relationship between bacteria and algae based on carbon and phosphorus, calculated from log bacteria versus log chlorophyll regression given in Bird and Kalff (1984). Bacterial carbon calculated as 20 fg/cell; bacterial P assumes a C/P (molar) of 53:1. Algal carbon and phosphorus estimated assuming C/Chl = 50, and assuming a Redfield ratio for C/N/P (molar) of 106:16:1. (B) Carbon biomass ratio of bacteria to algae in oceanic samples. Modified from Cho and Azam (1990). (C) Ratio of log of bacterial abundance to chlorophyll over a tropic range of lakes. Generalized trend from Currie (1990).

during the summer in the Antarctic. He concluded that the bacterial abundance and production were responses to release of macromolecules after death and lysis of the phytoplankton.

Dissolved organic matter exuded or secreted by phytoplankton is an important source of high-quality nutrients for bacteria and these exudates can provide greater than 50% of the carbon requirements of bacteria (Cole *et al.*, 1982; Larsson and Hagström, 1982; Bell and Kuparinen, 1984; Brock and Clyne, 1984). The proportion of primary production that is secreted as dissolved compounds is low compared with that incorporated into algal biomass (Sharp, 1981). Thus, it would be expected that bacterial abundance and production would be higher when a phytoplankton bloom was decaying than when the phytoplankton were actively growing. This simple relationship is not always found; it is modified by other factors that cause algal death, e.g., heavy grazing on the algae by zooplankton. Exoenzyme activity and temperature are also important factors governing bacterial growth in such situations. Bacterial communities can adapt to decreasing temperatures by alterations of the biochemical uptake mechanisms of particular species or by alterations of species composition. Thus, the overall growth rate of the heterotrophic bacterial community may not be severely restricted, provided there is an adequate supply of amino acids and sugars, except at very low temperatures. If, however, the bacteria are dependent on the hydrolysis of polymers by exoenzymes, then temperature could become a major factor limiting growth and in effect causing starvation.

Over an entire year, temperature certainly constrains bacterial growth in temperate lakes, and there are good statistical correlations between temperature and bacterial production (e.g., Bell, 1986; Morris, and Lewis, 1992). In some lakes, changes in bacterial growth rates are most dramatic at lower temperatures, for example as shown by Scavia and Laird (1987) at less than 10°C in the water column of Lake Michigan and by Bell and Ahlgren (1987) in surface sediments.

The biochemical composition as well as the concentration of organic matter is important in regulating bacterial growth in the sea. In the sub-Arctic Pacific Ocean, where concentrations of inorganic nutrients (N and P) are high and probably not limiting, bacterial growth and productivity is stimulated more by amino acids than glucose (Kirchman, 1990). This suggests that the most actively growing species of bacteria in the water were adapted to rapid uptake of amino acids and apparently energy limited if other substrates were supplied. In natural environments, bacteria grow more slowly than in culture; e.g., Kirchman (1990) found that bacterial generation times in the sub-Arctic pacific were 2 to 20 days and the bacteria required a period of at least 9 h and up to 2 days to respond to added organic compounds.

In general, the concentration of amino acids is low in coastal seawater; e.g., a range of 1 to 15 nM was reported by Fuhrman and Ferguson (1986). Much organic matter from phytoplankton or macrophytes is released as polymeric organic compounds from degrading tissue, lysing cells, or as slime secreted during growth. Bacteria need to hydrolyze the polymers and trap the monomers that are released before they become diluted and lost. Mechanisms are needed for coupling hydrolysis to uptake of amino acids, sugar, etc. (Hoppe *et al.*, 1988). The mechanisms and rates of exoenzyme activity and the relationships of bacteria to particles and macromolecules are important factors governing bacterial growth in natural environments that require further study.

2.3. Bacterial Cell Sizes in Natural Environments

Most aquatic bacteria are small (generally around 0.2 to 0.5 μm in diameter and volumes of 0.05 to 0.5 μm^3) and pass through filters with pore sizes of 0.8 to 1 μm. At least, they appear small to microbiologists who use laboratory media with organic C concentrations around 1 to 3 mg/liter. In fact, the bacteria in laboratory media are unusually large; we would argue that the normal size for bacteria is that observed in natural environments that support growth. Natural bacteria grow with perhaps only one or two copies of the genome per cell. Many increase in size before dividing when incubated with organic substrates, which suggests that they are living at suboptimal concentrations of organic nutrients (e.g., see Morita, 1982; Cho and Azam, 1988; Kirchman *et al.*, 1991). These small bacteria contain a high proportion of the biomass in the oceans and are major agents in the biogeochemical cycling of elements (Cho and Azam, 1988, 1990).

The size of the bacterial cells is an indicator of the trophic status of the bacteria, because starved bacteria are smaller than well-fed bacteria. There is a consistent trend for the bacteria in an oligotrophic environment to be smaller than those in eutrophic environments, but the differences are not large (Fuhrman *et al.*, 1989; van Duyl *et al.*, 1990; van Duyl and Kop, 1988). Numbers by themselves are not a good indication of the trophic status of bacteria because a response to starvation by bacteria is rapid division and the formation of small cells (Morita, 1982). Thus, numbers of bacteria may increase during starvation and, conversely, they may decrease when food is plentiful because of grazing by animals.

Although the majority of bacteria in aquatic environments are small and appear to be starved, they are active metabolically and are synthesizing DNA. Thus, many are at least growing slowly. Tabor and Neihof (1983) showed that at least 73% of bacteria present in water in Chesapeake Bay in September took up thymidine and incorporated tritium into macromolecules (probably DNA) and up to 94% of the bacteria took up amino acids. Tritiated thymidine is taken up only in dividing bacteria from natural environments (Fuhrman and Azam, 1982; Mården *et al.*, 1988). Thus, there is no doubt that the small bacteria in natural environments are dividing, even though some may be experiencing nutrient limitation and the early phases of starvation. Synthesis of DNA occurs in marine bacteria during starvation, although after about 3 h of starvation the rate is low (1% of that during growth) and probably the result of repair (Mården *et al.*, 1988).

The size of bacteria is influenced by temperature as well as by nutrient concentration and quality. Wiebe *et al.* (1992) found that volumes of marine bacteria were greater at high nutrient concentrations and in complex media, than at low concentrations or in a medium with a single source of carbon and nitrogen (proline); they also found that cell volumes were greater at 0°C than at 10°C (Table 1). Herbert and Bell (1977) reported volumes of marine bacteria to be larger at low (0°C) than at higher temperatures.

At temperatures below 0°C, substrate concentration and composition have a marked effect on bacterial cell sizes (Wiebe *et al.*, 1992). This suggests that the small size of bacteria in natural waters could be an effect of organic matter composition as well as concentration.

2.4. *Effects of Nitrogen and Phosphorus*

Nitrogen is a principal factor controlling or limiting primary productivity in the sea. The relationships of C and N to both algae and bacteria have been modeled by Lancelot and Billen (1985). They showed that where the C/N ratio of the organic matter utilized by bacteria was greater than 10:1, bacteria competed with the phytoplankton for uptake of inorganic N compounds. If

Table 1
Influence of Organic Matter Concentration (mg C/liter),
Composition, and Temperature on Volumes of Marine Bacteria[a]

Medium		Temperature			
Composition	Concentration	0°C		10°C	
Peptone, yeast extract	1500	2.3	0.13	1.7	0.091
	1.5	0.17	0.008	0.12	0.009
	0.15	0.11	0.008	0.070	0.005
Proline	1000	0.2	0.014	0.088	0.006
	1.0	0.15	0.010	0.072	0.005
	0.1	0.18	0.011	0.064	0.004

[a]From Wiebe *et al.* (1992).

inorganic N was limiting and the detritus or slimes secreted by algae had high C/N ratios, the growth of the bacteria would be restricted and their efficiencies of C utilization would be low (Lancelot and Billen, 1985). Bacteria growing under such conditions would be expected to show physiological signs of starvation, even though C concentrations were high. There are reports of algae secreting compounds inhibitory to bacteria (e.g., Hellebust, 1974; Cole, 1982). Such compounds would provide a mechanism for algae to minimize competition for nutrients.

Phosphorus is a principal factor controlling or limiting primary production in lakes (e.g., Schindler, 1977). Consequently, the interactions of algae, bacteria, and phosphorus have received more attention from freshwater microbial ecologists.

Previously, the importance of exudates for bacterial production has been overemphasized (Baines and Pace, 1991). This apparent dependence of bacteria on compounds excreted by algae led to the common view that bacterial growth in aquatic systems is primarily limited by organic carbon. On the contrary, bacterioplankton have been considered to have sufficient P, based primarily on observations that bacteria have a higher affinity (low K_m) for P (Currie and Kalff, 1984). In lakes where the algal biomass is less than about 10 μg chlorophyll/liter, between 50 and 99% of radioactive phosphate added to the water appears in the bacterial size fraction (Currie *et al.*, 1986). Taken together, such observations led to the "paradox paradigm" of algal–bacterial interactions proposed by Bratbak and Thingstad (1985), where algae are generally P-limited and bacteria are limited by dissolved organic C compounds, which they obtain from algae. Yet bacteria, by taking up P more effectively, may actually worsen the algal P-limitation. This paradox can be resolved by considering the activity of grazers that regenerate inorganic nutrients. This model may be accurate in the short term in some lakes, but it is insufficient to explain the dynamics of algae and bacteria over seasons and trophic states (e.g., Currie, 1990).

Bacteria contain at least twice as much P per unit biomass as do algae (e.g., Jürgens and Güde, 1990; Vadstein and Olsen, 1989). Thus, while bacteria are superior at taking up phosphate at low concentrations, they also have a high demand for P. In a eutrophic Norwegian lake, bacteria during the growing season were subsaturated with respect to P, and, rather than releasing P through the mineralization of organic matter (the traditional role for bacteria), were net consumers of P (Vadstein *et al.*, 1988). Experiments with natural bacterial communities grown in chemostats have confirmed that bacteria can be P-limited (Vadstein and Olsen, 1989; Jürgens and Güde, 1990). Based on his literature analysis, Currie (1990) postulated that both bacterial and algal growth are P-limited in very oligotrophic lakes and that bacteria become simultaneously limited by C and P in richer lakes. Because bacteria are better competitors when phosphate is scarce, bacteria grow more rapidly than algae as total P increases from 2 to 10 μg/liter (Fig. 1C). Algae then obtain greater proportions of the P and at concentrations greater than about 20 μg P/liter, algal biomass increases more rapidly than bacterial abundance (Fig. 1C; Currie *et al.*, 1986). Morris and Lewis (1992) demonstrated that bacteria were usually P-limited and occasionally limited by both C and P in a series of oligotrophic to mesotrophic lakes in Colorado. In a slightly acidic Swedish oligotrophic lake (total P about 9 μ/liter), bacteria were limited by both C and inorganic nutrients (Bell *et al.*, 1993b). Because the percentage of primary production released as organic carbon is very low in very eutrophic lakes (Baines and Pace, 1991), bacteria may eventually become predominantly limited by C in eutrophic lakes, especially during periods when cyanobacteria dominate (e.g., Robarts and Wicks, 1990).

3. MARINE ENVIRONMENTS

3.1. Pelagic Bacterial Growth

In the oligotrophic Sargasso Sea, bacterial productivity ranged from 0.8 to 6 μg C/liter per day, whereas in summer in Chesapeake Bay, it ranged from about 8 to 70 and over tropical seagrass beds from 20 in winter to 200 in summer (Ducklow, 1982; Fuhrman *et al.*, 1989; Moriarty *et al.*, 1990). In the North Sea and Dutch Wadden Sea, bacterial productivity and numbers were lower in winter than in other seasons (Van Duyl and Kop, 1988; Van Duyl *et al.*, 1990). Joint and Pomroy (1987) concluded that a large proportion of bacterial productivity was correlated with temperature and not with primary production in the Celtic Sea. Lancelot and Billen (1984) have, in contrast, found that bacterial production was correlated directly with phytoplankton production and in particular the release of dissolved organic matter during the phytoplankton bloom in the North Sea. In water upwelling from the deep ocean, bacteria were small and as phytoplankton blooms developed, there was a rapid increase in productivity (Painting *et al.*, 1989).

Dissolved organic compounds from phytoplankton turn over rapidly in the surface waters of the north Atlantic during the spring bloom (Kirchman *et al.*, 1991). The growth efficiency or cell biomass yield was low (2 to 9%), indicating that abundance would not increase markedly and that the bacterial production was limited by other factors. Below the mixed layer in the ocean, which is usually at a depth of 100 to 200 m, bacteria are dependent on organic matter falling from the surface as detritus, e.g., fecal pellets, and in feces from migrating animals and, in some seas, on organic matter carried by water masses that have sunk from the surface at convergence zones.

There are major differences between water masses in the photic zone, the upper mixed layer, and in the deep ocean. After sinking from the surface at convergence zones, water masses circulating below the mixed layer travel for quite long periods during which bacteria gradually use up organic nutrients that sank with the water and eventually they must rely on the low rate of supply from particles sinking from the surface layers. In the Pacific Ocean for example, residence times for water masses are 2 to 100 years or so for the Antarctic Intermediate Water mass which sinks in the subantarctic and 1000 or more years for the deep water which sinks below the Antarctic convergence. The low temperature in the deep ocean slows rates of metabolism, so organic matter lasts longer than at the surface (Sorokin, 1972, 1978). Bacteria traveling with these water masses would need to have strategies for surviving long periods with little nutrient input. When one considers the volume of water in the deep ocean compared with the photic zone, it is clear that the concentration of organic matter that is available to bacteria must be low, because it is all synthesized in the photic zone, either immediately above or in a distant, polar ocean.

The extent to which bacteria in the ocean depths are supplied by organic matter from the surface varies considerably between seas. Cho and Azam (1988) estimated that 17% of primary production settled to depths of 1000 m from the photic zone in the north Pacific gyre and 40% in the Santa Monica basin. In some parts of the sea, feces from migrating zooplankton and fish may be the main source; in other parts, transport of organic matter from polar convergence zones may be more important.

In oceans south of the equator, Sorokin (1978) and Sorokin *et al.* (1985) found a zone of high bacterial activity at intermediate water depths and temperatures of 5 to 10°C. Bacterial growth rates increased when the water was warmed, indicating that temperatures were limiting bacterial activity, but it would be surprising if psychrophilic bacteria were inhibited at these temperatures if organic matter were readily available. This suggests that the rate-limiting process was enzymatic activity external to bacteria, i.e., the hydrolysis of polymeric compounds. As the water masses originated south of the subtropical convergence zone, polymeric compounds would survive for much longer than monomers such as amino acids or simple sugars if rates of exoenzyme activity were inhibited by the low temperatures. If the bacteria were dependent solely on a vertical flux of organic matter, then productivities would decrease continuously with depth; the substantially higher abundance and production of bacteria suggests that there is horizontal advection of organic matter from polar zones. Further work on mechanisms of particulate organic matter formation and decomposition is needed to determine whether organic matter can be concentrated at boundaries between water masses with different densities and thus whether sinking particles can contribute to the zones of high bacterial activity observed by Sorokin (1972).

It is not only in the deep ocean that bacterial growth may be limited by the rates of polymer hydrolysis. The principal substrates for bacterial growth in the photic zone of the oceans are the macromolecules from dead algae (Lancelot and Billen, 1984, 1985; Billen, 1990). Fecal pellets of zooplankton are another source of polymeric material providing substrates for bacteria. The rates at which such particles are degraded and the extent to which they contribute to bacterial growth within or below the photic zone vary with location and are subjects requiring much more research (e.g., Karl *et al.*, 1988; Simon *et al.*, 1990). Some particles sink very rapidly to the bottom of the deep ocean. Phytoplankton detritus, with intact or partially decomposed algal cells, has been found at a depth of 4500 m in the northeast Atlantic (Thiel *et al.*, 1988).

Fine amorphous aggregates of organic matter in the sea are called "marine snow." These particles are found throughout the surface waters of the oceans and are enriched in organic matter and microorganisms. Particles in the sea also originate from fragments of decaying phytoplankton, the fecal pellets of zooplankton, and mucus secreted by a wide variety of plants and animals. Bacteria attached to the surfaces of particles are generally larger than those free in the water column and are more active metabolically. This indicates that bacteria attached to particles are not as limited in food supply as those free in the water column (Kirchman, 1983; Simon *et al.*, 1990). It seems that only a small proportion of bacteria are attached firmly to particles, but sampling procedures that do not cause disruption of loosely associated bacteria are difficult to apply. The dynamics of suspended particles that are neutrally buoyant in water are not known in relation to time before decomposition or ingestion by filter feeders, nor in relation to the mechanism of decomposition, i.e., the role of exoenzyme digestion either at bacterial cell surfaces or by free enzymes. The great diversity of particles and variations between seasons in particle formation from decaying phytoplankton and other sources make this a difficult topic to study, yet more research is needed because it is very pertinent to studies on bacterial growth and starvation in the natural environment.

Many of the starved bacteria that are found free in the water column may be species that normally grow on or within organic aggregates, fecal pellets, and other particles. It might be expected that small invertebrates such as copepods, which have short passage times for food in the gut, would not have a distinct gut flora, but in fact they do (Sochard *et al.*, 1979). Perhaps many of the bacteria that grow slowly or not at all in the water column are species that grow rapidly in the guts of animals after being ingested together with the food of the animal. When excreted in the fecal pellet, these bacteria would have a continued advantage over those free in the water, because they would have a concentrated food supply of phytoplankton fragments. This might explain why many marine vibrios have potential maximum growth rates that are very fast:

this would enable them to respond rapidly to transient high concentrations of organic matter in fecal pellets or lysing algal cells.

In general, bacterial production in the water column is dependent on primary production (Cole *et al.*, 1988). The relationship varies over short time and space scales and so bacterial productivity is not necessarily directly proportional to primary productivity at a given time or place (Ducklow, 1984). The time scales for bacterial growth vary, depending on whether the source of organic matter is exudation from normal phytoplankton during photosynthesis, or pulses of organic matter from decaying or stressed phytoplankton, or release by excretion from zooplankton, or lysis of cells during feeding by zooplankton. Bacterial growth then becomes "unbalanced," i.e., the rate of DNA synthesis or cell division does not occur at the same rate as synthesis of other cellular constituents, e.g., protein (Chin-Leo and Kirchman, 1990).

3.2. Bacterial Growth and Survival in Sediments

Bacteria in marine sediments experience fluctuating nutrient concentrations on time scales varying from hours to seasons and perhaps longer in the deep ocean, and this is reflected in the rates at which they take up organic matter and thus their productivity. Seasonal and daily variations in bacterial productivity, growth rates, and numbers have been found in many environments (Moriarty, 1986).

As mentioned above, bacterial productivity is a better indicator than abundance of the availability of organic nutrients and thus whether bacteria are likely to be starved, because abundance is limited by grazing animals. In sediments, this is particularly evident at the surface where low numbers and rapid specific growth rates indicate grazing (Alongi, 1989; Moriarty *et al.*, 1985, 1990, 1991). These effects of animals were only apparent in summer, when bacterial productivities were high; in winter, bacterial abundance was apparently not affected by animals.

Deeper in sediments, population density and production are controlled by the supply of organic matter (Alongi, 1989; Moriarty *et al.*, 1991). In deep waters where the supply of organic matter is limited, it has been found that bacterial population density at the sediment surface was high, but productivity, and therefore specific growth rates, were very low, indicating that not only was grazing low, but that the bacteria were starved more than those in shallow water (Table 2). Growth was apparently unbalanced on the Australian continental shelf at a water depth of 150 m because rates of protein synthesis relative to DNA synthesis were considerably faster than at deeper water depths (Table 2).

Among the factors that regulate bacterial production in sediment is the supply of organic matter (Moriarty, 1989). Bacterial productivity in coral reef

Table 2

*Specific Growth Rates (h^{-1}) of Bacteria in Sediment and the Ratio
of Rates of Protein to DNA Synthesis (P/DNA, Converted
to Equivalent Units of Carbon Production) at Various Water Depths
on the Continental Margin of Eastern Australia[a]*

| | Water depth (m) | | | | | |
| | 150 | | 400 | | 1000 | |
Sediment depth (mm)		P/DNA		P/DNA		P/DNA
2	0.008	3.3	0.011	1.7	0.0016	2.7
10	0.010	7.3	0.010	1.2	0.0013	1.3
20	0.006	10	0.014	2.5	0.0016	1.2

[a]From Moriarty *et al.* (1991).

and sea grass sediments is closely correlated to the productivity of sea grasses and benthic algae on daily and seasonal scales (Moriarty *et al.*, 1985, 1990; Moriarty and Hansen, 1990; Moriarty and Pollard, 1982; Pollard and Moriarty, 1991). On diurnal time scales, bacterial productivity and specific growth rates in sea grass rhizosphere sediment at depths of 3 and 10 cm varied with photosynthetic rates of the primary producers, but below the rhizosphere at a depth of 20 cm, there was little variation (Fig. 2). Sea grasses exude organic matter during photosynthesis into the rhizosphere, which is therefore a zone of intense bacterial growth. Coastal sea grass communities are very productive and there is a correspondingly high productivity of bacteria in the sediments and shallow waters of sea grass meadows. For example, in northern Australia, benthic bacterial productivity varied from about 3 to 10 g C/m^2 per day, whereas

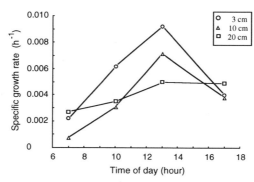

Figure 2. Diel variation in specific growth rate of bacteria at different depths in sediment in a tropical sea-grass sediment. Redrawn from Pollard and Moriarty (1991).

water column productivity was generally less than 1 g C/m^2 per day (Moriarty *et al.*, 1990). The productivity of bacteria in the water column was three to five times greater in summer than in winter and there was a corresponding seasonal variation in primary productivity in the same community. As this was a tropical environment, one would expect that if adequate nutrients had been available, bacterial productivity would not have decreased so markedly in winter, when temperatures were about 23°C. It seems likely that the seasonal variation was the result both of variation in the supply of organic nutrients as well as of temperature (Moriarty *et al.*, 1990).

Temperature is obviously an important factor controlling bacterial metabolic activity in sediments and thus growth rates and bacterial productivity, through its effect on exoenzyme hydrolysis of organic polymers, especially cellulose, chitin, and protein. This early phase in the decomposition of particulate organic matter, derived from dead plants as well as animals, seems to be the time-limiting one, but few ecological studies have been made of this process (Godschalk and Wetzel, 1977). Alongi (1988) found that bacterial production in tropical mangrove sediments was lower in winter and, furthermore, that when intertidal mud banks were exposed to air that was cooler than the water, bacterial growth rates decreased.

Bacterial productivity and primary productivity were correlated in coral reef sediments where a seasonal variation was also apparent (Moriarty *et al.*, 1985). It seems surprising that on coral reefs and to some extent tropical seagrass beds, the bacteria are apparently starved in winter, as specific growth rates are slow and productivity is low (Moriarty *et al.*, 1985, 1990; Hansen *et al.*, 1987). This could be caused by the lower rates of primary production and thus secretion of organic compounds from living cells and the slow rates of exoenzyme hydrolysis of plant and algal cell fragments.

These results indicate that bacteria in sediment are limited by both organic carbon substrate availability and temperature. There are periods varying from hours to months in surface sediments when organic substrate concentrations are so low that growth stops and the bacteria starve. In deep sediments, nutrients are permanently limiting, and the only bacteria that can survive are those that have developed strategies for maintenance and growth at very low concentrations of organic nutrients.

It is evident that once bacteria in sediment use up their sources of organic matter, most eventually die, because the abundance of bacteria below the zone of bioturbation decreases exponentially. Some bacteria are found deep within the sediments and this raises questions concerning starvation and survival, in particular for how long bacteria can survive, and whether they are able to grow and divide very slowly, using organic matter that was buried with them perhaps several thousand years ago, or whether they use dissolved organic compounds that diffuse very slowly from the surface. Bacteria buried deep in marine sediments with an age of 50,000 or more years have functioning biosynthetic pathways and the capacity for growth, even if they do not actually do so as inferred from the detection of DNA synthesis after a lag phase of about 12 h (Moriarty *et al.*, 1991). Bacteria in these very deep sediments may be simply surviving at a low metabolic rate. If they are growing, an important physiological question is: how is DNA synthesis regulated at very low and fluctuating concentrations of organic substrates?

4. FRESHWATER ENVIRONMENTS

4.1. Seasonal Dynamics of Bacterioplankton in Lakes

Lakes can be classified on the basis of mixing pattern (monomictic, dimictic, polymictic), temperature, and trophic status (oligotrophic, mesotrophic, eutrophic); consequently, there are about 15 generated patterns of seasonal phytoplankton profiles (Sandgren, 1988) concerning both biomass trends and successional patterns (e.g., Sommer, 1989). A few attempts have been made to produce generalized bacterial seasonal trends as well (Güde, 1989; Pedros-Alio, 1989), but real progress awaits the application of molecular biological techniques, such as 16S rRNA sequencing and total DNA hybridization (Pace *et al.*, 1986; Lee and Fuhrman, 1990).

We will first discuss the dynamics of planktonic bacteria during spring and early summer in a "typical" moderately eutrophic temperate lake, utilizing information primarily from Lake Erken, Sweden. Then the seasonal trends in several lakes with very different nutrient concentrations will be compared with respect to abundance, productivity, and specific growth rates.

There have been few studies of bacterial productivity under winter ice-cover, but population generation times of more than 50 days are typical (Bell, unpublished). After the ice melts in spring, algal biomass increases exponentially for several weeks. In Lake Erken, where the diatom *Stephanodiscus hantzchii* var. *pusillus* comprises 90% of the biomass, chlorophyll concentrations reach 20 μg/liter. During this period, bacterial abundance also increases and is strongly correlated with algal biomass (Bell and Kuparinen, 1984). Bacterial productivity increases toward the later phases of the diatom bloom, and bacterial numbers probably peak at least 1 week after the decline of the diatom bloom, as has been observed in marine systems (Billen, 1990). Bacteria frequently utilize more than 50% of the dissolved organic carbon compounds released by phytoplankton (Coveney, 1982; Sundh and Bell, 1992). Much of the dissolved organic compounds (DOC) excreted by phytoplankton during the spring bloom are nonetheless not utilized (Coveney and Wetzel, 1989). Scavia and Laird (1987) reported that during summer stratification in Lake Michigan, primary production was lower than bacterial production and suggested that the DOC accumulated during spring was subsequently utilized by the bacteria in the summer.

Several questions arise concerning bacterial activity in lakes: Is growth limited by temperature during spring? Are the dissolved organic carbon compounds excreted by phytoplankton mainly composed of high-molecular-weight material and does this organic matter contribute to the refractory dissolved organic carbon compounds in the lake or is it subsequently utilized by bacteria during summer? The algal excretion products during the end of the spring diatom bloom in May of 1988, when the water temperature was about 6 to 9°C, consisted of equal quantities of high-molecular weight ($> 10,000$ Da) and low-molecular-weight (<1000 Da) compounds. Sundh (1991) determined that about 80% of the low-molecular-weight material (sugars and amino acids) and about 50% of the high-molecular-weight material (carbohydrates) could be utilized by bacteria during 4-h incubations. Temperature must be the primary controlling factor: this high utilization of algal excretion products contrasts strongly with an average utilization of only 4% during 1983 when water temperatures were 3.5°C (Bell and Kuparinen, 1984). Thus, as suggested by Hobbie (1992), there is most likely a build-up during early spring of labile dissolved organic carbon compounds that can be rapidly utilized by bacteria at the end of spring blooms when water temperatures have risen.

Chrost (1989) demonstrated that the activity of bacterial extracellular enzymes increased during the decline of a spring bloom in a eutrophic lake when large amounts of high-molecular-weight DOC were released from dying algae. Thus, as temperatures approach 10°C, bacteria utilize the biologically available DOC that has been building up during the bloom and receive a new pulse of DOC from the dying algae. This pulse of bacterial activity in turn leads to an increase in abundance of heterotrophic flagellates, which by their grazing on bacteria, regenerate inorganic nutrients; this "microbial loop" occurs after the demise of the spring diatom bloom. It was at this time in oligotrophic Lake Dillon immediately after stratification in early June, when the water temperature was about 10°C, that the actual bacterial growth rate, determined from rates of thymidine incorporation into DNA, equaled the maximum specific growth rate (μ_{max}; about 0.05 h^{-1}), determined from bacterioplankton cultures supplemented with yeast extract (Morris and Lewis, 1992). Chlorophyll and particulate organic carbon concentrations were at their maximum. For the rest of summer the specific growth rate was much less than μ_{max}, suggesting bacteria were nutrient-limited.

During summer stratification, bacterial production is frequently an order of magnitude greater in the epilimnion than in the hypolimnion, at least in lakes with negligible external nutrient input during summer (Güde, 1989; Bell *et al.*, 1993a; Morris and Lewis, 1992). The release of phosphorus from surface sediments to the hypolimnetic waters can be significant (e.g., Pierson, 1991); thus, the lower bacterial production in the hypolimnion is likely the result of a combination of low temperature and lack of utilizable organic matter. Rates of organic matter sedimentation in lakes are greatest after the demise of the spring diatom bloom and after water mixing and circulation in autumn. During stratification, more than 70% of the primary production may be mineralized in the photic zone (e.g., see Tilzer, 1990).

The interactive effects of the nutrient and substrate supply and grazing pressure determine the abundance and temporal dynamics of bacteria (e.g., see Wright, 1988). Grazers enhance bacterial growth rates in two ways: by keeping bacteria at suboptimal abundances, allowing greater substrate concentration and thus more per bacterium, and by releasing dissolved organic and inorganic nutrients to the water (Lampert, 1978; Güde, 1985). The three seasonal trends shown in Fig. 3 are examples ranging from low substrate concentration with low grazing (Fig. 3A, oligotrophic Lake Njupfatet, Sweden), high substrate concentration with grazers present (Fig. 3B, mesotrophic Lake Erken, Sweden), and high substrate concentration with low grazing (Fig. 3C, hyper-eutrophic Lake Managua, Nicaragua).

Bacterial growth in the oligotrophic lake is constrained by nutrient limitation (Fig. 3A). There is, however, a potential microbial food web in the lake that keeps bacterial numbers down. When copepods, the dominant zooplankton which do not graze bacteria directly, were removed from enclosures in the lake, the biomass of ciliates increased by an order of magnitude and these ciliates feed on bacteria (Bell *et al.*, 1993b; Stensdotter *et al.*, in preparation). The dominant bacterial grazers in these nutrient-poor lakes are often mixotrophic flagellated algae (Chrysophyceae). Mixotrophic algae accounted for more than 75% of the total community grazing in two oligotrophic Canadian lakes (Bird and Kalff, 1984). This strategy is obvious: if nutrients are scarce, eat the competitors!

Bacterial productivity in Lake Erken (Fig. 3B) is bout ten times greater than in Lake Njupfatet, in proportion to the increase in algal biomass, but the bacterial standing stock is similar, thus mortality (mainly grazing) must balance production throughout the summer. The increased productivity is reflected in a faster growth rate (Fig. 3D). Consequently, nutrients are being recycled faster, and the algae are less nutrient stressed during this period (Istvanovics *et al.*, 1992). In contrast, in Lake Constance, which is similar in trophic status to lake Erken although much deeper (20 m versus 200 m), mean doubling times of epilimnetic bacteria exceeded 10 days during summer (Güde, 1990). Grazing pressure in this lake is low, and bacteria were a "nutrient sink" containing about half of the particulate P in the euphotic zone (Güde, 1991). Short periods of intensive growth were followed by long periods resembling stationary phases (Güde, 1990).

Lake Managua is a very eutrophic, polymictic (constantly mixing) tropi-

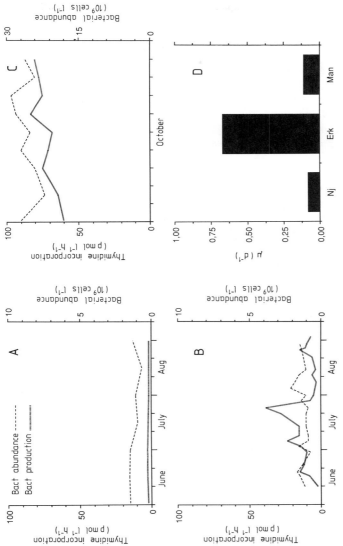

Figure 3. Seasonal trends of bacterial production (thymidine incorporation into DNA) and bacterial abundance. Trends shown are for the stratified period (June through August) in A and B. (A) An oligotrophic lake: Lake Njupfatet, Sweden (Bell *et al.*, unpublished). (B) A mesotrophic lake: Lake Erken, Sweden (Istvanovics *et al.*, 1992; Bell *et al.*, 1993a). (C) A hypereutrophic lake: Lake Managua, Nicaragua. The trend shown is from a 2-week period in October, 1988, when sampling was every 2–3 days (unpublished data); seasonal fluctuations are weak (Bell *et al.*, 1991). (D) Average specific growth rates (μ) in each lake.

cal lake at a low altitude and water temperature of $29 \pm 2°C$. The bacterial production in this lake has been channeled into bacterial biomass, which is constantly high (Fig. 3C), and thus the specific growth rate is not much faster than in an oligotrophic lake: the doubling time was about 10 to 13 days (Fig. 3D). Bacterial growth in this lake could be characterized as "permanent stationary phase". The dominant grazers in the lake are copepods (80% of grazer biomass); the flagellates and ciliates are apparently suppressed (Bell *et al.*, 1991; see also Table 2-1 in Sandgren, 1988). The high bacterial standing stock is at least in part the result of low community grazing. This pattern is not necessarily typical for other tropical lakes or eutrophic lakes in temperate regions. For example, the dominate grazer in Lake Tanganyika, a deep oligotrophic lake in Africa, is a mixotrophic ciliate (Hecky and Kling, 1981); significant communities of ciliates have been noted in other tropical lakes (e.g., see Gebre-Mariam and Taylor, 1989; Laybourn-Parry, 1992). Considering the strong relationship between μ_{max} and temperature (Morris and Lewis, 1992), we might predict the bacterial abundances would be significantly greater in the tropics, but despite some reports of high ($> 10^7$ ml^{-1}) abundances (Rai, 1979; Bell *et al.*, 1991), a comparative analysis of available data showed no significant difference between temperate and tropical regions (Pedros-Alio and Guerrero, 1991). However, there is still an extremely limited amount of information on the microbial ecology of tropical lakes.

4.2. *The Microbial Landscape in Lakes—Importance of Microenvironments*

The majority of bacteria in pelagic surface waters are free-living, and are actively growing and incorporating thymidine as determined by autoradiography (Simek, 1986). Nonetheless, if bacteria were randomly distributed in the water column, the high rates of bacterial production observed in surface waters would not be possible (Azam and Ammerman, 1984). The clustering of bacteria in microzones around active phytoplankton cells was termed the "phycosphere" (Bell and Mitchell, 1972) and the interactions of algae, bacteria, and protozoans have been integrated into a conceptual framework by Azam and Ammerman (1984) and Azam and Smith (1991).

The abundance of attached bacteria appears to be related to the amount of particulate matter (reviewed by Kirchman, 1993). Thus, it is not surprising that bacteria attached to particles may at times comprise up to 90% of the total bacterial abundance in eutrophic lakes. Attached bacteria are usually larger than free bacteria, but do not appear to generally have faster growth rates than free bacteria, as determined by rates of DNA synthesis (Kirchman, 1993). Attached bacteria do take up organic compounds such as glucose and amino acids at faster rates than free bacteria (Simon, 1985; Kirchman, 1993). Autoradiographic studies show that a large fraction of the organic carbon taken up by attached bacteria appears in the surrounding mucilage (Paerl, 1978). Azam and Smith (1991) suggested that attached bacteria were larger because they were, in a sense, packages of exoenzymes. Attached bacteria have a role in rapidly solubilizing particulate organic matter, which can then be utilized by the free-living bacteria surrounding the particle. During the initial stages of colonization, the bacteria thus compete with grazers for particles, but the presence of bacteria may make the particles more attractive to grazers.

The growth rate of bacteria on particles will depend on factors such as the age, nutritional status, and form (e.g., porous or dense) of the particle or aggregate, as well as the density of bacteria on the particle (e.g., Lewis and Gattie, 1990). Thus, it seems that simple definitions of "attached" bacteria as the bacteria not passing through a filter with a pore size of 1 μm, give few insights into the functioning of microbial communities. There are certainly aggregates in lakes that are similar to the various types of "flocs" that are termed "marine snow." Tranvik and Sieburth (1989) observed high abundances of bacteria and attached flagellates formed on flocculated humic matter.

Besides influencing bacterial growth rates and the rate of nutrient and substrate supply, the size and taxonomic composition of the grazer communities will also have a great impact on the resultant size structure of the bacterioplankton community. For example, grazing by large cladocerans such as *Daphnia* will produce bacterial populations comprising small, single-celled bacteria, whereas grazing by microflagellates, which select larger bacteria (e.g., Chrzanowski and Simek, 1990), may produce a size distribution comprising both small, free single-celled bacteria as well as filamentous, or other aggregates of bacteria that may escape grazing (Caron *et al.*, 1988; Güde, 1989). Bacterial aggregates, comprising 30% of total bacteria, developed on decomposing diatoms during a midsummer period of intensive grazing in Lake Erken; associated with the "diatom–bacteria aggregate" were high densities of attached heterotrophic flagellates (Bell *et al.*, 1993a). Bacterial production and specific growth rates were enhanced during this short period.

A substantial percentage of the bacteria in eutrophic lakes with colonial cyanobacteria are embedded in their gelatinous mucilage. Shallow, hypereutrophic lake Vallentunasjön, Sweden, is dominated by the cyanobacterium *Microcystis wesenbergii* and occasionally 50% to 90% of the total planktonic bacteria are embedded in its gelatinous sheath (Bell and Brunberg, unpublished). Microautoradiographic studies suggest that most of these bacteria are active (Bern, 1985). Their growth rate, however, is an order of magnitude lower than the growth rate of the free-living bacteria as observed for other aggregates (Lewis and Gattie, 1990; Brunberg and Bell, submitted).

4.3. Bacterial Productivity in the Surface Sediment of Lakes

Bacterial abundance is about three orders of magnitude higher in surface sediment (per unit volume) than in the overlying water (Jones, 1989). The cyanobacteria, especially *Microcystis* spp., are also major components in the surface sediment of Lake Vallentunasjön (Bell and Ahlgren, 1987; Brunberg and Boström, 1992). In the mid-1980s, *Microcystis* constituted about 80% of the total bacterial biomass and the bacteria bound in its mucilage were about 40% of the total heterotrophic bacteria in the upper 2 cm of sediment. Bacteria in mucilage grew two to nine times faster than other sedimentary heterotrophic bacteria (Brunberg and Bell, submitted). *Microcystis* support heterotrophic bacteria in the sediment by exuding organic matter while alive as well as by providing substrates for growth after they die. After 1985, *Microcystis* biomass decreased markedly in surface sediment (Fig. 4A), and by 1989, the mucilage bacteria were only 1 to 5% of the total in the sediment. The heterotrophic bacterial productivity declined at a similar rate (Fig. 4B), although numbers remained constant. There was also a strong covariation between cyanobacterial

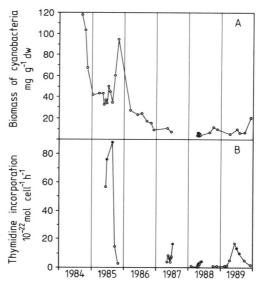

Figure 4. Microbial biomass in Lake Vallentunasjön, Sweden, 1984–1989. (A) Biomass of large cyanobacteria ($>$ 3 μm) in the surface sediment (0–2 cm) and (B) thymidine incorporation (heterotrophic bacterial production) in the surface sediment. Modified from Brunberg and Boström (1992). Darkened circles refer to occasions when the sediment temperature was 18–20°C.

biomass, heterotrophic bacterial activity, and the release of phosphorus from the sediments to the lake water (Brunberg and Boström, 1992).

The average growth rate of bacteria in lake sediments is often slower than that in the water column. For example, the fastest doubling times of bacteria in Lake Vallentunasjön were 2 to 4 days in July, 1985 (Bell and Ahlgren, 1987). By 1989, the fastest doubling times were about 14 days. The average generation time for bacteria in the euphotic zone of lakes is about 4 days in summer (calculated from data of Cole *et al.*, 1988). During the winter months, the average doubling times are 100 to 200 days, which indicates that most bacteria do not grow in winter. Boström and Törnblom (1990) reported that bacterial doubling times varied from thousands of days in winter to 100 to 200 days in summer in some shallow marine sediments and lakes, which indicates that a large portion of the bacteria were not growing and certainly starving even in summer.

5. CONCLUDING COMMENT

It is very rare to find bacteria in natural environments with nutrient concentrations around them that are greater than necessary to sustain maximum growth rates and with a cell size that is large and equivalent to that of cells in rich culture media. But, such microenvironments do occur, for example inside a diatom cell that has just lysed. In general, natural bacteria grow at slower rates than their maximum potential and may be limited by organic carbon, or nitrogen, phosphorus, etc., as well as by physicochemical factors such as temperature. These factors can vary rapidly and so, as bacteria experience varying impacts of these factors, growth rates vary. From our discussions above, it should be clear that we cannot arrive at simple, noncomplex generalizations about what controls bacterial growth and whether, where, or when bacteria are starved in natural environments without having to add qualifying statements. Having said that, we do wish to conclude with the comment that in general, aquatic bacterial populations are limited by nutrient availability more than by any other factor, as evidenced by the large diel and seasonal variations in growth that are related to availability of organic matter. Thus, they are not growing at maximum rates all the time, but at least some bacteria are growing at any given time in surface waters and sediments; i.e., they are limited by carbon or energy, but not starved. Individual bacterial and even all members of some species groups, however, may have no access to nutrients and thus be starved at some times and in some places. Slow growth rates and long periods in stationary phase and starvation are normal conditions for bacteria in the sea and lakes and for which they need strategies to survive.

ACKNOWLEDGMENTS. Research cited from the laboratory of R. Bell was supported by the Swedish Natural Science Research Council (NFR), the Swedish Environmental Research Board (SNV), and the Swedish Agency for Research Cooperation with Developing Countries (Sarec).

REFERENCES

Alongi, D. M., 1988, Bacterial productivity and microbial biomass in tropical mangrove sediments, *Microb. Ecol.* **15**:59–79.

Alongi, D. M., 1989, The fate of bacterial biomass and production in marine benthic food chains, in: *Recent Advances in Microbial Ecology* (T. Hattori, Y. Ishida, Y. Maruyama, R. Y. Morita, and A. Uchida, eds.), Japan Scientific Societies Press, Tokyo, pp. 355–359.

Azam, F., and Ammerman, J., 1984, Cycling of organic matter by bacterioplankton in pelagic marine ecosystems: Microenvironmental considerations, in: *Flows of Energy and Materials in Marine Ecosystems* (M. J. R. Fasham, ed.), Plenum Press, New York, pp. 345–360.

Azam, F., and Smith, D. C., 1991, Bacterial influence on the variability in the ocean's biogeochemical state: A mechanistic view, in: *Particle Analysis in Oceanography* (S. Demers, ed.), Springer-Verlag, Berlin, pp. 213–236.

Baines, S. B., and Pace, M. L., 1991, The production of dissolved organic matter by phytoplankton and its importance to bacteria: Patterns across marine and freshwater systems, *Limnol. Oceanogr.* **36**:1078–1090.

Bell, R. T., 1986, Thymidine incorporation as a measure of bacterial production in lakes, *Acta Universitatis Upsaliensis* 43 (Comprehensive Summaries of Uppsala Dissertations from the Faculty of Science).

Bell, R. T., 1993, Estimating production of heterotrophic bacterioplankton via incorporation of tritiated thymidine, in: *Handbook of Methods in Aquatic Microbial Ecology* (P. F. Kemp, B. F. Sherr, E. B. Sherr, and J. J. Cole, eds.), CRC Press, Boca Raton, FL, in press.

Bell, R. T., and Ahlgren, I., 1987, Thymidine incorporation and microbial respiration in the surface sediment of a hypereutrophic lake, *Limnol. Oceanogr.* **32**:476–482.

Bell, R. T., and Kuparinen, J., 1984, Assessing phytoplankton and bacterioplankton production during early spring in Lake Erken, Sweden, *Appl. Environ. Microbiol.* **48**:1221–1230.

Bell, R. T., and Tranvik, L., 1993, Impact of acidification and liming on microbial activity in lakes, *Ambio*, in press.

Bell, R. T., Erikson, R., Vammen, K., Vargas, M. H., and Zelaya, A., 1991, Heterotrophic bacterial production in Lake Xolotan (Managua) during 1988–1989, *Hydrobiol. Bull.* **25**: 145–149.

Bell, R. T., Stensdotter, U., Istvanovics, V., Pierson, D., and Pettersson, K., 1993a, Microbial dynamics and nutrient turnover in Lake Erken, Sweden, *J. Plank. Res.*, in press.

Bell, R. T., Vrede, K., Stensdotter, U., and Blomqvist, P., 1993b, Stimulation of the microbial food web in an oligotrophic slightly acidified lake, *Limnol. Oceanogr.*, in press.

Bell, W. H., and Mitchell, R., 1972, Chemotactic and growth responses of marine bacteria to algal extracellular products, *Biol. Bull.* **143**:265–277.

Bern, L., 1985, Autoradiographic studies on *methyl*-[³H]thymidine incorporation in a cyanobacterium (*Microcystis wesenbergii*)–bacterium association and in selected algae and bacteria, *Appl. Environ. Microbiol.* **49**:233–235.

Billen, G., 1990, Delayed development of bacterioplankton with respect to phytoplankton: A clue for understanding their trophic relationships, *Arch. Hydrobiol. Beih. Ergebn. Limnol.* **34**:191–201.

Billen, G., Servais, P., and Fontigny, A., 1988, Growth and mortality in bacterial population dynamics of aquatic environments, *Arch. Hydrobiol. Beih. Ergebn. Limnol.* **31**:173–183.

Bird, D. F., and Kalff, J., 1984, Empirical relationships between bacterial abundance and chlorophyll concentration in fresh and marine waters, *Can. J. Fish. Aquat. Sci.* **41**:1015–1023.

Boström, B., and Törnblom, E., 1990, Bacterial production, heat production, and ATP-turnover in shallow marine sediments, *Thermochim. Acta* **172**:147–156.

Bratbak, G., and Thingstad, T. F., 1985, Phytoplankton–bacteria interactions: An apparent paradox? Analysis of a model system with both competition and commensalism, *Mar. Ecol. Prog. Ser.* **63**:253–259.

Brock, T. D., and Clyne, J., 1984, Significance of algal excretion products for growth of epilimnetic bacteria. *Appl. Environ. Microbiol.* **47**:731–734.

Brunberg, A. K., and Bell, R. T., 1993, Contribution by the bacteria in the mucilage of the cyanobacterium *Microcystis* to benthic and pelagic bacterial production in a hypereutrophic lake, *Appl. Environ. Microbiol.*, submitted.

Brunberg, A. K., and Boström, B., 1992, Coupling between benthic biomass of *Microcystis* and phosphorus release from the sediments of a highly eutrophic lake, *Hydrobiologia* **235/236:** 375–385.

Caron, D. A., Goldman, J. C., and Dennet, M. R., 1988, Experimental demonstration of the roles of bacteria and bacterivorous protozoa in nutrient cycles, *Hydrobiologia* **159**:27–40.

Chin-Leo, G., and Kirchman, D. L., 1990, Unbalanced growth in natural assemblages of marine bacterioplankton, *Mar. Ecol. Prog. Ser.* **63**:1–8.

Cho, B. C., and Azam, F., 1988, Major role of bacteria in biogeochemical fluxes in the ocean's interior, *Nature* **332**:441–443.

Cho, B. C., and Azam, F., 1990, Biogeochemical significance of bacterial biomass in the ocean's euphotic zone, *Mar. Ecol. Prog. Ser.* **63**:253–259.

Chrost, R. J., 1989, Characterization and significance of β-glucosidase activity in lake water, *Limnol. Oceanogr.* **34**:660–672.

Chrzanowski, T. H., and Simek, K., 1990, Prey-size selection by freshwater flagellated protozoa, *Limnol. Oceanogr.* **35**:1429–1436.

Cole, J. J., 1982, Interactions between algae and bacteria in aquatic systems, *Annu. Rev. Ecol. Syst.* **13**:291–314.

Cole, J. J., Likens, G. E., and Strayer, D. L., 1982, Photosynthetically produced dissolved organic carbon: An important carbon source for planktonic bacteria, *Limnol. Oceanogr.* **27**:1080–1090.

Cole, J. J., Findlay, S., and Pace, M. L., 1988, Bacterial production in fresh and saltwater ecosystems: A cross-system overview, *Mar. Ecol. Prog. Ser.* **43**:1–10.

Coveney, M. F., 1982, Bacterial uptake of photosynthetic carbon from freshwater phytoplankton, *Oikos* **38**:8–20.

Coveney, M. F., and Wetzel, R. G., 1989, Bacterial metabolism of algal extracellular carbon, *Hydrobiologia* **173**:141–149.

Currie, D. J., 1990, Large-scale variability and interactions among phytoplankton, bacterioplankton and phosphorus, *Limnol. Oceanogr.* **35**:1437–1455.

Currie, D. J., and Kalff, J., 1984, A comparison of the abilities of freshwater algae and bacteria to acquire and retain phosphorus, *Limnol. Oceanogr.* **29**:298–310.

Currie, D. J., Bentzen, E., and Kalff, J., 1986, Does algal–bacterial phosphorus partitioning vary among lakes? A comparative study of orthophosphate activity in freshwater, *Can. J. Fish. Aquat. Sci.* **43**:311–318.

Ducklow, H. W., 1982, Chesapeake Bay nutrient and plankton dynamics. 1. Bacterial biomass and production during spring tidal destratification in the York River, Virginia, Estuary, *Limnol. Oceanogr.* **27**:651–659.

Ducklow, H. W., 1984, Geographical ecology of marine bacteria: Physical and biological variability at the mesoscale, in: *Current Perspectives in Microbial Ecology* (M. J. Klug and C. A. Reddy, eds.), American Society for Microbiology, Washington, D.C., pp. 22–31.

Fuhrman, J. A., and Azam, F., 1982, Thymidine incorporation as a measure of heterotrophic bacterioplankton production in marine surface waters: Evaluation and field results, *Mar. Biol.* **66:**109–120.

Fuhrman, J. A., and Ferguson, R. L., 1986, Nanomolar concentrations and rapid turnover of dissolved free amino acids in seawater, agreement between chemical and microbiological measurements, *Mar. Ecol. Prog. Ser.* **33:**327–342.

Fuhrman, J. A., Sleeter, T. D., Carlson, C. A., and Proctor, L. M., 1989, Dominance of bacterial biomass in the Sargasso Sea and its ecological implications, *Mar. Ecol. Prog. Ser.* **57:**207–217.

Gebre-Mariam, Z., and Taylor, W. T., 1989, Heterotrophic bacterioplankton production and grazing mortality rates in an Ethiopian rift-valley lake (Awassa), *Freshwater Biol.* **22:**369–381.

Godschalk, G. L., and Wetzel, R. G., 1977, Decomposition of macrophytes and the metabolism of organic matter in sediments, in: *Interactions between Sediment and Fresh Water* (H. R. Golterman, ed.), Junk, The Hague, pp. 258–264.

Güde, H., 1985, Influences of phagotrophic processes on the regeneration of nutrients in two-stage continuous culture systems, *Microb. Ecol.* **11:**193–204.

Güde, H., 1989, The role of grazing on bacteria in plankton succession, in: *Plankton Ecology: Succession in Plankton Communities* (U. Sommer, ed.), Springer-Verlag, Berlin, pp. 337–364.

Güde, H., 1990, Bacterial net production approaching zero—a frequent phenomenon in pelagic environments, *Arch. Hydrobiol. Beih. Ergebn. Limnol.* **34:**165–169.

Güde, H., 1991, Participation of bacterioplankton in epilimnetic phosphorus cycles of Lake Constance, *Int. Ver. Theor. Angew. Limnol. Verh.* **24:**16–20.

Hansen, J. A., Alongi, D. M., Moriarty, D. J. W., and Pollard, P. C., 1987, The dynamics of benthic microbial communities at Davies Reef, central Great Barrier Reef, *Coral Reefs* **6:**63–70.

Hecky, R. E., and Kling, H. J., 1981, The phytoplankton and protozooplankton of the euphotic zone of Lake Tanganyika: Species composition, biomass, chlorophyll content and spatiotemporal distribution, *Limnol. Oceanogr.* **26:**548–564.

Hellebust, J. A., 1974, Extracellular products, in: *Algal Physiology and Biochemistry* (W. D. P. Stewart, ed.), Blackwell, Oxford, pp. 838–863.

Herbert, R. A., and Bell, C. R., 1977, Growth characteristics of an obligately psychrophilic *Vibrio* sp., *Arch. Microbiol.* **113:**215–220.

Hessen, D. O., 1985, The relation between bacterial carbon and dissolved humic compounds in oligotrophic lakes, *FEMS Microbiol. Ecol.* **31:**215–223.

Hobbie, J. E., 1988, A comparison of the ecology of planktonic bacteria in fresh and salt water, *Limnol. Oceanogr.* **33:**750–764.

Hobbie, J. E., 1992, Microbial control of dissolved organic carbon in lakes: Research for the future, *Hydrobiologia* **229:**169–180.

Hobbie, J. E., and Cole, J. J., 1984, Response of a detrital foodweb to eutrophication, *Bull. Mar. Sci* **35:**357–363.

Hoppe, H. G., Kim, S. J., and Gocke, K., 1988, Microbial decomposition in aquatic environments: Combined process of extracellular enzyme activity and substrate uptake, *Appl. Environ. Microbiol.* **54:**784–790.

Istvanovics, V., Pettersson, K., Pierson, D., and Bell, R. T., 1992, An evaluation of phosphorus deficiency indicators for summer phytoplankton in Lake Erken, *Limnol. Oceanogr.,* **37:** 890–900.

Joint, I. R., and Pomroy, A. J., 1987, Activity of heterotrophic bacteria in the euphotic zone of the Celtic Sea, *Mar. Ecol. Prog. Ser.* **41:**155–165.

Jones, J. G., 1989, Bacterial populations in freshwater sediments: Factors affecting growth and their ultimate fate, in: *Recent Advances in Microbial Ecology* (T. Hattori, Y. Ishida, Y. Maruyama, R. Y. Morita, and A. Uchida, eds.), Scientific Societies Publ., Japan, pp. 343–348.

Jürgens, K., and Güde, H., 1990, Incorporation and release of phosphorus by planktonic bacteria and phagotrophic flagellates, *Mar. Ecol. Prog. Ser.* **59:**271–284.

Karl, D. M., Knauer, G. A., and Martin, J. H., 1988, Downward flux of particulate organic matter in the ocean: A particle paradox. *Nature* **332:**438–441.

Kirchman, D., 1983, The production of bacteria attached to particles suspended in a freshwater pond, *Limnol. Oceanogr.* **28:**858–872.

Kirchman, D. L., 1990, Limitation of bacterial growth by dissolved organic matter in the subarctic Pacific, *Mar. Ecol. Prog. Ser.* **62:**47–54.

Kirchman, D. L., 1993, Particulate detritus and bacteria in marine environments, in: *Aquatic Microbiology: An Ecological Approach* (T. Ford, ed.), Blackwell, Oxford, in press.

Kirchman, D. L., Suzuki, S., Garside, C., and Ducklow, H. W., 1991, High turnover rates of dissolved organic carbon during a spring phytoplankton bloom, *Nature* **352:**612–614.

Kjelleberg, S., Flärdh, K., Nyström, T., and Moriarty, D. J. W., 1993, Growth limitation and starvation of bacteria, in: *Aquatic Microbiology: an Ecological Approach* (T. Ford, ed.), Blackwell, Oxford, in press.

Lampert, W., 1978, Release of dissolved organic carbon by grazing zooplankton, *Limnol. Oceanogr.* **23:**195–198.

Lancelot, C., and Billen, G., 1984, Activity of heterotrophic bacteria and its coupling to primary production during the spring phytoplankton bloom in the southern bight of the North Sea, *Limnol. Oceanogr.* **29:**721–730.

Lancelot, C., and Billen, G., 1985, Carbon–nitrogen relationships in nutrient metabolism of coastal marine ecosystems, *Adv. Aquat. Microbiol.* **3:**263–321.

Larsson, U., and Hagström, Å., 1982, Fractionated primary production, exudate release, and bacterial production in a Baltic eutrophication gradient, *Mar. Biol.* **67:**57–70.

Laybourn-Parry, J., 1992, *Protozoan Plankton Ecology*, Chapman & Hall, London.

Lee, S., and Fuhrman, J. A., 1990, DNA hybridization to compare species compositions of natural bacterioplankton assemblages, *Appl. Environ. Microbiol.* **56:**739–746.

Lewis, D. L., and Gattie, D. K., 1990, Effects of cellular aggregation on the ecology of micro-organisms, *ASM News* **56:**263–268.

Mården, P., Hermansson, M., and Kjelleberg, S., 1988, Incorporation of tritiated thymidine by marine bacterial isolates when undergoing a starvation survival response, *Arch. Microbiol.* **149:**427–432.

Moriarty, D. J. W., 1986, Measurement of bacterial growth rates in aquatic systems from rates of nucleic acid synthesis, *Adv. Microb. Ecol.* **9:**245–292.

Moriarty, D. J. W., 1989, Relationships of bacterial biomass and production to primary production in marine sediments, in: *Recent Advances in Microbial Ecology* (T. Hattori, Y. Ishida, Y. Maruyama, R. Y. Morita, and A. Uchida, eds.), Japan Scientific Societies Press, Tokyo, pp. 349–354.

Moriarty, D. J. W., and Hansen, J. A., 1990, Productivity and growth rates of coral reef bacteria on hard calcareous substrates and in sandy sediments in summer, *Aust. J. Mar. Freshwater Res.* **41:**785–794.

Moriarty, D. J. W., and Pollard, P. C., 1982, Diel variation of bacterial productivity in seagrass *Zostera capricorni*, beds measured by rate of thymidine incorporation into DNA, *Mar. Biol.* **72:**165–173.

Moriarty, D. J. W., Pollard, P. C., Hunt, W. G., Moriarty, C. M., and Wassenberg, T. J., 1985, Productivity of bacteria and microalgae and the effect of grazing by holothurians in sediments on a coral reef flat, *Mar. Biol.* **85:**293–300.

Moriarty, D. J. W., Roberts, D. G., and Pollard, P. C. 1990, Primary and bacterial productivity of tropical seagrass communities in the Gulf of Carpentaria, Australia, *Mar. Ecol. Prog. Ser.* **61:** 145–157.

Moriarty, D. J. W., Skyring, G. W., O'Brien, G. W., and Heggie, D. T., 1991, Heterotrophic bacterial activity and growth rates in sediments of the continental margin of eastern Australia, *Deep-Sea Res.* **38:**693–712.

Morita, R. Y., 1982, Starvation-survival of heterotrophs in the marine environment, *Adv. Microb. Ecol.* **6:**117–198.

Morris, D. P., and Lewis, W. M., Jr., 1992, Nutrient limitation of bacterioplankton growth in Lake Dillon, Colorado, *Limnol. Oceanogr.* **37:**1179–1192.

Pace, N. R., Stahl, D. A., Lane, D. J., and Olsen, G. L., 1986, The analysis of natural microbial communities by ribosomal RNA sequences, *Adv. Microb. Ecol.* **9:**1–55.

Paerl, H. W., 1978, Microbial organic carbon recovery in aquatic systems, *Limnol. Oceanogr.* **23:**927–935.

Painting, S. J., Lucas, M. I., and Muir, D. G., 1989, Fluctuations in heterotrophic bacterial community structure, activity and production in response to development and decay of phytoplankton in a microcosm, *Mar. Ecol. Prog. Ser.* **53:**129–141.

Pedros-Alio, C., 1989, Toward an autecology of bacterioplankton, in: *Plankton Ecology: Succession in Plankton Communities* (U. Sommer, ed.), Springer-Verlag, Berlin, pp. 297–336.

Pedros-Alio, C., and Guerrero, R., 1991, Abundance and activity of bacteria in warm lakes, *Int. Ver. Theor. Agnew. Limnol. Verh.* **24:**1212–1219.

Pierson, D., 1991, Effects of vertical mixing on phytoplankton photosynthesis and phosphorus deficiency, Uppsala University, Dept. of Hydrology, Report Series A, No. 51

Pollard, P. C., and Moriarty, D. J. W., 1991, Organic carbon decomposition, primary and bacterial productivity, and sulphate reduction, in tropical seagrass beds of the Gulf of Carpentaria, Australia, *Mar. Ecol. Prog. Ser.* **69:**149–159.

Rai, H., 1979, Microbiology of Central Amazon lakes, *Amazoniana* **6:**583–599.

Robarts, R. D., and Wicks, R. J., 1990, Heterotrophic bacterial production and its dependence on autotrophic production in a hypereutrophic African reservoir, *Can. J. Fish. Aquat. Sci.* **47:** 1027–1037.

Sandgren, C. D., 1988, The ecology of chrysophyte flagellates: Their growth and perennation strategies as freshwater phytoplankton, in: *Growth and Reproductive Strategies of Freshwater Phytoplankton* (C. D. Sandgren, ed.), Cambridge University Press, London, pp. 9–104.

Scavia, D., and Laird, G. A., 1987, Bacterioplankton in Lake Michigan: Dynamics, controls, and significance to carbon flux, *Limnol. Oceanogr.* **32:**1017–1033.

Schindler, D. W., 1977, The evolution of phosphorus limitation in lakes, *Science* **195:**260–262.

Sharp, J. H., 1981, Inputs into microbial food chains, in: *Heterotrophic Activity in the Sea* (J. E. Hobbie and P. J. Williams, eds.), Plenum, New York, pp. 101–120.

Simek, K., 1986, Bacterial activity in a reservoir determined by autoradiography and its relation to phyto- and zooplankton, *Int. Rev. Gesamten Hydrobiol.* **71:**593–612.

Simon, M., 1985, Specific uptake rates of amino acids by attached and free-living bacteria in a mesotrophic lake, *Appl. Environ. Microbiol.* **49:**1254–1259.

Simon, M., Alldredge, A. L., and Azam, F., 1990, Bacterial carbon dynamics on marine snow, *Mar. Ecol. Prog. Ser.* **65:**205–211.

Sochard, M. R., Wilson, D. F., Austin, B., and Colwell, R. R., 1979, Bacteria associated with the surface and gut of marine copepods, *Appl. Environ. Microb.* **37:**750–759.

Sommer, U. (ed.), 1989, *Plankton Ecology: Succession in Plankton Communities*, Springer-Verlag, Berlin.

Sorokin, Y. I., 1972, Data on biological productivity in the western Pacific ocean, *Mar. Biol.* **20:**177–196.

Sorokin, Y. I., 1978, Decomposition of organic matter and nutrient regeneration, in: *Marine Ecology, Vol. IV* (O. Kinne, ed.), Wiley, New York, pp. 501–516.

Sorokin, Y. I., Kopylov, A. I., and Mamaeva, N. V., 1985, Abundance and dynamics of microplankton in the central tropical Indian Ocean, *Mar. Ecol. Prog. Ser.* **24**:27–41.

Sundh, I., 1991, The dissolved organic carbon released from phytoplankton: Biochemical composition and bacterial utilization, *Acta Universitatis Upsaliensis* 324 (Comprehensive Summaries of Uppsala Dissertations from the Faculty of Science).

Sundh, I., and Bell, R. T., 1992, Extracellular organic carbon released from phytoplankton as a source of carbon for heterotrophic bacteria in lakes of different humic content, *Hydrobiologia* **229**:9.

Tabor, P. S., and Neihof, R. A., 1983, Improved microautoradiographic method to determine individual microorganisms active in substrate uptake in natural waters, *Appl. Environ. Microbiol.* **44**:945–953.

Thiel, H., Pfannkuche, O., Schriever, G., Lochte, K., Gooday, A. J., Hemleben, C., Mantoura, R. F. C., Turley, C. M., Patching, J. W., and Riemann, F., 1989, Phytodetritus on the deep-sea floor in a central region of the northeast Atlantic, *Biol. Oceanogr.* **6**:203–239.

Tilzer, M., 1990, Environmental and physiological control of phytoplankton productivity in large lakes, in: *Large Lakes: Ecological Structure and Function* (M. Tilzer and C. Serruya, eds.), Springer-Verlag, Berlin, pp. 339–367.

Tranvik, L., 1988, Availability of dissolved organic carbon for planktonic bacteria in oligotrophic lakes of different humic content, *Microb. Ecol.* **16**:311–322.

Tranvik, L., 1989, Bacterioplankton growth, grazing mortality and quantitative relationship to primary production in a humic and a clearwater lake, *J. Plankton Res.* **11**:985–1000.

Tranvik, L., and Hölfe, M. G., 1987, Bacterial growth in mixed cultures on dissolved organic carbon from humic and clear waters, *Appl. Environ. Microbiol.* **53**:482–488.

Tranvik, L., and Sieburth, J. M. N., 1989, Effects of flocculated humic matter on free and attached pelagic microorganisms, *Limnol. Oceanogr.* **34**:688–699.

Vadstein, O., and Olsen, Y., 1989, Chemical composition and phosphate uptake kinetics of limnetic bacterial communities cultured in chemostats under phosphorus limitation, *Limnol. Oceanogr.* **34**:939–946.

Vadstein, O., Jensen, A., Olsen, Y., and Reinersten, H., 1988, Growth and phosphorus status of limnetic phytoplankton and bacteria, *Limnol. Oceanogr.* **33**:489–503.

Van Duyl, F. C., and Kop, A. J., 1988, Temporal and lateral fluctuations in production and biomass of bacterioplankton in the western Dutch Wadden Sea, *Neth. J. Sea Res.* **22**:51–68.

Van Duyl, F. C., Bak, R. P. M., Kop, A. J., and Nieuwland, G., 1990, Bacteria, auto- and heterotrophic nanoflagellates, and their relations in mixed, frontal and stratified waters of the North Sea, *Neth. J. Sea Res.* **26**:97–109.

Wetzel, R. G., 1992, Gradient-dominated ecosystems: Sources and regulatory functions of dissolved organic matter in freshwater ecosystems, *Hydrobiologia* **229**:181–198.

Wiebe, W. J., Sheldon, W. M., and Pomeroy, L. R., 1992, Bacterial growth in the cold: Evidence for an enhanced substrate requirement, *Appl. Environ. Microbiol.* **58**:359–364.

Wright, R. T., 1988, Methods for evaluating the interaction of substrate and grazing as factors controlling planktonic bacteria, *Arch. Hydrobiol. Beih. Ergebn. Limnol.* **31**:229–242.

Wright, R. T., and Coffin, R. B., 1984, Factors affecting bacterioplankton density and productivity in salt marsh estuaries, in: *Current Perspectives in Microbial Ecology* (M. J. Klug and C. A. Reddy, eds.), American Society for Microbiology, Washington, D.C., pp. 485–494.

3

Bacterial Responses to Soil Stimuli

J. D. van Elsas and L. S. van Overbeek

1. INTRODUCTION

Bacteria in the environment are subjected to many different stress factors such as nutrient, oxygen, or water limitations, temperature and pH extremes, UV irradiation, etc., which affect their physiological states. In particular, nutrient limitation and fluctuating nutrient availability are major stress factors in an environment such as soil. Hence, bacterial cells in soil may experience long periods of nongrowth next to sparse periods of growth. In fact, nongrowth may be the rule rather than the exception for cells in soil (Matin *et al.*, 1989).

Since bacterial cells in soil face nutrient and other stresses, which fluctuate in time, soil bacteria probably developed varying survival strategies enabling them to cope with such conditions. One such strategy obviously is the capacity of certain soil bacteria, e.g., bacilli, clostridia, and azospirilli, to form respectively spores or cysts, resistant survival forms with minimal metabolic activity. However, a substantial part of the soil bacteria, e.g., the soil pseudomonads, are not capable of producing such morphologically differentiated forms. These bacteria might have capabilities to cope with soil stresses which are akin to the stress responses of the well-studied marine *Vibrio* spp. or *Escherichia coli* (Groat and Matin, 1986; Tormo *et al.*, 1990; Östling *et al.*, 1991; Matin, 1992).

Recently, primarily the response to starvation has begun to become unraveled in *E. coli* and *Vibrio* spp. in physiological and molecular terms (e.g., Matin *et al.*, 1989; Matin, 1992; Östling *et al.*, this volume). Cellular physiology upon starvation is drastically affected in that many new proteins appear,

J. D. van Elsas and L. S. van Overbeek • Institute for Soil Fertility Research, 6700AA Wageningen, The Netherlands.

Starvation in Bacteria, edited by Staffan Kjelleberg. Plenum Press, New York, 1993.

and the cells enter a more stress-resistant form. Several key genes, e.g., $dnaK$, $katF$, have been identified which play a role in the starvation response (Lange and Hengge-Aronis, 1991a; Matin, 1992). Some gene products, such as DnaK and GroEL, were originally identified as heat shock proteins, suggesting overlaps between bacterial responses to different stress factors (Matin, 1992). Further, the bacterial response to both starvation and heat stress was revealed to be complex in terms of the number and relative abundance of new proteins induced. The shift from exponential growth to the stationary phase, in which starvation for essential nutrients takes place, induced the new proteins (Groat and Matin, 1986; Matin, 1992). Whereas some of the new proteins were shared between different stress factors, others were unique for each factor (Östling *et al.*, 1991). A detailed account of these data can be found elsewhere in this book (see chapters by Hengge-Aronis; Nyström, Östling *et al.*, this volume).

It is tempting to speculate that the starvation survival response found in *E. coli* and *Vibrio* is representative for the physiological processes bacteria undergo when present in or entering the soil environment. Importance differences, however, are likely to be found. First, the studies leading to the identification of the genes and gene products have been performed in laboratory systems, i.e., in well-mixed largely homogeneous aqueous phases which *a priori* are not representative for soil. Second, in addition to nutrient stress, soil may affect bacteria via other stress factors. Moreover, data on starvation survival so far have been obtained with organisms not indigenous in soil.

This chapter will discuss current knowledge on the response of soil bacteria to various stress factors. After describing the soil environment and the indigenous soil bacteria according to their most striking characteristics, the occurrence and putative effect of several stress and other stimuli will be treated. Description of the bacterial responses in a somewhat speculative manner has been unavoidable because of the current lack of detailed information on *in situ* stress in soil and responses of soil bacteria. Current approaches aimed at unraveling the molecular biology of responses to stress by soil bacteria, including some recent results obtained with soil pseudomonads, are also presented.

2. THE SOIL ENVIRONMENT

2.1. Observations on Bacterial Populations in Soil

Soil represents a highly structured environment, which is composed of solid, liquid, and gaseous phases. The solid phase dominates the system and is relatively static, as opposed to the dynamic conditions in the liquid and gaseous phases. All three phases are heterogeneous in terms of the distribution of different compounds within them (Kilbertus, 1980; Smiles, 1988). The components of the solid phase, i.e., inorganic substances such as clay, silt, and sand, and organic matter (humic substances), are distributed unevenly (van Veen and van Elsas, 1986). Moreover, complexes of these compounds are often present. These varying complexes of inorganic and/or organic matter form the so-called soil aggregates, with sizes in the order of micrometers to millimeters. Aggregates important for soil microorganisms are the clay–organic matter complexes, because of their charged surfaces and increased nutrient availability (Hattori and Hattori, 1976; Smiles, 1988). The solid phase is interspersed with the soil pore network (soil void), in which the liquid and gaseous phases are present. The relative abundance of pores with different neck sizes, governed by soil structure and texture, determines the water retention capacity of soil, and thereby to a large extent the water availability for soil microbes. The intricacies of the soil environment as a habitat for bacteria will not be treated in detail here; relevant reviews can be found in Foster (1988) and Smiles (1988).

Bacterial inhabitants of soil are located in soil pores of varying sizes and compositions. To analyze the effects of soil factors on bacterial life, conditions should be judged at the level of these soil pores, rather than at the overall soil level (Smiles, 1988). However, even though analytical methods have progressed in recent years, it is still difficult to describe soil in terms of the conditions at the pore level (Foster, 1988). These considerations show that soil is an extremely difficult environment to simulate; a corollary of this is the difficulty of meaningfully describing the responses to stimuli realistic for soil.

To simplify the concept of stimuli in soil, soil can be divided into two compartments: soil not directly influenced by plant roots (bulk soil), and soil directly affected by roots (rhizosphere soil). This separation permits the description, in simple factorial terms, of responses to simulated stimuli. The most striking difference between the two compartments is the flux and availability of organic carbon. Even though the overall carbon content of bulk soil can be considerable, most of the carbon is present in a recalcitrant form and soil can be regarded as an oligotrophic environment, i.e., an environment relatively poor in readily available organic carbon (Williams, 1985). The recalcitrant compounds, e.g., different humics, lignins, etc., may also complex available compounds, as argued by Morita (this volume). The overall soluble organic carbon content of four soils was between 70 and 200 μg/ml soil water in soil at field capacity, and, typically, a loamy sand soil contained 100 μg/ml soil water (van Ginkel, personal communication). Although this may seem still rather high, much of this carbon may be in a recalcitrant form, and localized in soil sites hard to reach for soil bacteria. The resulting low amount of available carbon in soil generally precludes bacterial growth. For instance, Gray and Williams (1971) and Shields *et al.* (1973) estimated the energy input into two soils overall to be sufficient for just a few cellular divisions of the soil microorganisms per year, assuming part of the energy was needed for maintenance. [See also Morita (this volume) for information on generation times in soil.] Nutrients may, however, become available locally, e.g., in decaying material of plant or animal origin, and trigger microbial growth and activity. Therefore, organic carbon may sometimes be locally sufficient for bacterial growth, if other key components (e.g., N, P, Fe, and micronutrients) are present. Bulk soil can thus be best described as an environment in which nutrients are very sparsely and infrequently available at discrete sites, so-called "hot spots"; bacterial growth in these hot spots might be empherally comparable with that in batch culture.

Plant roots represent one of the major sites of carbon input into soil. Both soluble compounds like root exudates and insoluble ones like remnants of root cortex cells are released in the rhizosphere. Between 5 and 50% of carbon of the standing root may end up in the rhizosphere in the form of rhizodeposits (Newman, 1985; Trofymow *et al.*, 1987). Thus, for a typical cereal root of 100 m/g dry weight (Atkinson, 1990) with a rhizosphere radius of 1 mm and a carbon concentration of 40%, between roughly 0.3 and 3 mg carbon/ml soil water might be released in a loamy sand soil at bulk density 1.3 and at field capacity (18% moisture content). Therefore, the ratio between the amount of soluble organic carbon in the rhizosphere and bulk soil might be roughly between 3 and 30. This carbon released from roots is also likely to be more available for soil bacteria than that present in the bulk soil because of its lower degree of recalcitrance. Most of the root-released carbon is liberated near the root tip during initial root development (Trofymow *et al.*, 1987). Thus, the rhizosphere can be characterized as a region in soil with a transient high availability of carbon; later, structural cellular material provides a more recalcitrant carbon source. Nutrient influx at a certain site in the rhizosphere thus changes in time because of root growth; the rate of release and quality of the compounds released depend on the growth stage of the plant (Keith *et al.*, 1986; Gregory and Atwell, 1991), on the soil water content (Martin, 1977), and on environmental conditions (Kraffczyck *et al.*, 1984). Rhizosphere bacteria often show increased growth and activity because of the increased availability of organic carbon in the rhizosphere (Foster, 1988). However, even in the relatively nutrient-rich rhizosphere, bacterial growth and activity are generally limited. For more details on the rhizosphere as a habitat for soil bacteria, the reader is referred to Curl and Truelove (1986) and Lynch and Whipps (1990).

2.1.1. Indigenous Populations

Bacterial populations in soil have been studied using both plating and direct microscopical observation techniques (e.g., Bae *et al.*, 1972; Alexander, 1977). Most indigenous bacteria in soil are closely associated with the soil solid phase, i.e., they are adsorbed to soil particles (Hattori and Hattori, 1976). Often, the cells are preferentially adsorbed to the organic or organic/clay complex. For instance, Hissett and Gray (1976) found that 64% of the bacteria in a sandy soil were associated with the organic fraction. Physical space in the soil pore network is probably not a limiting factor to soil microbes; using electron microscopy, Paul and Clark (1989) showed that only a minor fraction (1%) of the soil pore space was occupied by bacteria. On the basis of simple calculations, Postma and van Veen (1990) reached a similar conclusion.

Classically, bacteria in soil, at least those of the culturable fraction, can be divided into oligotrophic and copiotrophic forms (Poindexter, 1981). Oligotrophic bacteria such as *Arthrobacter* and *Caulobacter* spp. reveal low growth rates even under conditions of excess nutrients, and show physiological as well as morphological adaptations, like reduction of cell size, to reduced nutrient supply. Striking physiological characteristics are their capacity to rapidly accumulate reserve materials in periods of nutrient excess and to utilize this upon starvation. Further, they are able to reduce their endogenous metabolism to an extremely low level. It has long been assumed that oligotrophic bacteria, besides resting forms such as *Bacillus* spores, were the predominant bacterial inhabitants of bulk soil (Alexander, 1977; Poindexter, 1981). On the other hand, copiotrophic bacteria such as pseudomonads are geared to respond rapidly to increased availability of organic carbon, showing high growth rates and reduced storage of reserve materials. Bacteria with typical copiotrophic characteristics such as *Pseudomonas* spp. have been suggested to generally dominate bacterial populations in the rhizosphere. However, recently this concept has been challenged, and the occurrence of a higher percentage of oligotrophic bacteria like coryneforms has been shown by the use of more dilute media and lower incubation temperatures (Nijhuis *et al.*, 1993). For instance, a great proportion of bacteria isolated from the grass rhizosphere were coryneforms, which are classically believed to be oligotrophic in nature (Nijhuis *et al.*, 1993). Moreover, the concept of dividing bacteria in oligotrophic and copiotrophic forms may be too strict to describe bacteria in soil; bacteria may exist which have characteristics of both forms, being able to cope with both high and low nutrient availability (Morita, this volume).

A striking characteristic of indigenous bacteria in soil is the lack of culturability of the majority of cells; only a minor percentage of the soil bacteria observable via direct microscopy are generally capable of growth on common laboratory media. The use of dilute media and incubation temperatures and times realistic for soil have not led to a drastically enhanced culturability of these cells (Bakken, personal communication). This may indicate either that this majority is really silent or that the proper conditions for their culturing have not yet been found, possibly because of substrate toxicity. Using transmission electron microscopy, Bae *et al.* (1972) revealed that a major proportion of the bacterial cells in soil appeared as "dwarf" forms (ultramicrocells; Morita, this volume) with diameters below 0.3 μm, much like the *Vibrio* cells obtained after starvation-induced fragmentation (Novitsky and Morita, 1976; Morita, 1988). Besides spores and cysts, which amounted to 27% of forms observed, cellular forms were predominantly present. About 25% of the cells had a periplasmic space. Bakken and Olsen (1989) further showed that dwarfs with diameters below 0.4 μm had a very low percentage (0.2%) of culturable cells, whereas a much greater percentage (40%) of cells with diameters > 0.6 μm were culturable. However, the question of initial substrate toxicity was not addressed, since only full-strength medium was used. Dwarfs, as evidenced by extraction and detection of DNA via staining with Hoechst dye 33258, contained between 1.6 and 2.4 fg DNA per cell, indicating that the majority of them possessed at least 1 genome equivalent. Dwarfs in soil thus may represent a sporelike or "somnicell" stage of non-spore-forming bacteria, as proposed by Roszak and Colwell (1987). In this state, cells may be able to survive longer maintaining an extremely low metabolic rate. Three hypotheses might be put forward on the nature of dwarfs:

1. Dwarfs are minute forms of "common" normal-sized bacteria which as a response to nutrient starvation reduced their cell size and metabolic rate, and became efficient nutrient scavengers.
2. Dwarfs represent taxonomically quite different group(s) of bacteria, which are small-sized as a rule, since they are exclusively tuned to low nutrient conditions.
3. Dwarfs are moribund forms which originated from one of these two cellular forms; they are nonviable, nonrevivable, and possibly escaped predation or biodegradation because of physical protection.

Microbiological studies on dwarfs have been hampered by the difficulties of culturing them. They seem to occur abundantly in seawater and freshwater environments (Gottschal, 1992), and most data on them have been obtained in these environments. Pioneering studies of Torrella and Morita (1981) on dwarfs of seawater revealed that dwarfs may be a heterogeneous group encompassing all three possibilities. Although dwarfs smaller than 0.2 μm were missing in this study, it showed that some of the small cells were minute forms of common bacteria and some were possibly "authentic" oligotrophic dwarfs. The nature of the third, largest group remained unknown because of the lack of culturability. Finally, a dwarf strain obtained from oligotrophic lake water was culturable in very dilute media (0.5 mg trypticase and 0.05 mg yeast extract per liter lake water) maintaining its small cell size (0.7 μm long and 0.15 μm wide), showing that indeed part of the dwarf population in nature might be genuine small-sized viable oligotrophic bacteria (Ishida and Kadota, 1981). It is unknown to what extent the physiological state of soil dwarfs resembles that of the dwarfs and/or viable but nonculturable forms in seawater, i.e., whether they are viable, moribund, or dead. Another question pertains to the role of dwarfs in the ecosystem. Assuming that part of the dwarf population, the supposed oligotrophs extremely apt at scavenging scarce nutrients, are viable and metabolically active, nutrient cycling in nature, viz. soil, may take largely place through dwarfs. Recent evidence obtained with bacteria in the ocean's interior suggested that most of the carbon flux in this system was through forms of less than 0.1 μm^3 in volume, i.e., below 0.6 μm for cocci (Cho and Azam, 1988); 98% of the bacteria passed through a 1-μm filter. In soil, dwarfs also often show a numerical predominance over other bacterial forms. However, information is lacking on their functional prevalence given the known aptitude of many of the other bacteria to quickly respond to nutrients. Assuming bacterial cells in bulk soil are generally in a state of nutrient deprivation, both the dwarfs and other oligotrophic forms might be functionally predominant in processes under these generic conditions. However, in rhizosphere soil, a larger proportion of bacterial cells may be in a state of enhanced activity, as suggested by the greater percentage of bigger cells detected via electron microscopy (Foster, 1988). Nevertheless, dwarfs were also detected in the rhizosphere. Under these conditions, nondwarfs might be functionally more predominant.

2.1.2. Introduced Populations

The generally observed decline of bacterial numbers following introduction into soil has often hampered the effectiveness of bacterial inoculants. In particular, introduced fluorescent pseudomonads have been shown to be poor survivors. For instance, Tn5-tagged *Pseudomonas fluorescens* populations introduced into a loamy sand soil showed a progressive decline (Fig. 1; van Elsas *et al.*, 1991c), as evidenced from both cfu and immunofluorescence counts. Samples taken after prolonged incubation showed progressively more cells detectable via specific immunofluorescence than via a selective plate count, suggesting a possible conversion of part of the population into nonculturable forms. A comparison of the dynamics of different fluorescent pseudomonads in different soils (Table 1) further showed that all introduced populations decayed to low numbers in time spans between several weeks to months. Overall decay rates varied from ca. 0.2 to 1.1 (Table 1), and depended on the strain and on the soil used. Similar responses have often been found for other introduced bacteria such as *Salmonella typhimurium*, *Klebsiella pneumoniae* (Liang *et al.*, 1982), *Flavobacterium*, and *Alcaligenes* spp. (Nijhuis *et al.*, 1993; Thompson *et al.*, 1990). They might be characteristic for these copiotrophic organisms when exposed to soil.

It is difficult to pinpoint a single reason for the decline of the introduced

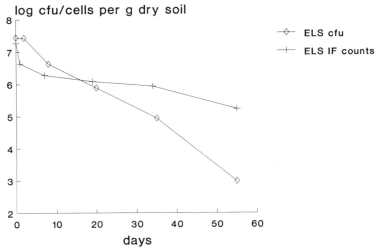

Figure 1. Total specific counts (immunofluorescence technique) and selective plate counts of *Pseudomonas fluorescens* introduced into soil microcosms. ELS, Ede loamy sand. Soil was kept at 15°C and 18% moisture content. Variation (S.D.) was within symbol dimensions.

cell numbers. Introduced bacteria, particularly shortly after release, are likely to be affected by the same adverse soil conditions mentioned for the indigenous bacteria. Given the relative immobility of bacterial cells in soil, the localization of introduced bacteria following introduction affects their ultimate fate, i.e., whether they are successful in colonizing soil and developing a cellular form resistant to soil stress or whether they will die out. Bacteria, after introduction into soil, are probably localized in more open spaces than the indigenous population, to a greater extent being subjected to soil factors. The local conditions for the individual cells determine the type of stress they endure. One major stress factor in soil, predation by protozoa (Habte and Alexander, 1975; Heijnen *et al.*, 1988), was suggested to be diminished via the enhanced localization of bacteria in soil pores with neck diameters below 3–6 μm, where protozoa do not easily enter due to their cellular dimensions (van Elsas *et al.*, 1991c). Other factors such as substrate and water availability and pH are also locally determined, but cannot be easily assessed because of soil heterogeneity.

The physiological responses of laboratory-grown bacteria exposed to the stresses of soil are likely to be manifold, depending on the prevailing conditions at the sites in soil where each cell is localized. It is difficult to study these because of the heterogeneous nature of soil. For instance, studies performed in soil extracts at best will provide a rough overall picture of cellular physiological responses. Nevertheless, it is this type of approach which may shed light on the

Table 1

Decay Rates of Different Fluorescent Pseudomonads Introduced into Soils in Soil Microcosms or in Field Microplots[a]

Soil	Experimental system	Introduced strain/marker	Decay rate[b]	Period (days)	Reference
Loamy sand	Field microplot	*P. fluorescens* (chr::Tn5)[c]	0.5–0.6	60	Van Elsas *et al.* (1986)
Silt loam	Field microplot	*P. fluorescens* (chr::Tn5)	0.3	60	Van Elsas *et al.* (1986)
Loamy sand (wheat rhizosphere)	Field microplot	*P. fluorescens* (chr::Tn5)	0.8	60	Van Elsas *et al.* (1986)
Silt loam (wheat rhizosphere)	Field microplot	*P. fluorescens* (chr::Tn5)	0.2	60	Van Elsas *et al.* (1986)
Loamy sand	Microcosm	*P. fluorescens* (chr::Tn5)	0.8	55	Heijnen *et al.* (1993)
L. sand + bentonite	Microcosm	*P. fluorescens* (chr::Tn5)	0.2	55	Heijnen *et al.* (1993)
Loamy sand	Microcosm	*P. fluorescens* (RP4)	0.9	60	Van Elsas and Trevors (1990)
L. sand + bentonite	Microcosm	*P. fluorescens* (RP4)	0.4	60	Van Elsas and Trevors (1990)
Sandy loam	Microcosm	*P. fluorescens* Pfl-2 Rp^R	0.2–0.4	36	Compeau *et al.* (1988)
Sandy loam	Microcosm	*P. fluorescens* Pfl-8 Rp^R	0.8	30	Compeau *et al.* (1988)
Silt loam	Microcosm	*P. fluorescens* R1 Rp^R	1.1	29	Wessendorf and Lingens (1989)
Sandy loam	Microcosm	*P. putida* Pp1-2	0.7	30	Compeau *et al.* (1988)
Clay loam	Field microplot	*P. putida* N-1R Rp^R	0.5	60	Dupler and Baker (1984)
Sandy loam	Field microplot	*P. putida* N-1R Rp^R	0.5	60	Dupler and Baker (1984)
Silty clay loam	Microcosm	*P. aeruginosa*	0.8	49	Zechman and Casida (1982)

[a]Initial cell numbers added were on the order of 10^7/g soil in most cases. Cells were added from washed fresh cultures without using carrier materials.
[b]Defined as the overall log decline in cfu counts per 10 days (calculated over the experimental period).
[c]chr::Tn5, chromosomal insertion of transposon Tn5 encoding kanamycin resistance; RP4, plasmid encoding kanamycin, tetracycline, and ampicillin resistance; Rp^R, resistant to rifampicin.

importance of global regulatory responses for bacteria exposed to soil. Moreover, the approach will also lead to possibilities for application of stress-induced genes for environmental engineering purposes.

3. STIMULI IN THE SOIL ENVIRONMENT

Bacteria located in soil or entering this environment, experience the conditions present at the level of their localization. Since bacteria commonly reside in soil in association with soil particles inside pores, the local stimuli are of obvious relevance to bacterial physiology and fate at the level of the individual cell. Because of stress or other stimuli in soil at the local level, specific proteins induced in bacterial cells change cellular physiology, possibly resulting in adaptation to stress. Specific promoters in the bacterial genome are likely to be involved. Adaptation of the bacterial cell to stress conditions at the pore level might enable it to subsequently colonize this niche given favorable conditions.

Whereas temperature may be regarded as characteristic over a large section of soil, other overall soil characteristics such as nutrient levels, water availability, pH, and the presence of specific compounds able to trigger responses, are only of relative importance for conditions at the level of individual cells, since local conditions may differ. Unfortunately, it is impossible to take full account of soil heterogeneity when considering triggers of bacterial physiological responses. Therefore, overall factors of prime importance for bacterial life in soil will be treated here. Although a strict division over the two soil compartments cannot be made, each stimulus will be set in relation to the compartment where it is considered to be prevalent. Figure 2 shows a conceptualization of the prevalence of selected stimuli in the compartments.

As indicated, a major abiotic factor affecting the physiology of bacteria in soil is nutrient limitation, i.e., the often limited availability of organic carbon, nitrogen, phosphorus, and other nutrients. The fate of both introduced and indigenous bacterial populations is governed by nutrient availability. In the absence of a continuous flux of nutrients, in particular of carbon, conditions in soil are frequently limiting and bacterial carbon starvation ensues. In the rhizosphere, the influx of root-released nutrients is a major factor influencing bacteria. Competition for nutrients in the rhizosphere may, however, also lead to limiting conditions for some species. For instance, Fe may be limiting and bacteria possess specialized systems based on siderophores to capture this element in competition with other microbes and the plant (Leong, 1986).

Further, low water activity affects bacteria in soil via osmotic and matric stress. Reduced water availability may be caused by a high concentration of soluble compounds in water, resulting in a low osmotic potential. Both

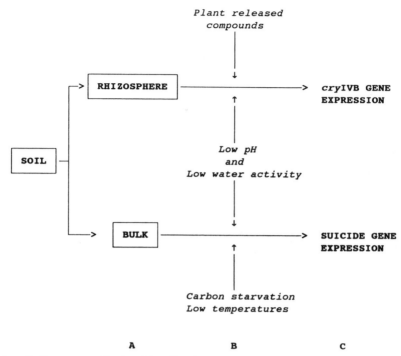

A   B   C

Figure 2. Compartmentalization of the soil environment into bulk and rhizosphere soil (A). Factors which influence physiological and molecular responses of soil bacteria related to either of the compartments (B) are presented. Promoters from both *Pseudomonas* spp. responding to these stimuli may be of use for the construction of environmentally regulated *cry*IVB and suicide genes (C).

evaporation and transpiration by plants affect the concentration of solutes in water-filled soil pores. Evaporation, which depends on atmospheric conditions, is the main process governing soil water availability (Smiles, 1988). It leads to reduced soil water content, reducing the water volume in soil pores and the water films surrounding soil particles. In the rhizosphere, transpiration by plants also plays a role (Smiles, 1988). The matric potential controls the amount of available water in soil; soil colloid surfaces attract water and considerable force has to be applied to remove this (Smiles, 1988). Water retention is further greater in pores with smaller neck diameters because of capillary forces maintaining water in pores. Consequently, the localization of bacteria in the soil pore network is a prime factor determining the degree of water stress experienced by a soil microorganism. In the rhizosphere, bacteria are subjected to fluctuations in water availability resulting from water fluxes induced by the plant roots.

The occurrence of temperature extremes in soil is dependent on atmospheric conditions. Whereas high temperature, sometimes exceeding 50°C, is typical for the top 1 cm of soil receiving solar influx, low temperatures of 10°C and less are quite realistic in top layers of soils in temperate climates during large parts of the year. For instance, the temperature at a depth of 10 cm in a field microplot in April/May of 1985 was around 10°C and temperature further fluctuated between 15 and 20°C during the summer months (van Elsas *et al.*, 1986). The lowest temperature registered was 6°C.

Soil acidity is another stimulus of importance. Acidification of rhizosphere soil takes place because of proton efflux by roots to exchange absorbed ions. In this way, a membrane potential is established at the plant root surface and the plant is able to take up nutrients via H^+ cotransport mechanisms. In wheat, the elongation zone is the most profound site of proton efflux (Bashan and Levanony, 1989). Local pH in the rhizosphere may transiently drop to a value as low as 3.5 depending on soil type. Finally, acid rain entering soils in many industrialized countries has led to overall pH decreases.

Other stimuli, such as anaerobiosis and UV irradiation, will also affect the physiology of bacteria in soil. These stimuli will not be treated further here. Stimuli relating to biotic factors (predators, competitors or antagonists and plants) also elicit bacterial responses, thereby determining bacterial fate in soil. For instance, the effect of predation by protozoa in reducing bacterial population sizes in soil is well known (e.g., Heijnen *et al.*, 1988); however, data on triggering of bacterial responses are now known. As discussed, triggers from plant roots, which are related to the increased availability of either general or specific substrates or to proton efflux, are factors of particular importance to rhizosphere bacteria such as commensalic, symbiotic, or plant pathogenic bacteria (e.g., Rahme *et al.*, 1992).

3.1. Responses to Stimuli of Bulk Soil

Recently, studies on the molecular biology of bacterial responses to soil and root triggers have started in different laboratories (Bhagwat and Keister, 1992; van Rhijn *et al.*, 1990; Lam *et al.*, 1991). The aim of these studies is to increase our understanding about how bacteria in soil cope with different conditions and which genetic systems are needed. For instance, Bhagwat and Keister (1992) used subtractive RNA hybridization with *Bradyrhizobium japonicum* to identify genes which were specifically induced as a response to soil extract. This black-box approach facilitates the detection of soil-induced genes. However, identification of the inducing compounds in soil extract is notoriously laborious. Van Rhijn *et al.* (1990) used promoterless Tn5::*lacZ* insertions in *Azospirillum brasilense* to obtain behavioral mutants, i.e., fla$^-$ or chemotaxis$^-$ mutants. This approach is also of use in the identification of stress- or plant-induced genes. Further, the promoters of some of the genes which are specifically induced under soil conditions, e.g., under conditions of reduced nutrient availability, may be utilizable in the design of modified organisms carrying specific genes for release to soil. This approach has already been followed by Little *et al.* (1991), who placed *tmo* (toluene monooxygenase) genes under the control of specific starvation-induced promoters in *E. coli*. They showed in pure culture studies that the degradation of phenol by the recombinant cells under nutrient limitation far exceeded that of cells which carried the gene under the influence of the *tac* promoter.

The experimental approach chosen has often been the use of transposon-based promoter probes, with a promoterless reporter gene such as *lacZ* contained within the truncated arms of the transposon (Simon *et al.*, 1989). Insertion of such elements at certain loci in the genome leads to transcriptional fusions and will thus identify genes which are expressed under selected conditions of stress. Screening for expression under the influence of the trigger and no expression without the trigger leads to the identification of differentially expressed genes and may identify the promoter sequences involved.

3.1.1. Starvation

Emphasis in this section will be on starvation for organic carbon; although important differences may prove to exist between carbon, nitrogen, and phosphorus starvation, all three seem to elicit a global cellular response. It may therefore be a generality that depriving a bacterial cell of a key element for growth leads to a global response resulting in a more "adapted" or resistant cell form. Whereas most of our knowledge comes from starvation studies with *E. coli* (e.g., Matin, 1992), starvation studies have also been performed with soil bacteria such as *Arthrobacter*, *Nocardia*, *Pseudomonas*, and *Bacillus* spp. (Boylen and Ensign, 1970; Boylen and Mulks, 1978; Nelson and Parkinson, 1978; Robertson and Batt, 1973). Briefly, *E. coli* has been shown to respond to carbon starvation by the formation of a more resistant cell form, reflected in both cellular morphology (cells become smaller and spherical; Lange and Hengge-Aronis, 1991b) and physiology (McCann *et al.*, 1991; Matin, 1992). The physiological response was characterized by the programmed degradation of polymeric cell constituents such as carbohydrates, proteins, ribosomal and messenger RNA, and ultimately DNA (down to one genome equivalent). Results with the soil bacteria also generally showed a programmed decrease of the amount of cellular macromolecules such as proteins, carbohydrates, poly-hydroxybutyric acid, and RNA. Thus, Nelson and Parkinson (1978) found an initial fast degradation phase, followed by a slower one for three bacteria obtained from arctic soil. In *Nocardia corallina* and *Arthrobacter* sp., carbon-containing cellular polymers were first degraded, and protein-N was initially conserved. In a second phase, proteins and RNA were degraded. Except for the reserve material polyhydroxybutyric acid, lipids were almost not degraded. Besides degradation of cellular materials and a redirection of cellular metabolism, an overall "shutdown" of cellular metabolism was notable. In fact, the capacity to remain viable at a very low level of endogenous metabolism was invoked as a reason for the better survival of *N. corallina* as compared with *E. coli* (Robertson and Batt, 1973). Moreover, a decrease of cellular dry weight resulting from decreased cell size was noted for the starved *Nocardia* (Robertson and Batt, 1973). Decreasing cell dimensions as a response to starvation has been known for some time, and in fact seems to be a common starvation response (Gottschal, 1992). It has been shown for soil-isolated pseudomonads and rhizobia in our laboratory (Postma *et al.*, 1988, and unpublished).

Further, *Pseudomonas fluorescens* R2f cells introduced into agricultural drainage water revealed conversion into nonculturable forms after 1 year (van Overbeek *et al.*, 1990). This reflects the behavior of many bacteria introduced into oligotrophic aquatic environments (Roszak and Colwell, 1987; Colwell *et al.*, 1985) and may represent adaptation of such cells to low nutrient conditions. No drastic size reduction was observed for these soil pseudomonads (van Overbeek *et al.*, 1990). On the other hand, culturable cells were only detected in sterilized drainage water and in nonsterilized water supplemented with nutrients, suggesting that the introduced pseudomonads may retain their culturability for a longer period in the presence of nutrients or in the absence of fierce competition for nutrients. The starvation response may be characterized as a general shift of metabolism from one tuned to rapid growth on easily degradable substrate in the presence of all other goodies needed ("feast" existence) toward one characterized by adaptation to hard times ("famine" existence), resulting in a cell of diminished size more resistant to various types of stresses (Matin, 1992; Östling *et al.*, this volume). Molecularly, this shift is brought about by a concerted drastic change in the relative abundance of many cellular macromolecules as discussed by Morita (this volume). The overall decrease in cellular constituents, however, is accompanied by the appearance of new materials. For instance, carbon starvation in *E. coli* resulted in the induction of at least 30 polypeptides (Groat and Matin, 1986; Matin, 1992). Starvation for other essential nutrients, e.g., nitrogen and phosphorus, also resulted in the induction of new proteins.

Two key classes of gene products are induced upon carbon starvation. The first class, encompassing the heat shock proteins DnaK and GroEL, is also induced by exposure to heat or ethanol. The genes coding for these proteins are apparently highly conserved (Matin *et al.*, 1989); for instance, the *dnaK* gene has recently been characterized in *Clostridium acetobutylicum* (Narberhaus *et al.*, 1992). The second class is characterized by the occurrence of a new sigma factor, σ^s, characteristic for the stationary phase. It is represented by the *rpo*S (*kat*F) gene product. RpoS/KatF acts as a central regulator of starvation-induced general resistance, including changes of cell morphology, in *E. coli*, by the regulation of transcription in stationary phase (McCann *et al.*, 1991). A detailed discussion of the role of KatF and other proteins in global responses can be found in Hengge-Aronis (this volume).

The possible occurrence of a global response to starvation and other stresses in soil pseudomonads and other bacteria is currently being tackled. Preliminary data obtained from two-dimensional gel electrophoresis of cellular proteins by the group of S. Molin (Copenhagen) revealed that starvation of the soil organism *Pseudomonas putida*, much like in *E. coli*, induced an estimated 40–50 new proteins. The possible involvement of KatF, DnaK, GroEL, and other proteins in the response is as yet unknown, but might be anticipated, since these proteins seem to be widespread (Narberhaus *et al.*, 1992; Östling *et al.*, this volume). We are currently assessing the possible involvement of these gene products in starvation survival of *P. fluorescens* and *P. cepacia*.

3.1.2. Low Temperature and Low Water Activity

Low temperature is a prime factor affecting bacterial physiology in soil, since metabolic and uptake rates are affected. Soil bacteria, e.g., pseudomonads, have been shown to survive better at lower than at higher temperatures in soil (e.g., van Elsas *et al.*, 1991a). However, only limited physiological data on the low-temperature response are available. Such data have been obtained with *Vibrio* and *E. coli* (see Oliver, this volume). By analogy, low temperature in soil may also trigger the expression of new proteins in bacteria possessing cold-inducible genes. Some of our preliminary data (unpublished) on possible cold-induced promoters and genes in soil pseudomonads indicate that they exist. It is hard to predict whether cold shock-regulated promoters will be sufficiently expressed under natural soil conditions given their potential dependence on substrate availability or other factors.

Osmotic and/or matric stresses are dominantly present in soil, next to starvation. Both stress types are therefore ecologically intertwined in soil. For instance, it might be possible that the starvation state induces relative indifference to osmotic or matric stress in nondifferentiating soil bacteria, much like in endospores. Starvation might, however, also impair cellular responses to osmotic pressure in soil, as exemplified by the putative blocking of uptake systems of compatible solutes in *E. coli* in nutrient-free seawater (Munro *et al.*, 1989). Conversely, much like *E. coli* released into seawater (Munro *et al.*, 1989), bacteria preadapted or adaptable to osmotic stress might be able to cope better with starvation conditions under osmotic pressure, a situation likely to occur in soil, than unadapted bacteria.

Osmotic or matric stress elicits a physiological response in bacterial cells leading to adaptation and enhanced survival. The response, osmoregulation, is molecularly complex. One mechanism involved might be the alteration of the topological state of DNA, which is also brought about by other stress types (Dorman, 1991; Mizuno and Mizushima, 1990). These mechanisms are obviously of potential importance for bacterial cells surviving in soil pores with their fluctuating and often high osmolarities. However, it is unknown to what extent responses to osmotic and matric stresses are operational in cells closely associated with soil particles and coping with nutrient stress. This will remain a trigger for future research.

3.2. Responses to Stimuli of Rhizosphere Soil

Conditions in the rhizosphere are different from those in bulk soil because of (1) the release of compounds, (2) pH changes, and (3) water fluxes. Bacterial population sizes are often enhanced and specific groups are favored. For instance, in a study on the competition between a rhizosphere organism, *Pseudomonas*, and a bulk soil organism, *Arthrobacter citreus* (Chan *et al.*, 1962), the *Pseudomonas* strain dominated in root extract, but not in soil extract. Because of the effect of roots on rhizosphere microbes, an increase of grazers can also be observed in the rhizosphere (Griffiths, 1990). Although seemingly more homogeneous than bulk soil, the rhizosphere represents a highly complex and heterogeneous environment for its bacterial inhabitants.

3.2.1. Plant Root-Derived Compounds

First, many root-derived compounds may serve as easily degradable nutrients for microorganisms in the rhizosphere. This was clearly shown in a model root chamber, in which the diffusion gradient of wheat root exudates was reduced in the presence of soil organisms (Beck and Gilmour, 1983). The bacterial response to nutrients in the rhizosphere can be characterized as a revival of the starvation condition characteristic for microorganisms in soil (nutrient shift-up). That starved bacteria are often capable of a rapid response to the addition of nutrients has been shown primarily for marine isolates (Morita, this volume). The enzymatic machinery needed to quickly utilize new carbon and energy sources may have been preserved during starvation survival. Also, enhanced expression of genes involved in key metabolic activities in this phase are thought to be important (Östling *et al.*, this volume).

Second, compounds present in root exudates or in sloughed-off cells may specifically induce bacterial genes without serving as a carbon or energy source. Induction of bacterial gene expression by plant signals or compounds does not represent a stress response *strictu sensu* or involve a global regulatory network. Rather, a specific set of genes (operon) is induced which is involved in the bacterial response to the presence of a plant. The bacterial response ultimately leads to plant colonization or invasion, in a pathogenic or symbiotic process. The mechanisms involved in the induction of bacterial responses are complex. They may be exemplified by the well-described *Rhizobium*–plant and *Agrobacterium tumefaciens*–plant interactions at the infection site (e.g., Mulligan and Long, 1985; Winans, 1990). In both, specific plant phenolic compounds act as inducers of the genes involved, respectively the *nod* genes in rhizobial nodulation and the *vir* genes in agrobacterial virulence. Flavonoid compounds present in minute amounts in exudates induce rhizobial *nod* genes, whereas acetosyringone from plant wounds induces agrobacterial *vir* genes. Such compounds might also trigger responses in other rhizosphere inhabitants. For instance, flavonoids which induced nodulation by *R. meliloti*, enhanced the growth rate of a *P. putida* strain (Hartwig *et al.*, 1991). Also, homology to *nod*D was observed in *P. syringae* cv. *tomato* (Roche *et al.*, 1991) and in *P. aeruginosa* (Chang *et al.*, 1989), suggesting these strains might possess a regulatory apparatus able to react on plant signals similar to the *nod* system, or remnants thereof.

Root colonization by rhizosphere bacteria probably involves chemotaxis (Bauer and Caetano-Anollés, 1991). Chemotaxis and motility (flagellar movement) are triggered by exudates. Many compounds, including amino acids, sugars, and flavonoids, are able to elicit a chemotactic response in rhizosphere bacteria, as shown for instance in Chet *et al.* (1973), van Rhijn *et al.* (1990), and Dharmatilake and Bauer (1992). That flagella are involved in root colonization by *P. fluorescens* was convincingly shown by de Weger *et al.* (1987). Expression of flagellar genes is probably controlled by alternative sigma factors of the σ^{28} class in *Bacillus subtilis* (Wigs *et al.*, 1981), *E. coli* and *S. typhimurium* (Helman and Chamberlin, 1987), and *P. aeruginosa* (Starnbach and Lory, 1992). In *Caulobacter crescentus*, genes responding to cell cycle signals controlled genes lower in the hierarchy of flagellar transcription (Dingwall *et al.*, 1992). Cascades in flagellar gene expression, ranging from class I (highest in hierarchy) to class III genes (lowest in hierarchy and regulated by gene products of the class I and II genes), have been proposed in a chemotaxis model of enteric bacteria (Helman, 1991).

Thus, many soil bacteria which evolved in close association with plants have probably developed genetic systems which enable them to quickly escape the starvation state in soil and colonize the plant (root). In particular, bacteria which can be regarded as copiotrophs (e.g., pseudomonads, rhizobia, agrobacteria) have been found apt in this process. The life cycle of such bacteria on annual crops can be regarded as a genuinely regular feast and famine existence.

Lam *et al*. (1991) tested *P. fluorescens* promoterless *lac*Z insertion mutants for their reaction to wheat root exudates. Three mutants showed high induction after exposure to root exudates and after isolation from gnotobiotic wheat roots. In the same study, five mutants which contained a constitutively expressed *lac*Z gene were impaired in root colonization as compared with the wild-type strain. However, the function of these genes in root colonization is still unclear. To identify environmentally induced genes and promoters in the root-colonizing bacteria *P. fluorescens* R2f and *P. cepacia* P2, we also used promoterless *lac*Z fusions (Simon *et al.*, 1989) and screened for specific responses to starvation, low temperature, and wheat root exudate obtained from axenically grown *Triticum aestivum* cv *Sicco*. To obtain carbon starvation-induced operon fusion mutants, a pool of mutants was comparatively screened on X-Gal-amended minimal medium with a low glucose concentration (0.01%) or a high glucose concentration (0.1%). Mutants positively responding to the low glucose concentration were further screened for response to both a nitrogen- and a phosphorus-limited minimal medium. One *P. fluorescens* carbon starvation mutant, RA92, only responded weakly to nitrogen starvation. However, no response to nitrogen and phosphorus starvation was observed with the *P. cepacia* P2 carbon starvation mutant PC3. Further, a low-temperature-induced mutant of P2, PD19, was obtained by screening for a differential response on minimal medium with the standard amount of glucose (0.1%), incubated at 27 or 10°C. No mutants of R2f which responded to low temperature were obtained under these conditions. Finally, for *P. fluorescens* R2f and *P. cepacia* P2, respectively four and two mutants responding to wheat root exudate were obtained. The mutants were selected by a differential response to X-Gal-amended root exudate on agar. They were screened for a response to 95 different carbon sources. One mutant of R2f, RIWE8, and one of P2, PEWE2, showed a specific response to an amino acid (proline for RIWE8 and leucine for PEWE2).

The responses of these mutants to the respective triggers as well as their response under uninduced conditions were quantified (Miller, 1972), providing the calculated ratio of expression under induced and uninduced conditions (Fig. 3). The promoters responding to root exudates gave a higher differential response than those of the stress-induced operon fusion mutants. This implies that promoters involved in utilization of compounds from root exudate may be more powerful than promoters involved in stress survival. We are involved in the construction of organisms obtained from soil carrying genes for the control of insect larvae, which damage grass and wheat roots (van Elsas *et al.*, 1991b; Waalwijk *et al.*, 1991). These genes might be controlled by environmental (e.g., soil) or plant factors. The gene for biological control, *cry*IVB (crystal protein; Waalwijk *et al.*, 1991), should be expressed near or on the plant roots, whereas for biosafety reasons a host-killing or suicide gene should be expressed were

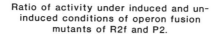

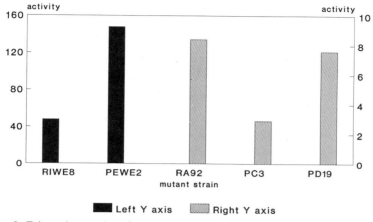

Figure 3. Enhanced expression of genes under starvation and other stress conditions in *Pseudomonas* spp. The ratio between expression levels (Miller units) under inducing and noninducing conditions is indicated. RIWE8, PEWE2: root exudate mutants; RA92, PC3: carbon starvation mutants; PD19: low temperature mutant.

escape of the organism from the rhizosphere to occur. An outline of this approach, including the soil and plant triggers of potential use, is given in Fig. 2. For the construction of the conditionally inducible genes *cry*IVB and a suicide gene, root exudate promoters may therefore be favored. Whereas the *cry*IVB gene could be placed directly under the influence of such promoter, the suicide gene might be controlled via a negatively regulated loop like the one used by Contreras *et al.* (1991).

4. CONCLUDING REMARKS

It is obvious from the foregoing that soil, because of its heterogeneity, poses a myriad of stresses and other influences on bacterial cells. A major effect certainly is nutrient limitation, leading to starvation. This nutrient scarcity is caused by the unavailability of carbon present in soil. The very small bacterial cells in soil (dwarfs) may represent a group of microbes well adapted to the often extremely low amounts of organic carbon locally available. Bacteria introduced into soil will be subjected to the same stress factors indigenous ones have been adapted to. In addition to biotic factors such as predation by protozoa, nutrient, water, low pH, and low temperature may be major stress factors affecting their physiology. Physiological studies performed on some soil isolates have shown that their physiological response to starvation resembles that of *E. coli* and *Vibrio* species in that there was a general programmed decline of cellular macromolecules, a concomitant reduction in cell size, and a metabolic arrest. However, the starvation response of soil bacteria has not been elucidated in molecular terms. On the other hand, the availability of both easily available nutrients and specific compounds in the rhizosphere signaling the presence of roots has been shown to trigger specific adaptive responses in rhizosphere bacteria.

Recent experimental approaches have used promoterless *lacZ* operon fusions to assess specific gene induction as a response to environmental triggers. Although *in vitro* evidence as to possibilities for specific induction, e.g., due to starvation, to low temperature or low pH, and to plant compounds, is accumulating, the ecological significance of these responses should be obtained after extensive testing of strains containing the inserts in soil microcosms. However, quantitative tests of promoter activity with *lacZ* operon fusions in soil, using spectrophotometrical methods, are obscured by the presence of (brown) impurities from soil in extracts. Therefore, there is a need to develop methodology which would enable detection.

Should modified organisms with soil- or rhizosphere-triggered genes become available, this would represent an example of "clean" environmental engineering, since no intervention following release is needed for the induction of expression of the beneficial or the suicide gene. Recently, suicide genes had to be induced in soil by addition of the trigger substance (Contreras *et al.*, 1991).

ACKNOWLEDGMENTS. We thank E. Bremer and E. Smit for critically reading the manuscript. We are also indebted to Soren Molin for use of some unpublished data, and for his helpful contributions to the manuscript.

REFERENCES

Alexander, M., 1972, *Introduction to Soil Microbiology*, 2nd ed., Wiley, New York.

Atkinson, D., 1990, Influence of root system morphology and development on the need for fertilizers and the efficiency of use, in: *Crops as Enhancers of Nutrient Use* (V. C. Balizer and R. R. Durean, eds.), Academic Press, New York, pp. 411–451.

Bae, H. C., Cota-Robles, E. H., and Casida, L. E., Jr., 1972, Microflora of soil as viewed by transmission electron microscopy, *Appl. Microbiol.* **23**:637–648.

Bakken, L. R., and Olsen, R. A., 1989, DNA content of soil bacteria of different cell size, *Soil Biol. Biochem.* **21**:789–793.

Bashan, Y., and Levanovy, H., 1989, Effect of the root environment on proton efflux in wheat roots. *Plant Soil* **119**:191–197.

Bauer, W. D., and Caetano-Anollés, G., 1991, Chemotaxis, induced gene expression and competitiveness in the rhizosphere, in: *The Rhizosphere and Plant Growth* (D. L. Keister and P. B. Cregan, eds.), Kluwer, The Netherlands, pp. 155–162.

Beck, S. M., and Gilmour, C. M., 1983, Role of wheat root exudates in associative nitrogen fixation, *Soil Biol. Biochem.* **15**:33–38.

Bhagwat, A. A., and Keister, D. L., 1992, Identification and cloning of *Bradyrhizobium japonicum* genes expressed strain selectively in soil and rhizosphere, *Appl. Environ. Microbiol.* **58**: 1490–1495.

Boylen, C. W., and Ensign, J. C., 1970, Intracellular substrates for endogenous metabolism during long-term survival of rod and spherical cells of *Arthrobacter crystallopoietes*, *J. Bacteriol.* **103**:578–587.

Boylen, C. W., and Mulks, M. H., 1978, The survival of coryneform bacteria during periods of prolonged nutrient starvation, *J. Gen. Microbiol.* **105**:323–334.

Chan, E. C. S., Katznelson, H., and Rouatt, J. W., 1962, The influence of soil and root extracts on the associative growth of selected soil bacteria, *Can. J. Microbiol.* **9**:187–197.

Chang, M., Hadero, A., and Crawford, I. P., 1989, Sequence of the *Pseudomonas aeruginosa trp*I activator gene and relatedness of *trp*I to other prokaryotic regulatory genes, *J. Bacteriol.* **171**: 172–183.

Chet, I., Zilberstein, Y., and Henis, Y., 1973, Chemotaxis of *Pseudomonas lachrymans* to plant extracts and to water droplets collected from the leaf surfaces of resistant and susceptible plants, *Physiol. Plant Pathol.* **3**:473–479.

Cho, B. C., and Azam, F., 1988, Major role of bacteria in biogeochemical fluxes in the ocean's interior, *Nature* **332**:441–443.

Colwell, R. R., Brayton, P. R., Grimes, D. J., Roszak, D. B., Huq, S. A., and Palmer, L. M., 1985, Viable but non-culturable *Vibrio cholerae* and related pathogens in the environment: Implications for release of genetically engineered microorganisms, *Bio/Technology* **3**:817–820.

Compeau, G., Jadoun Al-Achi, B., Platsouka, E., and Levy, S. B., 1988, Survival of rifampin-resistant mutants of *Pseudomonas fluorescens* and *Pseudomonas putida* in soil systems, *Appl. Environ. Microbiol.* **54**:2432–2438.

Contreras, A., Molin, S., and Ramos, J., 1991, Conditional-suicide containment system for bacteria which mineralize aromatics, *Appl. Environ. Microbiol.* **57**:1504–1508.

Curl, A. E., and Truelove, B., 1986, *The Rhizosphere*, Springer-Verlag, Berlin.

De Weger, L. A., van der Vlugt, C. I. M., Wijfjes, A. H. M., Bakker, P. A. H. M., Schippers, B., and Lugtenberg, B., 1987, Flagella of a plant-growth-stimulating *Pseudomonas fluorescens* strain are required for colonization of potato roots, *J. Bacteriol.* **169**:2769–2773.

Dharmatilake, A. J., and Bauer, W. D., 1992, Chemotaxis of *Rhizobium meliloti* towards nodulation gene-inducing compounds from Alfalfa roots, *Appl. Environ. Microbiol.* **58**:1153–1158.

Dingwall, A., Zhuang, W. Y., Quon, K., and Shapiro, L., 1992, Expression of an early gene in the flagellar regulatory hierarchy is sensitive to an interruption in DNA replication, *J. Bacteriol.* **174**:1760–1768.

Dorman, C. J., 1991, DNA supercoiling and environmental regulation of gene expression in pathogenic bacteria, *Infect. Immun.* **59**:745–749.

Dupler, M., and Baker, R., 1984, Survival of *Pseudomonas putida*, a biological control agent, in soil. *Phytopathol.* **74**:195–200.

Foster, R. C., 1988, Microenvironments of soil microorganisms. *Biol. Fertil. Soils* **6**: 189–203.

Gottschal, J. C., 1992, Substrate capturing and growth in various ecosystems, *J. Appl. Bacteriol. Symp. Suppl.* **73**:39S–48S.

Gray, T. R. G., and Williams, S. T., 1971, Microbial productivity in soil, *Symp. Soc. Gen. Microbiol.* **21**:255–286.

Gregory, P. J., and Atwell, B. J., 1991, The fate of carbon in pulse-labelled crops of barley and wheat, *Plant Soil* **136**:205–213.

Griffiths, B. S., 1990, A comparison of microbial-feeding nematodes and protozoa in the rhizosphere of different plants, *Biol. Fertil. Soils* **9**:83–88.

Groat, R. G., and Matin, A., 1986, Synthesis of unique proteins at the onset of carbon starvation in *Escherichia coli*, *J. Ind. Microbiol.* **1**:69–73.

Habte, M., and Alexander, M., 1975, Protozoa as agents responsible for the decline of *Xanthomonas campestris* in soil, *Appl. Microbiol.* **29**:159–164.

Hartwig, U. A., Joseph, C. M., and Phillips, D. A., 1991, Flavonoids released naturally from alfalfa seeds enhance growth rate of *Rhizobium meliloti*, *Plant Physiol.* **95**:797–803.

Hattori, T., and Hattori, R., 1976, The physical environment in soil microbiology: An attempt to extend principles of microbiology to soil microorganisms, *CRC Crit. Rev. Microbiol.* **4**: 423–461.

Heijnen, C. E., Hok-A-Hin, C. H., and van Elsas, J. D., 1993, Root colonization by *Pseudomonas fluorescens* introduced into soil amended with bentonite. *Soil Biol. Biochem.* **25**:239–246.

Heijnen, C. E., van Elsas, J. D., Kuikman, P. J., and van Veen, J. A., 1988, Dynamics of *Rhizobium leguminosarum* biovar *trifolii* introduced into soil; the effect of bentonite clay on predation by protozoa, *Soil Biol. Biochem.* **20**:483–488.

Helman, J. D., 1991, Alternative sigma factors and the regulation of flagellar gene expression, *Mol. Microbiol.* **5**:2875–2882.

Helman, J. D., and Chamberlin, M. J., 1987, DNA sequence analysis suggests that expression of flagellar and chemotaxis genes in *Escherichia coli* and *Salmonella typhimurium* is controlled by an alternative σ factor, *Proc. Natl. Acad. Sci. USA* **84**:6422–6424.

Hissett, R., and Gray, T. R. G., 1976, Microsites and time changes in soil microbe ecology, in: *The Role of Terrestrial and Aquatic Organisms in Decomposition Processes* (J. M. Anderson and A. MacFadyen, eds.), Oxford University Press, London, pp. 23–39.

Ishida, Y., and Kadota, H., 1981, Growth patterns and substrate requirements of naturally occurring obligate oligotrophs, *Microb. Ecol.* **7**:123–130.

Keith, H., Oades, J. M., and Martin, J. K., 1986, Input of carbon to soil from wheat plants, *Soil Biol. Biochem.* **18**:445–449.

Kilbertus, G., 1980, Etude des microhabitats contenus dans les agregats du sol; leur reaction avec la biomasse bacterienne et la taille des procaryotes presents, *Rev. Ecol. Biol. Sol* **17**:543–557.

Kraffczyck, I., Trolldenier, G., and Beringer, H., 1984, Soluble root exudates of maize: Influence of potassium supply and rhizosphere microorganisms, *Soil Biol. Biochem.* **16**:315–322.

Lam, S. T., Ellis, D. M., and Ligon, J. M., 1991, Genetic approaches for studying rhizosphere colonization, in: *The Rhizosphere and Plant Growth* (D. L. Kleister and P. B. Cregan, eds.), Kluwer, The Netherlands, pp. 43–50.

Lange, R., and Hengge-Aronis, R., 1991a, Identification of a central regulator of stationary-phase gene expression in *Escherichia coli*, *Molec. Microbiol.* **5**:49–59.

Leong, J., 1986, Siderophores: their biochemistry and possible role in the biocontrol of plant pathogens. *Ann. Rev. Phytopathol.* **24**:187–209.

Liang, L. N., Sinclair, J. L., Mallory, L. M., and Alexander, M., 1982, Fate in model ecosystems of microbial species of potential use in genetic engineering, *Appl. Environ. Microbiol.* **44**: 708–714.

Little, C., Fraley, C., McCann, M., and Matin, A., 1991, Use of bacterial stress promoters to induce biodegradation under conditions of environmental stress, in: *Proceedings of In Situ and On-site Bioreclamation*, Battelle Symposium, San Diego (R. E. Hichee, ed.), Butterworths, London, pp. 493–498.

Lynch, J. M., and Whipps, J. M., 1990, Substrate flow in the rhizosphere, *Plant Soil* **129**:1–10.

McCann, M. P., Kidwell, J. P., and Matin, A., 1991, The putative σ factor KatF has a central role in development of starvation-mediated general resistance in *Escherichia coli*, *J. Bacteriol.* **173:**4188–4194.

Martin, J. K., 1977, Effect of soil moisture on the release of organic carbon from wheat roots, *Soil Biol. Biochem.* **9:**303–304.

Matin, A., 1992, Physiology, molecular biology and applications of the bacterial starvation response, *J. Appl. Bacteriol. Symp. Suppl.* **73:**49S–57S.

Matin, A., Auger, E. A., Blum, P. H., and Schultz, J. E., 1989, Genetic basis of starvation survival in nondifferentiating bacteria, *Annu. Rev. Microbiol.* **43:**293–316.

Miller, J. H., 1972, *Experiments in Molecular Genetics*, Cold Spring Harbor Laboratory, Cold Spring Harbor, N.Y.

Mizuno, T., and Mizushima, S., 1990, Signal transduction and gene regulation through the phosphorylation of two regulatory components: The molecular basis for the osmotic regulation of the porin genes, *Mol. Microbiol.* **4:**1077–1082.

Morita, R. Y., 1988, Bioavailability of energy and its relationship to growth and starvation survival in nature, *Can J. Microbiol.* **34:**436–441.

Mulligan, J. T., and Long, S. R., 1985, Induction of *Rhizobium nod*C expression by plant exudate requires *nod*D, *Proc. Natl. Acad. Sci. USA* **82:**6609–6613.

Munro, P. M., Gauthier, M. J., Breittmayer, V. A., and Bongiovanni, J., 1989, Influence of osmoregulation processes on starvation survival of *Escherichia coli* in seawater, *Appl. Environ. Microbiol.* **55:**2017–2024.

Narberhaus, F., Giebeler, K., and Bahl, H., 1992, Molecular characterization of the *dna*K gene region of *Clostridium acetobutylicum*, including *grp*E, *dna*J, and a new heat shock gene, *J. Bacteriol.* **174:**3290–3299.

Nelson, L. M., and Parkinson, D., 1978, Effect of starvation on survival of three bacterial isolates from an arctic soil, *Can. J. Microbiol.* **24:**1460–1467.

Newman, E. I., 1985, The rhizosphere: Carbon sources and microbial populations, in: *Ecological Interactions in Soil* (A. H. Fitter, ed.), Blackwell, Oxford, pp. 107–121.

Nijhuis, E. H., Maat, M. J., Zeegers, I., Waalwijk, C., and van Veen, J. A., 1992, Selection of bacteria suitable for introduction into the rhizosphere of grass, *Soil Biol. Biochem.*, in press.

Novitsky, J. A., and Morita, R. Y., 1976, Morphological characterization of small cells resulting from nutrient starvation of a psychrophilic marine *Vibrio*, *Appl. Environ. Microbiol.* **32:**616–622.

Östling, J., Goodman, A., and Kjelleberg, S., 1991, Behaviour of IncP-1 plasmids and a miniMu transposon in a marine *Vibrio* sp.: Isolation of starvation inducible *lac* operon fusions, *FEMS Microbiol. Ecol.* **86:**83–94.

Paul, E. A., and Clark, F. E., 1989, *Soil Biology and Biochemistry*, Academic Press, New York.

Poindexter, J. S., 1981, Oligotrophy, *Adv. Microb. Ecol.* **5:**63–89.

Postma, J., and van Veen, J. A., 1990, Habitable pore space and survival of *Rhizobium leguminosarum* biovar *trifolii* introduced into soil. *Microb. Ecol.* **19:**149–161.

Postma, J., van Elsas, J. D., Govaert, J. M., and van Veen, J. A., 1988, The dynamics of *Rhizobium leguminosarum* biovar *trifolii* introduced into soil as determined by immunofluorescence and selective plating techniques, *FEMS Microbiol. Ecol.* **53:**251–260.

Rahme, L. G., Mindrinos, M. N., and Panopoulos, N. J., 1992, Plant and environmental sensory signals control the expression of *hrp* genes in *Pseudomonas syringae* pv phaseolicola, *J. Bacteriol.* **174:**3499–3507.

Robertson, J. G., and Batt, R. D., 1973, Survival of *Nocardia corallina* and degradation of constituents during starvation, *J. Gen. Microbiol.* **78:**109–117.

Roche, P., Debellé, F., Maillet, F., Lerouge, P., Faucher, C., Truchet, G., Dénarié, J., and Promé, J., 1991, Molecular basis of symbiotic host specificity in *Rhizobium meliloti*: *nod*H and *nod*PQ genes encode the sulfation of lipo-oligosaccharides signals, *Cell* **67**:1131–1143.

Roszak, D. B., and Colwell, R. R., 1987, Survival strategies of bacteria in the natural environment, *Microbiol. Rev.* **51**:365–379.

Shields, J. A., Paul, E. A., Lowe, W. E., and Parkinson, D., 1973, Turnover of microbial tissue in soil under field conditions, *Soil Biol. Biochem.* **5**:753–764.

Simon, R., Quandt, J., and Klipp, W., 1989, New derivatives of transposon Tn5 suitable for mobilization of replicons, generation of operon fusions and induction of genes in gram-negative bacteria, *Gene* **80**:161–169.

Smiles, D. E., 1988, Aspects of the physical environment of soil organisms, *Biol. Fertil. Soils* **6**:204–215.

Starnbach, M. N., and Lory, S., 1992, The *fli*A (*rpo*F) gene of *Pseudomonas aeruginosa* encodes an alternative sigma factor required for flagellin synthesis, *Mol. Microbiol.* **6**:459–469.

Thompson, I. P., Young, C. S., Cook, K. A., Lethbridge, G., and Burns, R. G., 1990, Survival of two ecologically distinct bacteria (*Flavobacterium* and *Arthrobacter*) in unplanted and rhizosphere soils, *Soil Biol. Biochem.* **22**:1029–1037.

Torrella, F., and Morita, R. Y., 1981, Microcultural study of bacterial size changes and microcolony and ultramicrocolony formation by heterotrophic bacteria in seawater, *Appl. Environ. Microbiol.* **41**:518–527.

Tormo, A., Almiró, M., and Kolter, R., 1990, *sur*A, an *Escherichia coli* gene essential for survival in stationary phase, *J. Bacteriol.* **172**:4339–4347.

Trofymow, J. A., Coleman, D. C., and Cambardella, C., 1987, Rates of rhizodeposition and ammonium depletion in the rhizosphere of axenic oat roots, *Plant Soil* **97**:333–344.

Van Elsas, J. D., and Trevors, J. T., 1990, Plasmid transfer to indigenous bacteria in soil and rhizosphere: Problems and perspectives, in: *Bacterial Genetics in Natural Environments* (J. C. Fry and M. J. Day, eds.), Chapman & Hall, London, pp. 188–199.

Van Elsas, J. D., Dijkstra, A. F., Govaert, J. M., and van Veen, J. A., 1986, Survival of *Pseudomonas fluorescens* and *Bacillus subtilis* introduced into two soils of different texture in field microplots, *FEMS Microbiol. Ecol.* **38**:151–160.

Van Elsas, J. D., Trevors, J. T., and van Overbeek, L. S., 1991a, Influence of soil properties on the vertical movement of genetically-marked *Pseudomonas fluorescens* through large soil microcosms, *Biol. Fertil. Soils* **10**:249–255.

Van Elsas, J. D., van Overbeek, L. S., Feldman, A. M., Dullemans, A. M., and de Leeuw, O., 1991b, Survival of a genetically engineered *Pseudomonas fluorescens* in soil in competition with the parent strain, *FEMS Microbiol. Ecol.* **85**:53–64.

Van Elsas, J. D., Heijnen, C. E., and van Veen, J. A., 1991c, The fate of introduced genetically engineered microorganisms (GEMs) in soil, in microcosms and the field: Impact of soil textural aspects, in: *Biological Monitoring of Genetically Engineered Plants and Microbes* (D. R. MacKenzie and S. C. Henry, eds.), Agricultural Research Institute, Bethesda, pp. 67–79.

Van Overbeek, L. S., van Elsas, J. D., Trevors, J. T., and Starodub, M. E., 1990, Long-term survival of and plasmid stability in *Pseudomonas fluorescens* and *Klebsiella* species and appearance of nonculturable cells in agricultural drainage water, *Microb. Ecol.* **19**:239–249.

Van Rhijn, P., Vanstockem, M., van der Leyden, J., and de Mot, R., 1990, Isolation of behavioral mutants of *Azospirillum brasilense* by using Tn5 *lac*Z, *Appl. Environ. Microbiol.* **56**:990–996.

Van Veen, J. A., and van Elsas, J. D., 1986, Impact of soil structure and texture on activity and dynamics of the soil microbial population, in: *Perspectives in Microbial Ecology* (F. Megusar and M. Gantar, eds.), Slovena Society for Microbiology, Ljubljana, pp. 481–488.

Waalwijk, C., Dullemans, A., and Maat, C., 1991, Construction of a bioinsecticidal rhizosphere isolate of *Pseudomonas fluorescens, FEMS Microbiol. Lett.* **77**:257–264.

Wessendorf, J., and Lingens, F., 1989, Effect of culture and soil conditions on survival of *Pseudomonas fluorescens* R1 in soil, *Appl. Microbiol. Biotechnol.* **31**:97–102.

Wigs, J., Gilman, M., and Chamberlin, M. J., 1981, Heterogeneity of RNA polymerase in *Bacillus subtilis*: Evidence for an additional sigma factor in vegetative cells, *Proc. Natl. Acad. Sci. USA* **78**:2762–2766.

Williams, S. T., 1985, Oligotrophy in soil: Fact or fiction? in: *Bacteria in the Natural Environment: The Effect of Nutrient Conditions* (M. Fletcher and G. Floodgate, eds.), Academic Press, New York, pp. 81–110.

Winans, S. C., 1990, Transcriptional induction of an *Agrobacterium* regulatory gene at tandem promoters by plant-released phenolic compounds, phosphate starvation, and acidic growth medium, *J. Bacteriol.* **172**:2433–2438.

Zechman, J. M., and Casida, L. E., Jr., 1982, Death of *Pseudomonas aeruginosa* in soil, *Can. J. Microbiol.* **28**:788–794.

4

Dynamics of Microbial Growth in the Decelerating and Stationary Phase of Batch Culture

C. Anthony Mason and Thomas Egli

1. INTRODUCTION

One of the most striking features of microbes is their ability to rapidly grow in the presence of abundant nutrients in both the natural and laboratory environments. It is therefore not surprising that a large amount of microbiological research has focused on this phenomenon. Although this book is concerned with starvation in bacteria, it is important that those events which occur during the antithesis of starvation, i.e., during bacterial growth prior to the stationary phase, are also examined. It is well known that the composition and physiology of a bacterial cell are a function of its growth rate and environment (Herbert, 1961; Harder and Dijkhuizen, 1983; Bremer and Dennis, 1987). Consequently, the response to starvation will depend on the conditions under which the cell was growing prior to the onset of exhaustion of a particular essential nutrient. Whereas other chapters in this book have emphasized the important genetic and physiological changes that occur after the onset of starvation and compare these events with those found during unlimited growth, this chapter will specifically deal with the physiological changes that take place during the transition phase from unlimited exponential growth to the early starvation phase.

C. Anthony Mason and Thomas Egli • Swiss Federal Institute for Water Resources and Water Pollution Control (EAWAG), CH-8600 Dübendorf, Switzerland.

Starvation in Bacteria, edited by Staffan Kjelleberg. Plenum Press, New York, 1993.

The first quantitative investigations of microbial growth date back to the late 1890s (Ward, 1895; Müller, 1895) and the subject had already received its first review in 1918 by Buchanan, by which time the concept of the batch growth curve was already established. Buchanan was the first to describe a generalized pattern and the different growth phases for a carbon-terminated bacterial culture (Fig. 1) (the reason for this unfamiliar nomenclature is given in the next section). Unfortunately, this pattern for a batch growth curve has become almost too familiar to microbiologists ("familiarity breeds contempt"). Other patterns for the bacterial batch growth curve exist but are often considered to be abnormal. The tendency has been to assume that the phenomenon of abrupt cessation of exponential growth and rapid transition into stationary phase, valid after carbon exhaustion, is identical irrespective of the nature of the nutrient terminating growth. During the course of this chapter, it will become clear that this is not the case and that pronounced deviations from the generalized pattern can be seen in the decelerating phase. The length and rate of passage through the deceleration phase are determined on the one hand by the nature of the nutrient terminating growth and on the other by the culture density. It is this time course which determines the pattern of physiological events leading to the stationary (starvation) phase. We will show for carbon-terminated growth (at

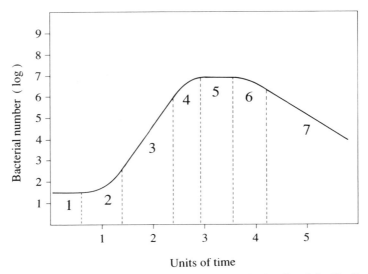

Units of time

Figure 1. Batch growth curve and the "life phases in a bacterial culture" as defined by Buchanan (1918). 1, initial stationary phase; 2, lag phase or phase of positive growth acceleration; 3, logarithmic growth phase; 4, phase of growth deceleration; 5, maximum stationary phase; 6, phase of accelerated death; 7, logarithmic death phase.

high cell densities) that the deceleration phase is almost undetectable whereas exhaustion, for example, of the phosphorus source results in an extensive deceleration phase. Moreover, the transition into the stationary phase is dependent on the culture density. In low-density cultures, the transition can occur over an extended period (up to hours) during which the cells are consistently exposed to a decreasing nutrient concentration. This is in contrast to cultures grown to high cell densities where the rapid exhaustion of the growth-terminating nutrient causes the cells to experience a change from feast to famine conditions within a fraction of a second. Unfortunately, especially with respect to the physiological alterations occurring in cells during this phase of growth, there exist very little data in the literature. One implication of this density effect concerns bacterial growth in the natural environment, where, since densities are frequently low, one would expect to see the gradual transition from exponential to stationary phase described above rather than a sudden change. One must expect that the nature of this transition period can greatly influence the long-term survival of natural bacterial populations under starvation conditions.

2. NUTRIENT LIMITATION AND NUTRIENT STARVATION

The term "nutrient limitation" has been used to describe two different phenomena, namely growth termination via nutrient exhaustion (or starvation) as encountered in batch culture, and nutrient-limited growth as is found in chemostat culture. The phenomenon of nutrient limitation in batch culture, i.e., where the specific growth rate of the culture decreases because of the restricted availability of a certain nutrient, is restricted to the deceleration phase where cells are only transiently experiencing these conditions. At low cell culture densities the cells pass slowly through this phase and have consequently more time to adapt to the decreasing nutrient availability, whereas in dense cultures the effect is hardly detectable. Strictly speaking, this means that the decelerating phase should be subdivided into two different phases. During the first, the growth rate becomes a function of the decreasing residual substrate concentration (e.g., according to the Monod equation). This phase is succeeded by a second during which growth continues as a result of mobilization and reallocation of the intracellular "reservoirs" of the exhausted nutrient but at a continuously decreasing rate and where the rate of growth is independent of the extracellular residual substrate concentration. This stage is followed by a prolonged period of starvation (the "true" stationary phase). In the absence of a renewed supply of the exhausted nutrient, these starvation conditions result in no net growth or when extensive cell death occurs, negative growth. In contrast, in continuous cultures, nutrient-limited conditions result in a growth rate which is dependent on the rate of supply of the growth-limiting nutrient. In this chapter we will use the term "nutrient-terminated" as opposed to "nutrient-limited" growth to differentiate between the situation as it occurs in batch culture and that in continuous culture.

3. GROWTH UNDER NUTRIENT-SUFFICIENT CONDITIONS

Nutrient-sufficient conditions are characteristic of the exponential phase during batch cultivation. This phase has been extensively studied and has resulted in the development of the concept of balanced growth (Schaechter *et al.*, 1958; Maaløe and Kjeldgaard, 1966). Most of the work reported in the literature has been carried out in media essentially composed of a single substrate although attention has also been focused recently on the use of substrate mixtures during exponential growth (Wanner and Egli, 1990). There are two factors from the exponential growth phase which influence the physiology of microbes in the stationary phase. First, it is the actual growth rate which determines the overall cell composition, i.e., content of ribosomes, DNA, protein, and so forth. And second, the selection of nutrients supplied essentially determines the biochemical potential. Both are of course not independent of each other. Medium composition effects on growth rate and therefore also on cell composition were extensively studied by various groups especially in the 1950s (e.g., Schaechter *et al.*, 1958). They found that in both batch and chemostat cultures of *Salmonella typhimurium*, the DNA and RNA content per cell as well as the cell volume increased in an exponential manner with growth rate, while the amounts of DNA and protein per cell mass were shown to decrease with increasing growth rate. A more detailed description of the relationship between growth rate and cell composition can be found in recent reviews on this topic (Bremer and Dennis, 1987; Neidhardt *et al.*, 1990).

4. EFFECTS OF EXHAUSTION OF DIFFERENT NUTRIENTS ON BATCH GROWTH CURVE

4.1. Carbon

Much of the early work describing microbial growth under batch conditions was concerned with the growth of heterotrophs. Usually these studies concentrated on the exponential growth phase with growth being ultimately limited by the availability of the carbon (and energy) source. Once the carbon source was exhausted, there was a very abrupt change in the growth rate with the bacterial concentration becoming immediately constant. This pattern was typically obtained when microorganisms were grown in a synthetic defined growth medium under controlled cultivation conditions (constant pH, adequate aeration). Such an example is shown in Fig. 2 for the growth of *Klebsiella pneumoniae* on a glucose minimal medium. The abrupt cessation of growth in experiments where growth was monitored by following biomass, has been confirmed for many different organisms and has been shown to be independent of the carbon source used (Niven *et al.*, 1977; Andersen and von Meyenburg, 1980; Nazly *et al.*, 1980; Wanner and Egli, 1990). However, frequently a gradual decline in the growth rate resulting in a distinct deceleration phase has been observed. This pattern has been commonly observed when media amended with complex components were used or where unexpected limitations (e.g., oxygen, trace elements) occurred at the end of the growth phase. For sporulating bacteria, e.g., *Bacillus* spp., deviations from this generalized pattern have been observed, where the biomass immediately declined following exhaustion of the carbon substrate (Al-Awadhi *et al.*, 1988).

4.2. Nitrogen

In contrast to the well-defined stationary phase observed for carbon exhaustion, no strict stationary phase occurs during the first few hours following exhaustion of the nitrogen source. An example of such a growth pattern is shown in Fig. 3, where an approximately linear increase in biomass is demonstrated for *K. pneumoniae*. As seen with carbon exhaustion, the deceleration phase is hardly detectable. As far as the authors are aware, there is no clear nomenclature which can accurately describe the growth during this phase, being distinct from both the decelerating and stationary phases. Frequently applied is the expression "linear growth phase," which is normally used to describe growth limited by gaseous substrates such as oxygen or methane. The continued increase in growth after exogenous nitrogen exhaustion is mainly the result of the intracellular accumulation of reserve carbonaceous polymers (Fig. 3). This will be discussed in more detail in a later section. The extent of continued growth is dependent on the type and concentration of the carbon substrate supplied and the type of reserve material accumulated. In addition, in some cases, the formation of extracellular slime layers, polysaccharides, and capsules has been reported to contribute a significant fraction to the continued biomass increase (Duguid and Wilkinson, 1953; Sutherland, 1982).

4.3. Phosphate

The growth curve obtained when phosphate is the nutrient whose exhaustion terminates growth is completely different from those already described.

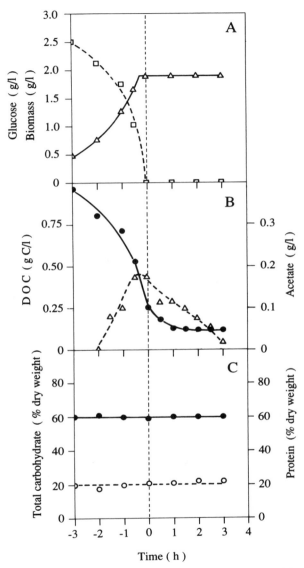

Figure 2. Batch growth of Klebsiella pneumoniae in a synthetic medium with glucose as the growth-limiting substrate. $T = 30°C$, pH = 7.0. (A) Growth given as dry biomass produced ($\triangle$); glucose ($\square$). (B) Dissolved organic carbon (DOC; $\bullet$); acetate produced ($\triangle$). (C) Cellular content of total carbohydrates ($\bullet$) and protein ($\circ$). With permission from Wanner and Egli (1990).

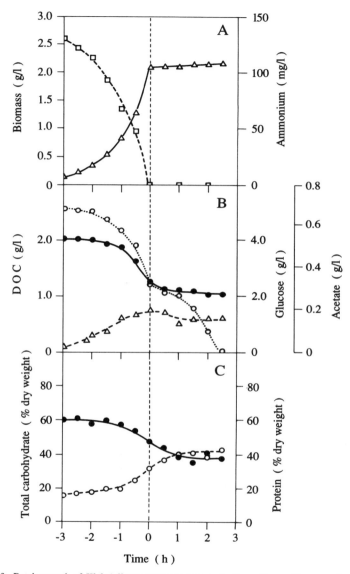

Figure 3. Batch growth of *Klebsiella pneumoniae* in a synthetic medium with ammonium as the growth-limiting substrate. $T = 30°C$, pH $= 7.0$. (A) Growth given as dry biomass produced ($\triangle$); ammonium ($\square$). (B) Dissolved organic carbon (DOC; $\bullet$); glucose ($\circ$); acetate produced ($\triangle$). (C) Cellular content of total carbohydrates ($\circ$) and protein ($\bullet$). With permission from Wanner and Egli (1990).

An example of such a growth curve is shown in Fig. 4, where the growth pattern is characterized by an extended deceleration phase. Biomass increases as much as sixfold have been observed using *K. pneumoniae* when sufficient carbon and other nutrients were available (Wanner, 1986). A similar growth pattern was reported for phosphate-terminated batch growth of *Streptomyces tendae* with glucose (Mundry and Kuhn, 1991). The effects of phosphorus exhaustion are different from those of nitrogen exhaustion in that the increase in biomass which occurs subsequent to depletion of phosphorus from the growth medium is caused by continued cell growth, i.e., initiation of new rounds of DNA replication. Unfortunately, there are no data to indicate whether a stationary phase (or death phase) actually occurs under these conditions since exhaustion of another nutrient, typically the carbon source, is always ultimately responsible for terminating further growth.

4.4. Other Nutrients

The three previous examples represent those nutrients which are most frequently described in the literature as the ones which terminate growth in batch cultures. However, with respect to other nutrients, similar extended deceleration phases can be expected and in some cases, e.g., magnesium, have been documented (reviewed in Wanner and Egli, 1990). Potassium (and this is probably also true for other nutrients which are not covalently bound in the cell) does not fit into this pattern of a clear reduction in growth rate after the onset of nutrient-limitation. Exponential growth of a culture of *K. penumoniae* appeared to be unaffected by the exhaustion of potassium (Wanner and Egli, 1990) with no apparent change in the growth rate occurring at the point at which the potassium concentration in the medium reached "zero." Exactly the same results were found for potassium-terminated growth of *E. coli* (Mulder *et al.*, 1988). The only indication of an effect as a result of potassium exhaustion in both *E. coli* and *K. pneumoniae* (Mulder *et al.*, 1988; Wanner and Egli, 1990) was that the specific growth rate of the culture was approximately 30% lower than during exponential growth under potassium-sufficient conditions, as a consequence of the low potassium concentrations required in the experiments to ensure potassium became the first nutrient to be exhausted. The growth curve obtained where sulfur exhaustion terminates growth is also an exception in that for the few data available, no extended deceleration phase has been shown to occur (Wanner and Egli, 1990). The biomass increase which occurred following exhaustion of sulfur probably is a result of further growth combined with a redistribution of this nutrient among those proteins necessary for survival. The small increase is probably the result of the fact that the pool of available sulfur is small with the majority of the sulfur bound in proteins.

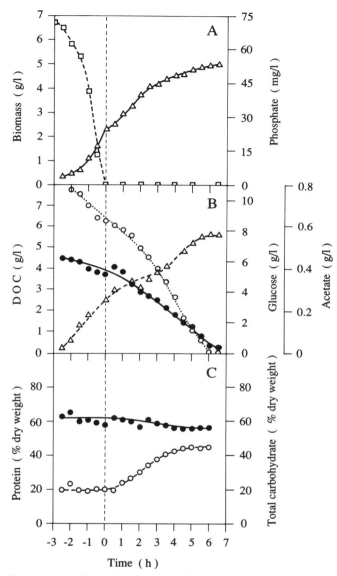

Time (h)

Figure 4. Batch growth of *Klebsiella pneumoniae* in a synthetic medium with phosphate as the growth-limiting substrate. $T = 30°C$, pH = 7.0. (A) Growth given as dry biomass produced ($\triangle$); phosphate ($\square$). (B) Dissolved organic carbon (DOC; $\bullet$); glucose ($\circ$; acetate produced ($\triangle$). (C) Cellular content of total carbohydrates ($\circ$) and protein ($\bullet$). With permission from Wanner and Egli (1990).

5. EFFECT OF NUTRIENT EXHAUSTION IN BATCH CULTURE ON PHYSIOLOGICAL PARAMETERS

5.1. Cell Number and Morphology

As described above, it is usual with heterotrophic bacteria for the stationary-phase biomass to remain effectively constant for several hours following carbon exhaustion. In contrast to this, the number of bacterial cells continues to increase during the first 1–2 h of the stationary phase (Maaløe and Kjeldgaard, 1966). This occurs because the cell is committed to complete already initiated rounds of DNA replication and is therefore a function of the growth rate in the exponential phase (Bremer and Dennis, 1987; Cooper, 1991). For very fast growing bacteria, as many as four additional rounds of cell division should occur subsequent to carbon source exhaustion. For example, an approximately twofold increase in cell number has been reported during the first part of the stationary phase for both *Beneckea natriegens* and *Escherichia coli* (Wade, 1952; Nazly *et al.*, 1980). The increase in cell number in the stationary phase consequently leads to a reduction in the cell size (Henrici, 1923; Huntington and Winslow, 1937; Hershey, 1938). In this phase it is also common for morphological changes to occur such as those observed in *Arthrobacter* spp. where the cells change from rod-shaped to coccoid forms (Luscombe and Gray, 1971, 1974). Similarly, a reduction in cell size with a concomitant change in shape from rods to cocci has been reported for many different gram-negative bacteria including *K. pneumoniae*, *E. coli*, *Vibrio* spp., or *Pseudomonas* spp. when subjected to starvation conditions (Marden *et al.*, 1985; Wrangstadh *et al.*, 1986; Lappin-Scott *et al.*, 1988; Malmcrona-Friberg *et al.*, 1986; Morita, this volume). Other morphological changes can also occur such as those found in prosthecate bacteria (*Caulobacter, Hyphomicrobium*) where it has been reported that the fraction of swarmer cells increased in the late exponential and stationary phase (Morgan and Dow, 1985).

In the case of nitrogen exhaustion, where biomass continues to increase, there are to our knowledge very little data available from experiments in which the number of cells has been directly measured during the transition from nutrient-sufficient to nutrient-exhausted conditions without removing the cells and putting them into a N-starvation medium. However, one would expect that fast-growing cells were potentially capable of up to four rounds of division in order to complete already initiated DNA replication, in an analogous manner to that described above for carbon exhaustion. For *Salmonella typhimurium* this has in fact been observed (Schaechter, 1961; von Meyenburg and Hansen, 1987). Nevertheless, it is also conceivable that, since carbon and energy are still sufficient and nitrogen can be remobilized from excess RNA, further DNA replication initiation (i.e., growth) can occur. However, this remains to be examined.

In all other cases the biomass produced after exhaustion of the nutrient increased by more than twofold (e.g., phosphorus, magnesium). Growth continued in the sense that new rounds of DNA replication were initiated and completed. Nevertheless, strict evidence for this is lacking since no data concerning changes in cell size or morphology in conjunction with good culture data seem to be available in the literature.

5.2. Product Formation and Utilization

During unrestricted exponential growth of *E. coli*, *K. pneumoniae*, or *B. natriegens* with glucose, a significant part (5–10%) of the carbon utilized is excreted in the form of acetate (Niven *et al.*, 1977; Andersen and von Meyenburg, 1980; Holms, 1986). After exhaustion of glucose this acetate is reutilized without a significant lag (Fig. 2) and the authors are unaware of cases where this has resulted in a second phase of growth. Although acetate has received most attention, other products are also released during exponential growth. The type of products excreted will depend on the carbon source, the microorganism, and the nature of the growth-terminating nutrient. For example, in addition to acetate, at low pH *K. pneumoniae* also produced pyruvate which was preferentially utilized after exhaustion of glucose (Lonsmann Iversen, 1987). In the experiment shown in Fig. 2, approximately 10% of the initial carbon supplied remained in the medium after utilization of acetate in the stationary phase. This indicates that other, less readily utilizable or nondegradable products may be excreted or, e.g., released as a result of cell lysis (Mason and Hamer, 1987).

At high initial substrate concentrations, significant concentrations of excretion products can be formed during exponential batch growth which can subsequently serve as a source of carbon and energy after exhaustion of the primary carbon source. In contrast, at low initial substrate concentrations the resulting concentrations of excreted products will be much lower and it remains to be seen if they have any impact on starvation survival.

In *K. pneumoniae* following nitrogen exhaustion when glucose is still present in excess, acetate production appears to stop (Fig. 3) whereas in the case of sulfur, potassium, and phosphate (Fig. 4) exhaustion, acetate production continued as long as the primary carbon source was available (Wanner and Egli, 1990). It is possible that when other primary carbon substrates are used, effecting less severe catabolite repression or supporting lower growth rates, metabolic overflow products might not be formed.

5.3. Accumulation of Storage Compounds

Very few organisms accumulate carbonaceous storage compounds such as polyhydroxyalkanoates (PHA) or glycogen during exponential growth in the presence of a large excess of all nutrients. Consequently, after exhaustion of carbon there is generally no available endogenous source of mobilizable carbon storage compounds to support those metabolic processes which occur in the stationary phase. One exception concerns *Arthrobacter crystallopoietes* which can accumulate a significant amount of a glycogen-like polysaccharide, which could be immediately consumed after carbon in the medium becomes exhausted (Boylen and Ensign, 1970). Similarly, *Vibrio* species S14 accumulated PHA during growth and the utilization of this reserve compound appears to in part regulate the starvation response (Östling *et al.*, this volume). Regulation of glycogen and PHA synthesis and the role of carbonaceous storage compounds in survival have been the subject of various reviews (Dawes and Senior, 1973; Preiss, 1984; Dawes, 1985; Morita, 1985).

Accumulation of reserve polymers such as glycogen, lipids, PHA, and polyphosphates, either singly or in combination, has been widely reported during the response to nitrogen depletion in batch culture (Kay and Gronlund, 1969; Dawes and Senior, 1973; Zevenhuizen and Ebbink, 1974; Niven *et al.*, 1977; Nazly *et al.*, 1980). These storage polymers can account for a significant fraction of the increase in dry weight observed under this growth regime (Duguid and Wilkinson, 1953; Nazly *et al.*, 1980; Dawes, 1985). An increase in the amount of storage compounds will result in a dilution effect of the other cellular constituents. This is clearly demonstrated in Fig. 3 for *K. pneumoniae*, where it can be seen that the increase in intracellular polymer concentration begins during the late exponential growth phase, in response to the steadily decreasing availability of ammonium.

Extracellular polysaccharide excretion (slime or capsules) is a common phenomenon for many bacteria following exhaustion of the nitrogen supply under otherwise nutrient-sufficient conditions. Up to 60% of the dry weight of *K. pneumoniae* can, in fact, be made up of exopolysaccharide. The production of these extracellular polysaccharides is enhanced when oxygen is simultaneously limiting. Production of this polymer has also been described for conditions where growth was terminated or limited by the availability of phosphorus, sulfur, and potassium (Duguid and Wilkinson, 1953; MacKelvie *et al.*, 1968; Neijssel and Tempest, 1979; Sutherland, 1982).

5.4. Content of DNA

The amount of DNA present in a cell during exponential growth is primarily a function of growth rate which is itself a function of the medium composition. As a result of the delay between the time needed for DNA replication and cell division, fast-growing cells can accumulate up to four DNA equivalents per cell (Bremer and Dennis, 1987; Cooper, 1991). As growth rate changes during the deceleration phase, the number of DNA equivalents is readjusted. At the onset of stationary phase the bacteria are still obliged to complete any already initiated rounds of DNA replication. The number of additional replication rounds will therefore be determined by the growth rate prior to entry into the stationary phase and thus dependent on the presence or duration of the deceleration phase. For carbon-terminated batch cultures such as in Fig. 2, DNA and cell replication are out of sequence and thus further cell divisions are inevitable, and occur within the first few hours after carbon depletion (Brdar *et al.*, 1965; Boylen and Ensign, 1970; Scherer and Boylen, 1977). As such, there is a requirement for both energy and building blocks to be supplied. In contrast, where an extended deceleration phase occurs, DNA replication initiation occurs less frequently and the number of chromosomes per cell approaches one. Therefore, by the time stationary phase is reached, no mobilization of energy or building blocks is necessary for this purpose. Even following carbon exhaustion, termination of DNA replication does not impose any excessive energetic demand on the cell because of low energy requirement and because energy and building blocks become available through the degradation of RNA (Képès, 1986; von Meyenburg and Hansen, 1987; Neidhardt *et al.*, 1990). Whereas earlier results have suggested that the number of DNA equivalents per cell generally diminished to one during the early stationary phase (resulting in an increase in the cell number), new data suggest that they can maintain, within the same cell, several DNA equivalents for an extended period of time in the stationary phase (Åkerlund *et al.*, 1992).

5.5. Content of RNA

The proportion of the different RNA fractions varies with the growth rate of the cell (Neidhardt *et al.*, 1990). The mRNA, which has a very high turnover rate and a half-life of approximately 4 min, contributes only very little (maximum 4%) to the total RNA. Its relevance in deceleration and stationary-phase processes is, therefore, negligible. As far as the stable forms of RNA (tRNA and rRNA) are concerned, their stability also varies with growth rate. At very slow growth rates, tRNA is more stable than rRNA which indicates that degradation of rRNA primarily provides energy and precursors for synthesis.

Immediately after carbon exhaustion, it has been shown for many different organisms that the degradation of RNA is initiated. It was found that extracellular products or reserve materials can retard the degradation of RNA. However, when such extracellular products are removed by washing, degradation of RNA starts immediately (Strange *et al.*, 1961; Dawes and Ribbons, 1965; Burleigh and Dawes, 1967; Boylen and Ensign, 1970; Scherer and Boylen, 1977; Nazly *et al.*, 1980). Extracellular accumulation of UV-adsorbing material (purines and pyrimidines) as well as inorganic phosphate has been reported to occur during the degradation of RNA (Strange *et al.*, 1961; Postgate and Hunter, 1962). Ribose serves as the main source of energy as a result of degradation of RNA. Most of the studies of RNA degradation have been carried out over a long period of time under starvation conditions, whereby up to 80% of the RNA has been found to be degraded (Burleigh and Dawes, 1967; Boylen and Ensign, 1970). Even during net degradation of RNA, synthesis continues in the cell at a rate of ca. 10% of that during exponential growth (Barner and Cohen, 1958; Mandelstam and Halvorson, 1960; Mandelstam, 1960). Recently, results have been published suggesting that ribosomal particles might stay intact for an extensive period of time during starvation (Östling *et al.*, this volume).

The authors are unaware of any good information concerning changes in RNA composition and concentration following nitrogen depletion. The only problem under such conditions is the availability of nitrogen for the synthesis of nucleic acids for DNA. However, since the growth rate is decreasing, the amount of RNA must also be adjusted and as such some RNA degradation should occur which would satisfy this requirement.

RNA serves as the major source of phosphorus following the exhaustion of the P source from the growth medium. RNA degradation occurs, after depletion of the small cellular orthophosphate pool, at a rapid rate (10–15% per hour) in cells exposed to phosphate starvation conditions (Horiuchi, 1959; Medveczky and Rosenberg, 1971). The transfer of as much as 60% of the P initially present in RNA to DNA could be demonstrated using labeled [^{32}P]-RNA. When polyphosphate is present in the form of volutin granules, these can also serve as a readily mobilizable source of P for nucleic acid synthesis (Harold, 1963, 1964). In gram-positive bacteria, P-containing cell wall components can also contribute to the mobilizable P pool (see Section 5.7).

5.6. Proteins

During the early stages following exhaustion of carbon, no net change in the concentration of proteins is usually observed. This has been confirmed for various organisms (e.g., see Fig. 2). Nevertheless, very significant alterations in the composition of the proteins present occur as indicated by the rapid increase in the turnover rate from 1–3% per hour to roughly 5–7% after carbon exhaustion (Mandelstam, 1960; Pine, 1972; Goldberg and Dice, 1974). Details of the proteins that are synthesized in response to carbon exhaustion as well as those induced as a result of exhaustion of other nutrients are discussed in detail in various other chapters of this book. The general degradation pattern of cellular constituents indicates that the breakdown of proteins only becomes a primary source of energy after other less essential components such as storage compounds and RNA have been exploited to their limit (Strange *et al.*, 1961; Postgate and Hunter, 1962; Burleigh and Dawes, 1967; Scherer and Boylen, 1977; Nazly *et al.*, 1980; Boyaval *et al.*, 1985). There is strong evidence to suggest that during the first hours of nitrogen depletion, there is no net degradation of protein and the observed decrease is in fact caused by a dilution effect linked to the incorporation of storage material (Holme and Palmstierna, 1956; Mandelstam and Halvorson, 1960). The turnover rate also increased from 1% per hour prior to nitrogen exhaustion to 5% subsequently (Schleissinger and Ben-Hamida, 1966; Willetts, 1967; Miller, 1975). Among those proteins specifically induced as a result of reduced availability of nitrogen are amino acid binding proteins and high-affinity ammonium assimilation systems (Kustu *et al.*, 1979; Magasanik and Neidhardt, 1987). There is unfortunately very little information regarding the dynamics of induction of such proteins subsequent to the exponential growth phase during batch cultivation, in contrast to the large amount of information available from starvation experiments carried out by transferring the cells into a N-deficient medium.

Little information exists on change in the protein content following exhaustion of other nutrients. However, after exhaustion of, e.g., sulfur, both the carbon and nitrogen sources should, by definition, still be in excess. The observed small increase of biomass following sulfur exhaustion is then probably a reflection of the fact that the available pool of mobilizable sulfur is very restricted.

5.7. Cell Envelope

Carbon starvation results in an increase in the degree of cross-linking in the murein layer in gram-negative bacteria (Schleifer *et al.*, 1976; Pisabarro *et al.*, 1985). An increase in the thickness of the peptidoglycan layer has also been demonstrated from two to three layers of peptidoglycan to four to five layers in *E. coli* (Leduc *et al.*, 1989). Turnover of the peptidoglycan layer has been especially studied in gram-positive bacteria. It was found that carbon starvation resulted in a reduction in the turnover of this layer from 30–50% per hour, as found during exponential growth, to negligible levels. This has important consequences for the susceptibility, especially of medically important bacteria, to cell-wall-active agents. Various other less well defined alterations have also been described such as formation of fibrillar structures, increased surface hydrophobicity, surface roughness, and formation of vesicles on the bacterial surfaces during long-term carbon starvation (Kefford *et al.*, 1982; Kjelleberg and Hermansson, 1984; Kjelleberg *et al.*, 1987).

Alterations to cell walls in gram-positive bacteria have also been recently studied using electron microscopy of cells of *Bacillus subtilis* (Archibald *et al.*,

1989; Merad *et al.*, 1989; Clark-Sturman *et al.*, 1989). These studies confirmed earlier data showing the replacement of phosphorus-containing teichoic acids by phosphorus-free teichuronic acids in the cell walls (Ellwood and Tempest, 1972; Neidhardt *et al.*, 1990). However, these results were obtained using continuously grown cells, and little if any direct information is available from batch experiments. It is known that gram-positive bacteria slough off a large amount of phosphorus-containing cell wall material during exponential growth (Wanner and Egli, 1990). In dense cultures this may serve as an additional source of phosphorus following its exhaustion. Also the content of lipoteichoic acids has been observed to decrease significantly in stationary-phase cells of *Streptococcus mutans* under carbon starvation (Fisher, 1988).

Various alterations to the inner and outer membrane have been observed following nutrient exhaustion. These include chemical alteration (saturation and epoxylation) of unsaturated fatty acids of phospholipids (Marr and Ingraham, 1962; Shaw and Ingraham, 1965; Cronan, 1968; Hood *et al.*, 1986; Guckert *et al.*, 1986), changes in the ratio of the various fatty acids present (White and Tucker, 1969; Cronan, 1968; Short and White, 1971; DeSiervo and Salton, 1973; Malmcrona-Friberg *et al.*, 1986), and even the disappearance of phospholipids (Oliver and Stringer, 1984; Hood *et al.*, 1986; Lonsmann Iversen, 1987). Other changes in membranes have been reported in the literature as a result of nutrient exhaustion (summarized in Wanner and Egli, 1990). However, it should be mentioned that most of these changes were investigated during long-term starvation and that no detailed information has been published where adaptations and alterations in the cell envelope occur during the first few hours after nutrient depletion.

6. CONCLUDING REMARKS

In this restricted overview we have tried to indicate the ways in which bacteria respond to diminished nutrient availability at the end of growth in batch culture. These responses are dependent on many variables, including the bacterium examined, the complexity of the growth medium, and the growth-terminating nutrient. We have tried to summarize the generally observed response patterns in Table 1 with respect to the major cellular constituents.

It is remarkable how little we really know about the batch growth curve despite its history and establishment in the doctrines of microbiology. The present concepts of bacterial growth curves are dominated by the pattern exhibited by carbon-terminated cultures, characterized by a sudden abrupt change from exponential growth into the stationary phase. The deceleration phase is usually poorly described or ignored. Here we have shown that the deceleration phase can be quite extensive and might be characteristic for

Table 1
Changes in Parameters of Batch Culture Cells Observed within the First Few Hours after Depletion of Various Nutrients[a]

Changes observed after substrate exhaustion	Substrate terminating batch growth				
	Carbon	Nitrogen	Phosphorus	Sulfur	Potassium
Biomass	0^b	+	+ +	+	+ +
Cell size	−	?	?	?	?
Cell number	+	+	+ +	+	+ +
Metabolites	−	+ +	+	+ +	+ +
Carbohydrates	0	+ +	+	+	0
Polyphosphate	0	+	−	+ +	0
DNA	+	?	?	?	?
RNA	−	−	−	(0)	?
Protein	0	−	0/−	0	0

[a]Adapted from Wanner and Egli (1990).
[b]+(+), (strong) increase; 0, no change; −, decrease; ?, no data available in the literature.

exhaustion of different nutrients. What is clear from many of the growth curves published in the literature is that despite the simplicity of appearance and familiarity of the bacterial batch growth curve, analysis of them is often superficial and the complexity of the processes occurring is then overlooked.

Interpretation of a lot of the data in the older biochemical literature is difficult because essential culture parameters were not described. In many cases, unfortunately, this is still true for more recent work. As we have tried to explain in this contribution, culture data, such as residual substrate and biomass concentrations and their time courses, are crucial for a detailed and thorough interpretation of cellular rearrangement processes in the early stationary phase. To illustrate this, a large number of experiments have been carried out where growth was terminated by exhaustion of a compound other than that described. Frequently, oxygen was a limiting factor. In experiments designed to examine, e.g., nitrogen-terminated growth, it is not uncommon for simultaneous exhaustion of carbon to also occur, albeit temporarily delayed. The experiments shown in Figs. 3 and 4 clearly demonstrate that carbon utilization does continue despite exhaustion of another essential nutrient. This feature must be taken into account during starvation experiments especially when one wants to study the effects of exhaustion of one particular nutrient only. In this respect it is also worth mentioning that the design or choice of the culture medium must also be carried out carefully to avoid omitting essential nutrients. One particularly good example where such a poor medium is frequently employed is M9. In its strict composition it contains no trace elements. In earlier times this medium was probably an excellent cheap minimal medium because all of the trace elements were provided by impurities in the retailed chemicals and from the use of tap water. Today, where everything has to be of the highest possible purity, these trace elements probably do not find their way into the culture medium by this source and it is not uncommon to see M9 supplemented with, e.g., yeast extract or peptone, to improve growth. *E. coli* can only grow for two transfers from complex medium through trace-element-free minimal M9 medium before growth becomes poor.

The concepts and results of a more detailed understanding of the dynamics of microbial growth in batch culture find an important application not only in environmental microbiology but also in biotechnology. Many technical applications require the optimal growth of microorganisms and some, e.g., secondary metabolite production, depend on the ability to control the dynamics of growth and metabolism in the decelerating and stationary phases of growth.

In conclusion, it is now clear that a better understanding of the starvation process requires more comprehensive knowledge of the decelerating and early stationary phases of growth. The dynamics of these processes are complex but can provide essential information on the pattern of responses one can expect as a result of nutrient exhaustion.

REFERENCES

Åkerlund, T., Bernader, R., and Nordström, K., 1992, DNA content and cell size distribution in exponentially growing and stationary phase cultures of *Escherichia coli*, Abstr. Meeting on Kinetics, Dynamics and Physiology of Microbial Growth. Rüschlikon, Switzerland.

Al-Awadhi, N., Egli, T., and Hamer, G., 1988, Growth characteristics of a thermotolerant methylotrophic *Bacillus* sp. (NCIB 12522) in batch culture, *Appl. Microbiol. Biotechnol.* **29:**485–493.

Andersen, K. B., and von Meyenburg, K., 1980, Are growth rates of *Escherichia coli* in batch cultures limited by respiration? *J. Bacteriol.* **144:**114–123.

Archibald, A. R., Glassey, K., Green, R. S., and Lang, W. K., 1989, Cell wall composition and surface properties in *Bacillus subtilis*: Anomalous effect of incubation temperature on the phage-binding properties of bacteria containing varied amounts of teichoic acids, *J. Gen. Microbiol.* **135:**667–673.

Barner, H. D., and Cohen, S. C., 1958, Protein synthesis and RNA turnover in a pyrimidine-deficient bacterium, *Biochim. Biophys. Acta* **30:**12–20.

Boyaval, P., Boyaval, E., and Desmatseaud, M. J., 1985, Survival of *Brevibacterium linens* during nutrient starvation and intracellular changes, *Arch. Microbiol.* **141:**128–132.

Boylen, C. W., and Ensign, J. C., 1970, Intracellular substrates for endogenous metabolism during long-term starvation of rod and spherical cells of *Arthrobacter crystallopoietes*, *J. Bacteriol.* **103:**578–587.

Brdar, B., Kos, E., and Drakulic, M., 1965, Metabolism of nucleic acids and protein in starving bacteria, *Nature* **16:**303–304.

Bremer, H., and Dennis, P. P., 1987, Modulation of chemical composition and other parameters of the cell by growth rate, in: *Escherichia coli* and *Salmonella typhimurium* (F. C. Neidhardt, J.

L. Ingraham, K. B. Low, B. Magasanik, M. Schaechter, and H. E. Umbarger, eds.), American Society for Microbiology, Washington, D.C., pp. 1527–1542.

Buchanan, R. E., 1918, Life phases in a bacterial culture, *J. Infect. Dis.* **23**:109–125.

Burleigh, I. G., and Dawes, E. A., 1967, Studies on the endogenous metabolism and senescence of starved *Sarcina lutea*, *Biochem. J.* **102**:236–250.

Clark-Sturman, A. J., Archibald, A. R., Hancock, I. C., Harwood, C. R., Merad, T., and Hobot, J. A., 1989, Cell wall assembly in *Bacillus subtilis*: Partial conservation of polar wall material and the effect of growth conditions on the pattern of incorporation of new material at the polar caps, *J. Gen. Microbiol.* **135**:657–665.

Cooper, S., 1991, *Bacterial Growth and Division: Biochemistry and Regulation of Prokaryotic and Eukaryotic Division Cycles*, Academic Press, New York.

Cronan, J. E., 1968, Phospholipid alterations during growth of *Escherichia coli*, *J. Bacteriol.* **95**:2054–2061.

Dawes, E. A., 1985, Starvation, survival and energy reserves, in: *Bacteria in Their Natural Environments* (M. Fletcher and G. D. Floodgate, eds.), Academic Press, New York, pp. 43–79.

Dawes, E. A., and Ribbons, D. W., 1965, Studies on the endogenous metabolism of *Escherichia coli*, *Biochem. J.* **95**:332–343.

Dawes, E. A., and Senior, P. J., 1973, The role and regulation of energy reserve polymers in microorganisms, *Adv. Microb. Physiol.* **10**:135–266.

DeSiervo, A. J., and Salton, M. R. J., 1973, Changes in phospholipid composition of *Micrococcus lysodeikticus* during growth, *Microbios* **8**:73–78.

Duguid, J. P., and Wilkinson, J. F., 1953. The influence of cultural conditions on polysaccharide production by *Aerobacter aerogenes*, *J. Gen. Microbiol.* **9**:174–189.

Ellwood, D. C., and Tempest, D. W., 1972, Effects of environment on bacterial wall content and composition, *Adv. Microb. Physiol.* **7**:83–117.

Fisher, W., 1988, Physiology of lipoteichoic acids in bacteria, *Adv. Microb. Physiol.* **29**:233–302.

Goldberg, A. L., and Dice, J. F., 1974, Intracellular protein degradation in mammalian and bacterial cells, *Annu. Rev. Biochem.* **43**:835–869.

Guckert, J. B., Hood, M. A., and White, D. C., 1986, Phospholipid ester-linked fatty acid profile changes during nutrient deprivation of *Vibrio cholerae*: Increases in the *trans/cis* ratio and proportions of cyclopropyl fatty acids, *Appl. Environ. Microbiol.* **52**:749–801.

Harder, W., and Dijkhuizen, L., 1983, Physiological responses to nutrient limitation, *Annu. Rev. Microbiol.* **37**:1–23.

Harold, F. M., 1963, Accumulation of inorganic polyphosphate in *Aerobacter aerogens*. I. Relationship to growth and nucleic acid synthesis, *J. Bacteriol.* **86**:216–221.

Harold, F. M., 1964, Enzymatic and genetic control of polyphosphate accumulation in *Aerobacter aerogenes*, *J. Gen. Microbiol.* **35**:81–90.

Henrici, A. T., 1923, Influence of age of parent culture on size of cells of *Bacillus megatherium*, *Proc. Soc. Exp. Biol. Med.* **21**:343–346.

Herbert, D., 1961, The chemical composition of micro-organisms as a function of their environment, *Symp. Soc. Gen. Microbiol.* **11**:391–416.

Hershey, A. D., 1938, Factors limiting bacterial growth. II. Growth without lag in *Bacterium coli* cultures, *Proc. Soc. Exp. Biol. Med.* **38**:127–129.

Holme, T., and Palmstierna, H., 1956, Changes in glycogen and nitrogen-containing compounds in *Escherichia coli* B during growth in deficient media. I. Nitrogen and carbon starvation, *Acta Chem. Scand.* **10**:578–586.

Holms, W. H., 1986, The central metabolic pathways of *Escherichia coli*: Relationship between flux and control at a branch point, efficiency of conversion to biomass, and excretion of acetate, *Curr. Top. Cell. Regul.* **28**:69–105.

Hood, M. A., Guckert, J. B., White, D. C., and Deck, F., 1986, Effect of nutrient deprivation on lipid, carbohydrate, DNA, RNA, and protein levels in *Vibrio cholerae*, *Appl. Environ. Microbiol.* **52:**788–793.

Horiuchi, T., 1959, RNA degradation and DNA and protein synthesis of *E. coli* B in a phosphate-deficient medium, *J. Biochem.* **46:**1467–1480.

Huntington, E., and Winslow, C. E. A., 1937, Cell size and metabolic activity at various phases of the bacterial culture cycle, *J. Bacteriol.* **33:**123–144.

Kay, W. W., and Gronlund, A. F., 1969, Influence of carbon or nitrogen starvation on amino acid transport in *Pseudomonas aeruginosa*, *J. Bacteriol.* **100:**276–282.

Kefford, B., Kjelleberg, S., and Marshall, K. C., 1982, Bacterial scavenging: Utilization of fatty acids localized at a solid–liquid interface, *Arch. Microbiol.* **133:**257–260.

Képès, F., 1986, The cell cycle of Escherichia coli and some of its regulatory systems, *FEMS Microbiol. Rev.* **32:**225–246.

Kjelleberg, S., and Hermansson, M., 1984, Starvation effects on bacterial surface characteristics, *Appl. Environ. Microbiol.* **48:**497–503.

Kjelleberg, S., Hermansson, M., and Marden, P., 1987, The transient phase between growth and nongrowth of heterotrophic bacteria, with emphasis on the marine environment, *Annu. Rev. Microbiol.* **41:**25–49.

Kustu, S. G., McFarland, N. C., Hui, S. P., Esmon, B., and Ames, G. F.-L., 1979, Nitrogen control in *Salmonella typhimurium*: Co-regulation of synthesis of glutamine synthetase and amino acid transport system, *J. Bacteriol.* **138:**218–234.

Lappin-Scott, H. M., Cusack, F., MacLeod, A., and Costerton, J. W., 1988, Starvation and nutrient resuscitation of *Klebsiella pneumoniae* isolated from oil well waters, *J. Appl. Bacteriol.* **64:**541–549.

Leduc, M., Fréhel, C., Siegel, E., and Van Heijenoort, J., 1989, Multilayered distribution of peptidoglycan in the periplasmic space of *Escherichia coli*, *J. Gen. Microbiol.* **135:**1243–1254.

Lonsmann Iversen, J. J., 1987, The pH mediated effects of initial glucose concentration on the transitory occurrence of extracellular metabolites, gas exchange and growth yields of aerobic batch cultures of *Klebsiella pneumoniae*, *Biotechnol. Bioeng.* **30:**352–362.

Luscombe, B. M., and Gray, T. R. G., 1971, Effect of varying growth rate on the morphology of *Arthrobacter*, *J. Gen. Microbiol.* **69:**433–434.

Luscombe, B. M., and Gray, T. R. G., 1974, Characteristics of Arthrobacter grown in continuous culture, *J. Gen. Microbiol.* **82:**213–222.

Maaløe, O., and Kjeldgaard, N. O., 1966, *Control of Macromolecular Synthesis*, Benjamin, New York.

MacKelvie, R. M., Campbell, J. J. R., and Gronlund, A. F., 1968, Absence of storage products in cultures of *Pseudomonas aeruginosa* grown with excess carbon or nitrogen, *Can. J. Microbiol.* **14:**627–631.

Magasanik, B., and Neidhardt, F. C., 1987, Regulation of carbon and nitrogen utilization, in: *Escherichia coli and Salmonella typhimurium* (F. C. Neidhardt, J. L. Ingraham, K. B. Low, B. Magasanik, M. Schaechter, and H. E. Umbarger, eds.), American Society for Microbiology, Washington, D. C., pp. 1318–1325.

Malmcrona-Friberg, K., Tunlid, A., Marden, P., Kjelleberg, S., and Odham, G., 1986, Chemical changes in cell envelope and poly-β-hydroxybutyrate during short term starvation of a marine bacterial isolate, *Arch. Microbiol.* **144:**340–345.

Mandelstam, J., 1960, The intracellular turnover of protein and nucleic acids and its role in biochemical differentiation, *Bacteriol. Rev.* **24:**289–308.

Mandelstam, J., and Halvorson, H., 1960, Turnover of protein and nucleic acid in soluble ribosome fractions of non-growing *Escherichia coli*, *Biochim. Biophys. Acta* **40:**43–49.

Marden, P., Tunlid, A., Malmcrona-Friberg, K., Odham, G., and Kjelleberg, S., 1985, Physiological and morphological changes during short-term starvation of marine bacterial isolates, *Arch. Microbiol.* **142**:326–332.

Marr, A. G., and Ingraham, J. L., 1962, Effect of temperature on the composition of fatty acids in *Escherichia coli, J. Bacteriol.* **84**:1260–1267.

Mason, C. A., and Hamer, G., 1987, Cryptic growth in *Klebsiella pneumoniae, Appl. Microbiol. Biotechnol.* **25**:577–584.

Medveczky, N., and Rosenberg, H., 1971, Phosphate transport in *Escherichia coli, Biochim. Biophys. Acta* **241**:494–506.

Merad, T., Archibald, A. R., Hancock, I. C., Harwood, C. R., and Hobot, J. A., 1989, Cell wall assembly in *Bacillus subtilis*: Visualization of old and new wall material by electron microscopic examination of samples stained selectively for teichoic acid and teichuronic acid, *J. Gen. Microbiol.* **135**:645–655.

Miller, C. G., 1975, Peptidases and proteases of *Escherichia coli* and *Salmonella typhimurium, Annu. Rev. Microbiol.* **29**:485–504.

Morgan, P., and Dow, C. S., 1985, Environmental control of cell-type expression in prosthecate bacteria, in: *Bacteria in Their Natural Environments* (M. Fletcher and G. D. Floodgate, eds.), Academic Pres, New York, pp. 131–169.

Morita, R. Y., 1985, Starvation and miniaturization of heterotrophs with special emphasis on maintenance of the starved viable state, in: *Bacteria in Their Natural Environments* (M. Fletcher and G. D. Floodgate, eds.), Academic Press, New York, pp. 111–130.

Mulder, M. M., van der Gulden, H. M. L., Postma, P. W., and Van Dam, K., 1988, Continued growth of *Escherichia coli* after stopping medium addition to a potassium-limited chemostat culture, *J. Gen. Microbiol.* **137**:777–783.

Müller, M., 1895, Ueber den Einfluss von Fiebertemperaturen auf die Wachstumsgeschwindigkeit und die Virulenz des Typhus-Bacillus, *Z. Hyg. Infektionskrankh.* **20**:245–280.

Mundry, C., and Kuhn, K-P., 1991, Modelling and parameter identification for batch fermentations with *Streptomyces tendae* under phosphate limitation. *Appl. Microbiol. Biotechnol.* **35**:306–311.

Nazly, N., Carter, I. S., and Knowles, C. J., 1980, Adenine nucleotide pools during starvation of *Beneckea natriegens, J. Gen Microbiol.* **116**:295–303.

Neidhardt, F. C., Ingraham, J. L., and Schaechter, M., 1990, *Physiology of the Bacterial Cell: A Molecular Approach*, Sinauer, Sunderland, Mass.

Neijssel, O. M., and Tempest, D. W., 1979, The physiology of metabolite overproduction, in: *Microbial Technology* (A. T. Bull and D. C. Ellwood, eds.), Cambridge University Press, London, pp. 53–82.

Niven, D. F., Collins, P. A., and Knowles, C. J., 1977, Adenylate energy charge during batch culture of *Beneckea natriegens, J. Gen. Microbiol.* **98**:95–108.

Oliver, J. D., and Stringer, W. F., 1984, Lipid composition of a psychrophilic marine *Vibrio* sp. during starvation-induced morphogenesis, *Appl. Environ. Microbiol.* **47**:461–466.

Pine, M. J., 1972, Turnover of intracellular proteins, *Annu. Rev. Microbiol.* **26**:103–126.

Pisabarro, A. G., de Pedro, M. A., and Vazquez, D., 1985, Structural modifications in the peptidoglycan of *Escherichia coli* associated with changes in the state of growth in the culture, *J. Bacteriol.* **161**:238–242.

Postgate, J. K., and Hunter, J. R., 1962, The survival of starved bacteria, *J. Gen. Microbiol.* **29**:233–263.

Preiss, J., 1984, Bacterial glycogen synthesis and its regulation, *Annu. Rev. Microbiol.* **38**:419–458.

Schaechter, M., 1961, Patterns of cellular control during unbalanced growth, *Cold Spring Harbor Symp. Quant. Biol.* **26**:53–62.

Schaechter, M., Maaløe, O., and Kjeldgaard, N. O., 1958, Dependency on medium and temperature of cell size and chemical composition during balanced growth of *Salmonella typhimurium*, *J. Gen. Microbiol.* **19:**592–606.

Scherer, C. G., and Boylen, C. W., 1977, Macromolecular synthesis and degradation in *Arthrobacter* during periods of nutrient deprivation, *J. Bacteriol.* **132:**584–589.

Schleifer, K. H., Hammes, W. P., and Kandler, O., 1976, Effect of endogenous and exogenous factors on the primary structures of bacterial peptidoglycan, *Adv. Microb. Physiol.* **13:**246–292.

Schleissinger, D., and Ben-Hamida, F., 1966, Turnover of protein in *Escherichia coli* starving for nitrogen, *Biochim. Biophys. Acta* **119:**171–182.

Shaw, M. A., and Ingraham, J. L., 1965, Fatty acid composition of *Escherichia coli* as a possible controlling factor of the minimal growth temperature, *J. Bacteriol.* **90:**141–146.

Short, S. A., and White, D. C., 1971, Metabolism of phosphatidylglycerol, lysylphosphatidylglycerol, and cardiolipin of *Staphylococcus aureus*, *J. Bacteriol.* **108:**219–226.

Strange, R. E., Dark, F. A., and Ness, A. G., 1961, The survival of stationary phase *Aerobacter aerogenes* stored in aqueous suspension, *J. Gen. Microbiol.* **25:**61–76.

Sutherland, I. W., 1982, Biosynthesis of microbial exopolysaccharides, *Adv. Microb. Physiol.* **23:**79–150.

von Meyenburg, K., and Hansen, F. G., 1987, Regulation of chromosome replication, in: *Escherichia coli and Salmonella typhimurium* (F. C. Neidhardt, J. L. Ingraham, K. B. Low, B. Magasanik, M. Schaechter, and H. E. Umbarger, eds.), American Society for Microbiology, Washington, D. C., pp. 1555–1577.

Wade, H. E., 1952, Observations on the growth phases of *Escherichia coli*, American type 'B', *J. Gen. Microbiol.* **7:**8–23.

Wanner, U., 1986, Auswirkung verschiedener Nährstoff-limitationen auf das Batch-Wachstum und die Zellzusammensetzung von *Klebsiella pneumoniae*, Diploma thesis, Institute for Water Resources and Water Technology, Federal Institutes of Technology, Zurich.

Wanner, U., and Egli, T., 1990, Dynamics of microbial growth and cell composition in batch culture, *FEMS Microbiol. Rev.* **75:**19–44.

Ward, H. M., 1895, On the biology of *Bacillus ramosus* (Fraenkel), a schizomycete of River Thames, *Proc. R. Soc. London* **58:**265–268.

White, D. C., and Tucker, A. N., 1969, Phospholipid metabolism during bacterial growth, *J. Lipid Res.* **10:**220–233.

Willetts, N. S., 1967, Intracellular protein breakdown in non-growing cells of *Escherichia coli*, *Biochem. J.* **103:**453–461.

Wrangstadh, M., Conway, P. L., and Kjelleberg, S., 1986, The production and release of an extracellular polysaccharide during starvation of a marine *Pseudomonas* sp. and the effect on adhesion, *Arch. Microbiol.* **145:**220–227.

Zevenhuizen, L. P. T. M., and Ebbink, A. G., 1974, Interrelation between glycogen, poly-β-hydroxy butyric acid and lipids during accumulation and subsequent utilization in a *Pseudomonas*, *Antonie van Leeuwenhoek* **40:**103–120.

5

Starvation and Recovery of Vibrio

Jörgen Östling, Louise Holmquist, Klas Flärdh, Björn Svenblad, Åsa Jouper-Jaan, and Staffan Kjelleberg

1. INTRODUCTION

Vibrio species are generally found in the cultivable fraction of marine bacteria. Their coexistence with the dominant pool of oligobacteria in carbon-limited marine waters (Button *et al.*, 1993) as well as their frequent appearance in association with higher organisms (see below) imply a successful life cycle of feast and famine. This chapter presents data on the emerging pattern of an elaborate starvation-induced "differentiation-like programme" in marine *Vibrio* species, the inducing conditions of this response and its specific, non-growth-associated physiology, and the routes to successful persistence as well as recovery of the starved ultramicrocell.

It is suggested that the *Vibrio* species that are examined in several laboratories today are greatly appropriate for studies of the starvation-induced response in nondifferentiating bacteria. The long-term starvation and stress resistance of these organisms allows for convenient and reliable studies of ultramicrocell formation, free from experimental uncertainties such as loss in viability, cryptic growth, and the development of a heterogeneous population with time of starvation.

Jörgen Östling, Louise Holmquist, Klas Flärdh, Björn Svenblad, Åsa Jouper-Jaan, and Staffan Kjelleberg • Department of General and Marine Microbiology, University of Göteborg, S-413 19 Göteborg, Sweden. *Present address of S. K.*: School of Microbiology and Immunology, University of New South Wales, Kensington, New South Wales 2033, Australia.

Starvation in Bacteria, edited by Staffan Kjelleberg. Plenum Press, New York, 1993.

While not further explored in this contribution, some applications of starvation-specific responses and the gene and protein responders that mediate the adaptation pathways are worth mentioning. These applications are likely to attract a wide interest of research. For example, novel and efficient means of controlling *in situ* bioremediation and constructing biological containment for safe deliberate release of bacteria into the environment are developing rapidly. In both cases, starvation-specific promoters are asked to control genes of interest for the specific application, such as those for degradation of pollutants and genes for killing of the host cell to prevent persistence in the ecosystem. A third example is the development of a new generation of biomarkers that monitor the physiological status of environmental bacteria in response to starvation and stress. These are directed against specific protein responders. Finally, there is a renewed interest for the use of nonpathogenic bacteria for production of recombinant proteins. The efficient secretion machinery of extracellular proteins in marine *Vibrio* species can be utilized for such production. The fact that the nongrowing cell can efficiently deliver the recombinant protein extracellularly allows the additional benefit of production that is independent of biomass increase.

This chapter includes an introductory presentation of *Vibrio* species in the environment and the current understanding of the taxonomy of this genus. The core of this contribution focuses on the physiology and molecular biology of the carbon starvation network. This discussion is introduced by summarizing the past few years of studies of the adaptation to multiple-nutrient limitation, which forms the basis for the present model of formation of starved ultramicrocells of *Vibrio*. The last section provides a brief comparison of the starvation-induced responses displayed by different *Vibrio* spp. Hence the diversity in the adaptation pathways among different species is highlighted.

2. VIBRIOS IN THE ENVIRONMENT

Members of the genus *Vibrio* appear to successfully proliferate in areas of high substrate availability and cell density, as well as to persist as free-living cells in pelagic waters. There are several examples of sites of colonization. Dominant culturable bacteria in kelp beds are vibrios (Davis, 1985). *Vibrio* spp. are also known to constitute a significant fraction of the gastrointestinal microflora of fish and invertebrates (Belas and Colwell, 1982; Onarheim, 1990; Ruby and Morin, 1979; Westerdahl *et al.*, 1991) (see Appendix 1). Moreover, the same species can be found as free-living cells as well as associated with higher organisms and at inanimate surfaces. For example, *V. fischeri* exists as free-living as well as symbionts in light organs of fish and squid (Ruby and McFall-Ngai, 1992; Ruby and Morin, 1979) and *V. parahaemolyticus* has been isolated from seawater, sediments, and plankton samples on numerous occasions (Chowdhury *et al.*, 1990; Kaneko and Colwell, 1973; Venkateswaran *et al.*, 1989). Long-term persistence of several *Vibrio* species in marine waters is commonly shown (see Morita, this volume) and members of the genus *Vibrio* are generally identified among marine psychrophilic isolates (Morita, 1975). It should be reinforced that many studies have demonstrated the ability of *Vibrio* species to form ultramicrocells (e.g., Davis, 1985, 1992; MacDonell and Hood, 1982; Tabor *et al.*, 1981). Of 29 isolated ultramicrobacteria, i.e., cells that passed through a 0.2-μm filter, from different sites in an estuarine, 62% belonged to the genus *Vibrio* (MacDonell and Hood, 1984).

While it is clear that *Vibrio* species are well represented in the cultivable fraction of bacterial communities in a variety of marine ecosystems (Bauman and Bauman, 1977; Belas and Colwell, 1982; Chowdhury *et al.*, 1990; Colwell, 1984; Corpe, 1970; Davis, 1985), it appears that they may also be considered as significant members of marine pelagic bacterial populations as deduced by 16 S rRNA sequences extracted from the whole community. This consists of largely unculturable bacteria. In an oligotrophic marine picoplankton community, it was found that several sequences were related to common marine isolates of the gamma subdivision of Proteobacteria, in addition to 16 S rRNA sequences from phylogenetic groups of bacteria that are not related to any known rRNA sequences from cultivated organisms (Schmidt *et al.*, 1991). Further, all barophilic bacteria identified in a recent study of phylogenetic relationships between deep-sea bacteria were grouped into one of five distinct lineages within the gamma subdivision and clustered around the genera *Vibrio/ Photobacterium*, *Shewanella*, *Colwellia*, and two as yet undescribed lineages (DeLong and Franks, 1992).

The opportunistic character of *Vibrio* is displayed in a variety of ways. For example, *Vibrio* strains possess a large number of exoenzymes that allow for efficient degradation of a variety of polymeric substances in the environment as well as host tissue of living surfaces (Buck *et al.*, 1991; Colwell, 1984; Davis, 1992; Deane *et al.*, 1987; Yu and Lee, 1987). As can be seen in Appendix 1, several species of *Vibrio* are pathogenic for man and others are pathogenic for marine vertebrates and invertebrates (e.g., Austin and Austin, 1987; Colwell, 1984; Oliver, this volume).

The family Vibrionaceae has undergone considerable taxonomic revisions during the last few years. The pattern emerging from an increasing number of 16 S rRNA sequences suggests that either there should be one genus, *Vibrio*, which also includes the species now assigned to the genus *Photobacterium*, or should the latter genus be kept, the genus *Vibrio* may be divided into five or six genera (Stackebrandt, personal communication). The marine bacterial strains S14 and DW1, for which detailed physiological and molecular studies of the starvation-induced differentiation program are presented in this chapter, have been 16 S rRNA sequenced and found to reside in the genus *Vibrio* (Zabkar *et al.*, to be published). The sequences of these strains clearly do not cluster with any of the *Vibrio* species sequenced to date and warrant species status.

3. ADAPTATION TO MULTIPLE-NUTRIENT LIMITATION

This section offers a brief overview of the results obtained from earlier studies of the adaptations displayed by *Vibrio* sp. S14 cells in response to multiple-nutrient starvation, i.e., simultaneous depletion of carbon, nitrogen, and phosphorus. For a more detailed description on multiple-nutrient limitation in *Vibrio*, see Morita (1985, 1988, this volume) and Kjelleberg *et al.* (1987, 1990, 1993). The model that emerged from these studies forms the basis for the present detailed research on carbon starvation physiology, addressed in the subsequent sections of this chapter.

3.1. Morphology and Physiology

Since a bacterial cell that becomes exposed to nutrient limitation may not be able to physically escape, an adjustment of the phenotype is a prerequisite for survival. This process can be observed microscopically as a process of reductive division where each rod-shaped cell transforms into a number of small coccoid cells (Fig. 1). This includes the degradation of endogenous material, while previously initiated genome replications are completed (Humphrey *et al.*, 1983; Nyström and Kjelleberg, 1989). Electron microscopy studies show that inclusion bodies made up of poly-β-hydroxybutyrate (PHB) and present during growth disappear from *Vibrio* sp. S14 cells as they gradually become smaller (Mårdén *et al.*, 1985).

In that multiple-nutrient starvation induces the formation of ultramicrocells, without the appearance of defined morphological stages, the phenotypic adaptation to nutrient depletion is better described in physiological terms. When *V.* S14 is transferred from a rich complex medium to an artificial seawater microcosm, depleted of phosphorus, nitrogen, and carbon, a series of physiological changes take place. On the basis of the macromolecular synthesis pattern, three phases can be distinguished (Fig. 1) (Nyström *et al.*, 1990b).

1. The first phase (stringent control phase, 0–30 min), which is initiated by a temporary accumulation of guanosine 3'-diphosphate 5'-diphosphate (ppGpp), is characterized by a rapid down-regulation in the synthesis of macromolecules like RNA, proteins, and peptidoglycan (Nyström *et al.*, 1990b; Nyström and Kjelleberg, 1989). In concert with this shutdown, the rate of intracellular proteolysis is increased (Nyström *et al.*, 1988). The cell also increases its nutrient scavenging capacity by inducing a high-affinity leucine

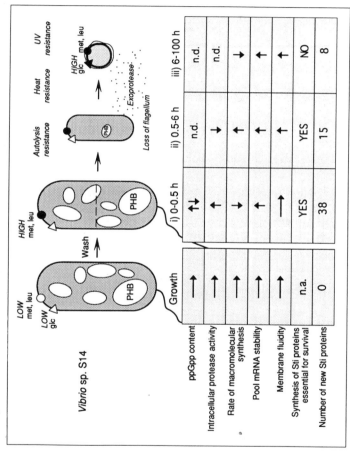

Figure 1. Schematic representation of some of the morphological and physiological alterations that *Vibrio* sp. S14 undergoes during exposure to multiple-nutrient starvation. The 100-h starvation period is divided into three morphological and physiological phases (i, ii, iii). The arrows indicate an increase ($\uparrow$) or decrease ($\downarrow$) in individual responses as compared with logarithmically growing ($\rightarrow$) cells. "*LOW*" and "*HIGH*" refer to the affinity of the uptake systems for glucose and leucine. n.d., not determined; Sti, starvation-induced; n.a., not applicable.

uptake system with a broad substrate specificity (Mårdén *et al.*, 1987). At the same time, the exoprotease activity is increased (Albertson *et al.*, 1990a).

2. The stringent response seems to be relieved after 20 min of starvation. Simultaneously with the decreased ppGpp content, a partial recovery of the rates of macromolecular synthesis is observed (Nyström *et al.*, 1990b; Nyström and Kjelleberg, 1989). This may be accomplished by the increased rate of proteolysis, decreased consumption of amino acids resulting from a drastic decrease in the total rate of protein synthesis, and also by *de novo* synthesis of amino acids, possibly fueled by the intracellular carbon and energy reserve PHB. This phase also includes a decrease in the total amount of fatty acids per cell, by at least a factor of 10 after 5 h, and a change in the fatty acid composition to shorter and more unsaturated fatty acids. It is suggested that this leads to an increased membrane fluidity, possibly in order to enhance nutrient transport (Malmcrona-Friberg *et al.*, 1986). When cells starved for different times were subjected to various stress treatments, maximal resistance, as deduced by plating efficiency, was developed toward ampicillin-induced autolysis and heat by 1 and 6 h, respectively (Nyström and Kjelleberg, 1989; Nyström *et al.*, 1992).

3. During the third phase (6–100 h), the macromolecular synthesis rate as well as the protein content decrease (Nyström *et al.*, 1986, 1988). It should be emphasized, however, that protein synthesis can be detected after more than 1 week of starvation. Protein synthesis during prolonged starvation, i.e., several months in marine *Vibrio* strains, has previously been reported (Davis, 1992). During this third phase, resistance toward UV light develops (Nyström *et al.*, 1992) and the affinity of the glucose uptake system is increased (Albertson *et al.*, 1990d). The mean mRNA half-life is increased during the first 24 h of starvation from 1.6 to 10.2 min, while for some starvation-specific transcripts the half-life can be as long as 71 min (Albertson *et al.*, 1990b).

3.2. Sequential Expression of Starvation-Induced Proteins

The first indication that a starved *Vibrio* cell may be considered as the product of a differentiation process was given by two-dimensional gel electrophoresis studies. These demonstrated that such cells have an altered protein composition, including some unique starvation-specific polypeptides (Amy and Morita, 1983; Jouper Jaan *et al.*, 1986). Recent detailed investigations of pulse-labeled *V.* S14 cells showed that the expression of some 60 proteins is increased in a time-dependent sequential pattern in response to multiple starvation (Nyström *et al.*, 1990b). The synthesis rates of at least 38 proteins are increased during the first 30 min of energy and nutrient starvation. During the second (0.5–6 h) and third (6–100 h) phases another at least 15 and 8 newly induced proteins can be detected. The synthesis of some of the starvation-induced proteins is transient, whereas for others the synthesis is permanent throughout the 100-h period studied. Inhibition of protein synthesis by chloramphenicol further showed that proteins synthesized during the first and second phases are essential for long-term starvation survival. This aspect is extensively discussed by Nyström (this volume) and Spector and Foster (this volume).

4. RESPONSES TO STARVATION FOR INDIVIDUAL NUTRIENTS

Based on results presented for other starvation-induced differentiation and differentiation-like responses such as those of *Bacillus* spp (Hecker and Völker, 1990) and *E. coli* (Matin *et al.*, 1989), it was reasoned that individual starvation for either carbon, nitrogen, or phosphorus would induce the formation of stress-resistant ultramicrocells of *Vibrio*. This would imply the presence of a core set of general gene and protein responders that is essential for the starvation-induced differentiation response. To examine whether this is the case or whether a specific individual starvation condition serves as the determinant for the essential starvation functions in *Vibrio* cells, starvation for either carbon, nitrogen, or phosphorus was analyzed by using a defined medium (Nyström *et al.*, 1992).

4.1. Viability

The viability of *V.* S14 cells during individual nutrient starvation conditions was evaluated by determination of colony forming units (cfu). Cells exposed to multiple-nutrient and carbon starvation remained at least 100% viable during one week of starvation, whereas a great loss of plateable cells was observed already after 2–3 days of nitrogen and phosphorus starvation. However, with prolonged incubation of the agar plates, microcolonies eventually appeared, in approximately the same numbers of cfu as at the onset of nitrogen or phosphorous starvation. Microscopic measurements of the INT-dehydrogenase activity in the different starvation regimens proved that the number of cells actively reducing INT, 2-(p-indophenyl)-3-(p-nitrophenyl)-5-phenyl tetrazolium chloride, were no less than 85% after one week of starvation (Holmquist and Kjelleberg, 1993a) (Fig. 2).

4.2. Morphology

Microscopic observations during starvation for the different individual nutrients revealed very distinct differences in the morphotypes (Holmquist and Kjelleberg, 1993a) (Fig. 2). Upon multiple-nutrient or carbon starvation, the rod-shaped motile cells become small, coccoid, and nonmotile, after as short

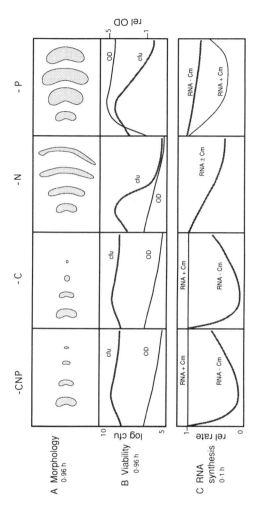

Time of starvation

Figure 2. Changes during starvation for multiple as well as for individual nutrients in *Vibrio* sp. S14. −CNP, multiple nutrients; −C, carbon; −N, nitrogen; −P, phosphorus. (A) Relative changes in morphology and cell size during 96 h of starvation; (B) relative changes in optical density (OD) and colony-forming units (cfu) during 96 h of starvation; (C) relative changes in the rate of RNA synthesis with or without the relaxing agent chloramphenicol (Cm) followed during the first hour of starvation. All relative values are related to the immediate onset of starvation.

a time as 3 h of starvation, by undergoing reductive division. Nitrogen starvation results in thin filaments, which may indicate that some late or intermediate stage of cell division is blocked in nitrogen-starved *V.* S14. Phosphorus starvation results in swollen elongated cells after prolonged starvation. This may be the result of the massive accumulation of the storage compound PHB, previously observed during phosphorus limitations by Malmcrona-Friberg *et al.*, (1986). These morphological alterations correlate well with the changes in optical density (OD). An immediate decrease in OD was observed for carbon, nitrogen, and multiple-nutrient starvation, whereas in the phosphorus-depleted medium a significant continuous increase in OD was observed during the first 48 h (Fig. 2) (Holmquist and Kjelleberg, 1993a).

4.3. RNA and Protein Synthesis

Carbon and multiple-nutrient starvation are the only conditions that at the onset of growth arrest provoke a rapid shutdown in overall protein and RNA synthesis, indicative of a stringent controllike response. This is supported by the prevention of the immediate decrease in RNA synthesis by addition of the relaxing agent chloramphenicol (Nyström *et al.*, 1992) (Fig. 2). Interestingly, the synthesis rate of proteins localized to the periplasmic space was shown to increase during the initial phase of carbon and multiple-nutrient starvation. Nitrogen- and phosphorus-starved cells display a slow decrease in the rate of periplasmic space protein synthesis and do not induce the synthesis of the specific Csp-responders (Holmquist and Kjelleberg, 1993b) (see 5.3.2.).

4.4. Overlap in Protein Synthesis during Starvation for Different Individual Nutrients

In terms of protein synthesis, it appears that the carbon starvation is not the major determinant for the multiple-nutrient starvation stimulon (Nyström *et al.*, 1992). The overlap between multiple-nutrient-starvation-inducible proteins, and those induced by individual starvation conditions is not significantly different. Comparing $-P$, $-N$, $-C$ and multiple-nutrient starvation as resolved by two-dimensional gel electrophoresis it becomes clear that very few proteins are unique to carbon starvation. Hence, it is possible that a limited number of proteins are essential for the starvation survival (Nyström *et al.*, 1992).

4.5. Stress Resistance

The starvation-induced development of resistance against stress conditions, such as temperature upshift, UV irradiation, and cadmium chloride exposure, appears to be identical for multiple-nutrient and carbon-starved cells (Nyström *et al.*, 1992). Maximal thermotolerance develops within 10 h of carbon or multiple-nutrient starvation, whereas cells depleted of nitrogen or phosphorus display less than 10% survival capacity subsequent to heat shock measured at 10 h of starvation. Sensitivity to UV irradiation is gradually lost during carbon or multiple-nutrient starvation. Only 10 and 50% of nitrogen- and phosphorus-starved cells respectively survive UV irradiation after prolonged starvation.

In summary, energy and/or carbon depletion appear to be the determinant for a series of essential functions related to nongrowth of the marine *Vibrio* sp. S14. This includes ultramicrocell formation, stringent control, and the development of starvation and stress resistance. Carbon starvation experiments with other *Vibrio* cells, i.e., comparative studies with *V.* DW1, *V. anguillarum*, and *V. vulnificus*, provide evidence that this may be a general feature of at least *Vibrio* species (unpublished observations).

5. CARBON STARVATION

5.1. Physiology of Carbon-Starved Cells

Vibrio sp. S14 cells enter a phase of gradually decreased metabolic activity after a few hours of carbon starvation. Such relatively long-term-starved *Vibrio* cells are not dormant, however, but survive for prolonged periods in a metabolically active state, as demonstrated by the low but easily detectable rate of leucine incorporation into protein (Flärdh *et al.*, 1992; Nyström *et al.*, 1986). In fact, the starved population appears to be primed for rapid and efficient recovery. High-affinity transport systems are maintained, and even induced (Albertson *et al.*, 1990d; Mårdén *et al.*, 1987). Perhaps most significantly, cells starved for several days respond instantaneously to substrate addition. No preconditioning or activation is required, as it is for optimal germination of *Bacillus* spores (Foster and Johnstone, 1989). Multiple-nutrient-starved *V.* S14 cells increase the rates of RNA and protein synthesis severalfold within the first minute when given fresh medium (Albertson *et al.*, 1990c). Therefore, in addition to the conservation of cellular and genetic integrity, the long-term-starved cells also retain an excess of the essential parts of their metabolic machinery, not for usage during starvation but to enable the efficient recovery process. Furthermore, the energetic status of the starved cells, i.e., ATP concentrations and/or membrane potential, must be high enough to allow highly efficient active transport to take place. *V. fluvialis* cells that had been starved for 48 h contained substantial concentrations of ATP, while the GTP and UTP concentrations were shown to be much lower than during growth (Smigielski *et al.*, 1989). We anticipate that some overall metabolic control mechanisms exist that tightly regulate catabolic and anabolic processes in the starved cells, and prevent wasteful metabolism and degradation of endogenous constituents from occurring. Such controls would be rapidly reversible, as demonstrated by the substrate-induced response by *V.* S14 (Albertson *et al.*, 1990d). Since the acceleration of protein synthesis is the first measurable manifestation of that response, the overall control of transcription and translation may be crucial features of starvation physiology.

5.1.1. Ribosomes and Protein Synthesis

The ribosome concentration of *E. coli* cells is closely correlated to growth rate (Neidhardt *et al.*, 1990), and extensive degradation and loss of ribosomes have repeatedly been observed during starvation conditions (e.g., Davis *et al.*, 1986; Kaplan and Apirion, 1975; Okamura *et al.*, 1973). Ribosome degradation has even been suggested as a major cause of cell death during starvation (Davis *et al.*, 1986). In *V.* S14, the rate of ribosome degradation is much lower (half-life 80 h) (Flärdh *et al.*, 1992). The possible correlation between the slow loss of viability and the ribosome content in this organism has not been investigated in detail. The most remarkable finding of this study is that the ribosomes exist in large excess over the apparent demand for translation in carbon-starved cells. For example, when more than 50% of the ribosomes are left, the total rate of protein synthesis is less than 1% of the value during growth. The ribosomal particles appear to be intact and no degradation intermediates could be observed (Flärdh *et al.*, 1992). Furthermore, the ribosomes in carbon-starved *V.* S14 were found to be as active as growth ribosomes in polyuridylic acid-directed polyphenylalanine synthesis in S30 extracts (unpublished results). Consequently, the availability of ribosomes appears not to limit protein synthesis in the starved cells, and there is a large pool of ribosomes that can be used in the upshift response.

Approximately ninety percent of the ribosome populations in carbon-starved *V.* S14 cells are 70S particles, while some 10% are free 30S and 50S subunits (Fig. 3) (Flärdh *et al.*, 1992), 100S ribosome dimers have been shown to accumulate in *E. coli* during the transition to stationary phase (Wada *et al.*, 1990) but no such particles could be detected in starved *V.* S14. Most of the protein synthesis during starvation occurs on the 70S monosomes, which indicates that protein synthesis may be limited at or close to the initiation step of translation. This does not contradict the above-mentioned observation of active ribosomes, since the poly(U)-directed system is independent of translational initiation, and involves only the elongation step. Using an *in vitro* translation

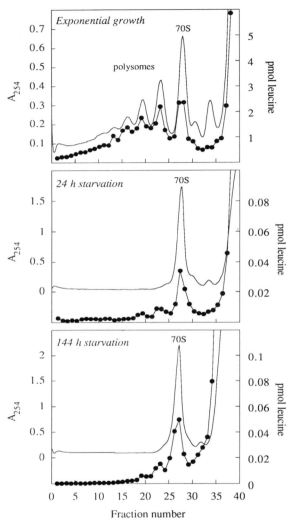

Figure 3. Polyribosome profiles and rate of protein synthesis in *Vibrio* sp. S14 during growth and carbon starvation (Flärdh *et al.*, 1992). The cells were pulse-labeled for 10 s with [³H]leucine prior to polysome extraction and sucrose density gradient analysis. Fractions for radioactivity determinations were collected starting from the bottom of the gradient. Absorbance at 254 nm and amount of leucine incorporated into growing polypeptide chains (●) are shown. (Reprinted with permission from the American Society for Microbiology)

system directed by phage R17 RNA, Davis *et al.* (1986) demonstrated that ribosomes from phosphate-starved *E. coli* are defective, and that the addition of purified initiation factors to such ribosomes restored the translational activity.

5.1.2. Regulation of Recovery-Related Protein Synthesis

An interesting feature of the starved *Vibrio* cell is the ability of the ultramicrocell to instantaneously respond to nutrient addition by increasing the rate of protein synthesis. The conservation of a functional protein synthesis machinery is seemingly independent of the length of the starvation period. However, the length of the subsequent lag or maturation phase before DNA synthesis, cell division, and regrowth commence, is dependent on the time of starvation before upshift (Albertson *et al.*, 1990c). An ordered and sequential pattern of protein synthesis, as revealed by two-dimensional gel electrophoresis of pulse-labeled cells, was found during this maturation phase. This pattern includes a set of proteins (Mat) that are exclusively synthesized during the maturation phase (see Nyström, this volume). An understanding of the fast response of the translation apparatus, and how it is initiated, may contribute to the elucidation of the sensing and signaling pathways involved in the induction of the Mat proteins.

The very first increase in the rate of protein synthesis is independent of *de novo* RNA synthesis as judged from rifampicin inhibitory studies (Albertson *et al.*, 1990c). As seen by two-dimensional gel electrophoresis of cell extracts labelled during the presence of rifampicin, the synthesis of at least some of the Mat protein seems to be independent of *de novo* RNA synthesis (Marouga, Flärdh & Kjelleberg, unpublished results). It is thus feasible that the synthesis of some of the first Mat proteins may begin at the transcriptional level prior to the upshift. Hence, in addition to the stable and active messengers (Albertson *et al.*, 1990b), there may be some stable but silent transcripts that are post-transcriptionally activated in response to nutritional upshifts. N-terminal amino acid sequencing of these proteins and reverse genetics should provide a tool for studying the modification and stability of upshift-specific transcripts and information on the means by which the cells regulate the induction of recovery and "maturation" proteins.

5.2. Physiological Basis for Stress Resistance of Starved Cells

The starvation resistance, as well as the cross-protection against other stresses that develops during carbon starvation, may be ascribed to some of the Sti (starvation-induced) proteins, the majority of which are synthesized during the initial hours of starvation (Nyström *et al.*, 1990b). Attempts are made to attribute the stress resistance to specific proteins or regulatory networks that are induced by starvation in *Vibrio* as well as in *E. coli* and other enterobacteria (Jenkins *et al.*, 1992; Hengge-Aronis, this volume; Lange and Hengge-Aronis, 1991; Nyström, this volume; Nyström *et al.*, 1992; Spector and Foster, this volume). It may also be fruitful to investigate the basis for cross-protection by examining physiological and physicochemical features of the cells. The thermoresistance of bacterial spores has for example been ascribed primarily to the dehydration of the protoplast (Gerhardt and Marquis, 1989; Murrell, 1988). Smigielski *et al.* (1989) reported a decreased water content of starved *V. fluvialis* cells. Strikingly, low water contents were also observed for the very small bacterial cells harvested from seawater (Simon and Azam, 1989).

The conservation and protection of the genome is essential for the persistence and recoverability of nongrowing bacterial cells. A class of small acid-soluble proteins (SASPs) has been demonstrated as essential for the marked UV-resistance of *Bacillus* spores (Setlow, 1992). Detailed information on the status of the DNA in starved *Vibrio* spp. is unfortunately lacking, except for some reports on decreased nucleoid volumes (Hood *et al.*, 1986; Moyer and Morita, 1989). The nonreplicative state of the genome may include nucleoid condensation, as demonstrated for swarmer cells of *Caulobacter crescentus* (Swoboda *et al.*, 1982) and *Rhodomicrobium vannielii* (Dow *et al.*, 1983). Clearly, the role of DNA binding proteins in protection and conformation of the DNA in starved cells needs further attention. The discovery of a starvation-induced histonelike protein in *E. coli*, the Dps protein, may prove to be very interesting in this respect (Almirón *et al.*; Siegele *et al.*, this volume).

5.3. Carbon Starvation Responders

The carbon starvation stimulon has recently been identified by two-dimensional gel electrophoresis of pulse-labeled cells and shown to comprise at least 20 unique proteins (Nyström *et al.*, 1992). The two main questions that need to be addressed in this context are: What are the function of the starvation-specific proteins and what are the pathways for the information of the carbon starvation stimulus to the member operons of this stimulon? Several carbon starvation-specific proteins are presently characterized in an attempt to address the first question (see Sections 5.3.2. and 5.3.3.). As outlined in the next section, an answer to the second question is being sought by isolation of mutants that fail to regulate genes belonging to the carbon starvation stimulon. Screening such regulatory mutants for deficiencies in starvation-related properties makes it possible to identify regulons that are part of the carbon starvation stimulon and which have an essential role in starvation survival.

5.3.1. Identification of a Putative Carbon Starvation Response Regulator in Vibrio sp. S14

The approach employed can be described as a two-step process in which the first step includes isolation of a *lac* fusion that responds exclusively to a desired stimulus. The second step involves the introduction of a secondary transposon and subsequent screening for either constitutively LacZ-positive or LacZ-negative phenotypes as a result of insertional inactivation of a gene product with regulatory impact on the primary target gene promoter.

A carbon starvation-induced *lac* fusion (*csi*A, formerly designated S141 M5) was isolated by screening a *V.* S14 miniMu(lac) transposon mutant bank on low-glucose minimal X-Gal plates (Östling *et al.*, 1991). The glucose concentration was adjusted so that it limited growth of the colonies within 48 h. Before the secondary mutant was isolated, the primary fusion was checked for overlap with various stress treatments and starvation conditions. Of several different individual nutrient depletions, this fusion responds exclusively to carbon starvation. In order to assign a specific role for a regulatory fusion, it is important that the *lac* fusion strain, used as a recipient for the secondary transposon, has a restricted induction pattern.

A miniTn10 transposon was delivered to the *V.* S14 *csi*A fusion strain by conjugational transfer (Östling and Kjelleberg, in press). The vector pLOF Km used for this transfer has a broad host range origin of transfer (mob RP4) and provides the transposase activity from a gene in *trans* which allows for the isolation of stable mutants. An R6K origin of replication ensures that no replication occurs in the recipient strain (de Lorenzo *et al.*, 1990). By plating the conjugation mixture directly onto both high- and low-glucose minimal plates, with the appropriate selective antibiotics and X-Gal, several constitutively LacZ-negative and -positive clones were found. One of the constitutively LacZ-negative double mutants (*csr*Il) isolated exhibits at least a 100-fold decrease in viability when starved for carbon as compared with the wildtype over a 4-day period. By applying pulse labeling and two-dimensional gel electrophoresis techniques, it was demonstrated that this regulatory mutant fails to synthesize several proteins (Östling and Kjelleberg, in press). Interestingly, the same proteins can be identified as carbon starvation-induced proteins in the wild type, which implies that they are directly regulated by the target gene product. The problem with this approach is that it is difficult to ascertain if the altered induction patterns of a protein are a direct or indirect effect of the inactivation of the regulatory gene. When interpreting induction pattern from mutants, one also has to bear in mind that the target gene products are not produced in active forms and hence may abolish possible feedback mechanisms.

Out of approximately 100,000 colonies that were screened, only six constitutively white and six constitutively blue colonies were found. Among the six constitutively LacZ-negative clones, three were inactivated at the same location as judged by the size of the miniTn10 hybridizing fragment. These results might reflect that there are only a few regulatory genes that are involved in the regulation of the *csi*A fusion and hence possibly the entire carbon starvation stimulon. Thorough discussions on regulatory genes, implicated to control global regulatory networks involved in the starvation response, can be found in several chapters in this volume (Hengge-Aronis; Nyström; Siegele *et al.*; and Spector and Foster).

The approach for the isolation of regulatory mutants and the subsequent identification of gene products and the regulatory gene itself is of general value. The transposon serves as an excellent marker gene for cloning since the transposon is physically linked to the gene of interest. The flanking sequences can be amplified by inverse PCR (Hulton *et al.*, 1990; Ochman *et al.*, 1988), using an oligonucleotide complementary to the ends of the miniTn10. The PCR product is then used in order to identify the intact gene in a wild-type library.

5.3.2. Periplasmic Space Proteins

The periplasmic binding proteins are involved in improved solute transport during starvation (Albertson *et al.*, 1990d; Mårdén *et al.*, 1987). Several studies have demonstrated starvation-specific alterations in chemotaxis toward amino acids and sugars, which possibly involve binding proteins for delivery of the attractant to the membrane-bound transducer (Malmcrona-Friberg *et al.*, 1990). Exoproteases are additional periplasmic space-localized starvation proteins. Increased exoprotease activity in response to starvation most likely stimulates the cell to induce new uptake systems that allow the cell to mobilize substrates otherwise inaccessible to them.

The carbon starvation stimulon in *V.* S14 includes at least three periplasmic space protein responders, *carbon starvation-induced periplasmic* space protein Csp1 (120 kDa), Csp5 (37 kDa), and Csp6 (30 kDa) (Holmquist and Kjelleberg, 1993b). Csp1 exhibits a more than 100-fold increase in synthesis at 24 h of starvation, and is synthesized during at least 96 h of carbon starvation, whereas Csp5 and Csp6 are transient responders with an eight- to tenfold induction within the first few hours of starvation. Although no or minor synthesis can be observed for the latter proteins during prolonged starvation, the amounts of these proteins, visualized by Western blotting using specific polyclonal antibodies, are as pronounced by 96 h as observed at 24 h of starvation.

None of these proteins are induced by other starvation conditions or stress exposures that have been tested. While no functions can yet be assigned to the three carbon-starvation specific responders, Csp5 and Csp6 are within the size range of most periplasmic binding proteins of gram-negative bacteria (Philips *et al.*, 1987). Csp1 has been identified as the product of a starvation-specific mRNA species with a functional half-life of 71 min after 24 h of starvation (Albertson *et al.*, 1990b). This reflects an unusual messenger stability and strongly suggests that starvation-induced modification and stability of transcripts may be an important level of regulation in the nonreplicating cell. N-terminal amino acid sequences of these proteins do not display significant homologies with any of the proteins in available data bases. The sequencing of the encoding genes of these proteins, which are induced at different times of the carbon starvation-induced program, allows us to identify regulatory regions and generate mutants to further understand the functions of the proteins, as well as their role in the differentiation program.

5.3.3. Stress Proteins

In support of the notion that carbon starvation is the determinant for development of resistance to various stress conditions, a significant overlap between carbon starvation-inducible proteins and those induced by specific stress conditions, such as heat, nalidixic acid, and cadmium exposure, has been observed (Nyström *et al.*, 1990a). Two of the dominant responders are the prokaryotic heat shock proteins DnaK and GroEL, members of the Hsp70 and Hsp60 families, respectively. By using antisera raised against *E. coli* DnaK and GroEL, it was demonstrated that several *Vibrio* species respond to starvation by induction of these proteins. Interestingly, Sis1 (*starvation-induced stress protein*), a new stress responder that is induced by carbon starvation, was also found to have homologous epitopes to DnaK (Holmquist *et al.*, 1993).

The induction and kinetics of synthesis of DnaK, GroEL, and Sis1 were investigated in the three *Vibrio* species *V.* S14, *V. vulnificus*, and *V.* DW1 (Holmquist *et al.*, 1993). These stress proteins were found to be induced in response to both nutrient limitation and temperature shifts. Interestingly, each species displays an individual induction and time-dependent synthesis pattern for all three proteins. For example, with respect to carbon starvation, DnaK is synthesized during prolonged starvation in *V.* S14 whereas for *V.* DW1 and *V. vulnificus* the induction is only transient. GroEL and Sis1 are transiently induced in *V. vulnificus* but remain synthesized during prolonged carbon starvation in *V.* S14. Both DnaK and Sis1 accumulate in *V. vulnificus*, as the concentration of these proteins increases during prolonged carbon starvation.

The differential induction and synthesis of the heat shock proteins DnaK and GroEL in each of the several different *Vibrio* species raise interesting questions with respect to the regulation of these proteins. For example, it is possible that the heat shock sigma factor RpoH is not the sole regulator of the starvation-induced expression of either or both DnaK and GroEL.

Carbon starvation induction of DnaK and GroEL has also been observed in *V. anguillarum*, again displaying a different induction pattern of synthesis (Nelson *et al.*, in press). As discussed in Section 6, it remains to be seen whether the synthesis of stress proteins such as DnaK, GroEL and Sis1 reflects the core program of carbon starvation.

DnaK has also been found in the periplasmic space in long-term-starved *V.* S14 cells and *V. anguillarum* (Holmquist *et al.*, 1993). It may be suggested that the DnaK chaperone participates in assisting export of starvation-specific proteins to the periplasmic space. In preliminary co-immunoprecipitation studies on whole cell extracts of starved cells, we have demonstrated that Csp5 coprecipitates with anti-DnaK–DnaK complexes (unpublished results). It is possible that the starved cell experiences denaturation of cellular proteins which would serve as a stimulus in the response pathway for synthesis of molecular chaperones.

6. THE DIVERSITY OF STARVATION RESPONSES IN THE GENUS VIBRIO

A comparison of the starvation survival process among the most studied *Vibrio* species, i.e., *V. cholerae*, *V. Vulnificus*, *V. anguillarum*, *V. fluvialis*, *Vibrio* sp. ANT 300, *Vibrio* sp. S14, and *Vibrio* sp. DW1, may allow the identification of general responses as well as responses that are unique for a given strain.

It is obvious, however, that the different research groups studying starvation survival of *Vibrio* species have focused their efforts on different parts or phases of the starvation survival program. As carefully explained by Mason and Egli, and Siegele *et al.* (this volume), the composition of the growth medium also greatly influences the outcome of the starvation response. A comparison of results obtained in different laboratories also requires a detailed analysis of how the data are presented. It is possible to arrive at very different interpretations, for example, depending on whether the data are calculated per total cell, per culturable cell, per biovolume, or per milliliter starvation culture. Nevertheless, some general traits of the starvation survival process of *Vibrio* are emerging, and perhaps more interestingly, it also appears to represent considerable diversity between species.

Electron micrograph analysis and size measurements of starved vibrios generally reveal that the bacteria become smaller and coccoid during starvation (Baker *et al.*, 1983; Mården *et al.*, 1985; Novitsky and Morita, 1978). The cell wall is assumed to become thicker and the composition of the cell membrane changes (Guckert *et al.*, 1986; Malmcrona-Friberg *et al.*, 1986; Oliver and Stringer, 1984; Smigielski *et al.*, 1989). An enlarged periplasmic space is generally found for ultramicrobacteria (Baker *et al.*, 1983; Hood *et al.*, 1986; Mårdén *et al.*, 1985). Several different survival strategies for bacteria have been identified. The most common scheme seems to be that the bacteria complete the ongoing rounds of replication and generate the maximum number of small cells possible containing at least one genome. Different degrees of survival are observed subsequent to this replicative phase at the onset of starvation (Morita, this volume).

Rather dramatic differences have also been observed between different strains. For example, the formation of surface structures such as fibrils and outer membrane vesicles or blebs has been reported to be unique for certain *Vibrio* strains (Dawson *et al.*, 1981; Mårdén *et al.*, 1985). Differences in the physicochemical characteristics of the cell surface and adhesion to inanimate surfaces have also been found for different organisms during nongrowth (Kjelleberg and Hermansson, 1984). Specific protein responders that act as signatures of starvation in a given *Vibrio* species are easy to establish by two-dimensional gel electrophoresis of pulse-labeled cells. It may in fact be suggested that the programmed survival exhibited by nondifferentiating bacteria in response to starvation-induced growth arrest is one of considerable diversity, and that we may not necessarily arrive at a general model that accurately describes the carbon starvation survival response in nondifferentiating bacteria. In the following, differences in three presumably essential features of the physiology of starvation survival are highlighted.

Addition of chloramphenicol to interrupt the protein synthesis during starvation revealed that *Vibrio* sp. S14 is dependent on protein synthesis for viability in the beginning of the starvation period (Nyström and Kjelleberg, 1989). In contrast, *V.* DW1 is dependent on protein synthesis during a starvation period as long as a week (Jouper-Jaan *et al.*, 1992). These results indicate that essential proteins for survival during starvation are synthesized at different times after the onset of the differentiation program in different *Vibrio* species. A similar induction pattern has been observed for *Pseudomonas* strains (S. Molin, personal communication).

At the onset of starvation, the rate of protein synthesis undergoes transient changes. After this initial short-term phase of starvation, the rate of protein synthesis decreases rapidly, such that by 24 h of starvation it remains at a very low level. This level varies between different *Vibrio* species. It appears that *V.* S14 displays a much lower rate of protein synthesis compared with *V.* DW1 (Jouper-Jaan *et al.*, 1992; Nyström *et al.*, 1986) and *V. vulnificus* (Weichart, personal communication) during the long-term starvation. The rate of protein synthesis might reflect the level of basal metabolism of the starved ultra-microcell. The question whether bacteria with a low rate of protein synthesis survive longer periods than bacteria with a higher rate of protein synthesis has not been explored in detail.

As discussed above, different *Vibrio* strains exhibit individual induction patterns of the stress proteins DnaK, GroEL, and Sis1 during carbon starvation (Holmquist *et al.*, 1993). It is obvious that homologous proteins can be differentially induced and presumably may also exhibit different functions in different *Vibrio* species. There also appears to be different induction patterns of the heat shock proteins DnaK and GroEL in the same species. This raises interesting questions with respect to the regulation of stress proteins, such as DnaK and GroEL, during starvation and the significance of these proteins for the starvation-induced differentiation response.

It is possible that bacteria that form ultramicrocells during prolonged starvation follow different routes and induce different sets of proteins, in order to achieve enhanced survival and stress resistance. Hence, the starvation-induced differentiation program would be different in different species. Should there be a common core pathway in the different species, however, we have probably not yet addressed the significant features of such a program. For example, the role of stress proteins may prove to be of little significance for the differentiation pathway of starved *Vibrio* cells. A detailed analysis of the characteristics that appear to be rather different for starved cells of different strains will most likely provide a means of addressing these questions.

7. APPENDIX

Species Currently Assigned as Vibrios, Their Site of Isolation, and Pathogenicity for Other Organisms

Species[a]	Sites of isolation[b]	Pathogenicity[c]
V.(L.) aestuarianus	W, I	
V. alginolyticus	W, F, I	HP, FP, IP
V.(L.) anguillarum	S, W, F	FP
V.(P.) angustum	W	
V. campbellii	W	
V. carchariae	F	FP
V. cholerae	W, F, I	HP, FP
V. cincinnatiensis	—	HP
V. costicola	—	
V.(L., P.) damsela	W, F, I	HP, FP
V. diazotrophicus	W, I	
V.(P.) fischeri	W, F, I	
V. fluvialis	W, I	HP
V. furnissii	W, I	
V. gazogenes	W	
V. harveyi	W, F	
V. hollisae	—	HP

(Continued)

Species^a	Sites of isolation^b	Pathogenicity^c
V.(P.) leiognathi	W, F, I	
V.(P.) logei	W, F, I	
V. mediterranei	W	
V. metschnikovii	W, I	HP
V. mimicus	W, I	HP
V. natriegens	W	
V. navarrensis	W	
V. nereis	W, I	
V. nigripulchritudo	W	
V.(L.) ordalii	F	FP
V. orientalis	W	
V. parahaemolyticus	S, F	HP, (FP)
V.(L.) pelagius biovar. I	W	
biovar. II	W	
V.(P.) phosphoreum	F	
V. proteolyticus	I	
V. salmonicida	S, W, F	FP
V. splendidus biovar. I	W	
biovar. II	W	
V.(L.) tubiashii	I	IP
V. vulnificus	S, W, F, I	HP, FP

^{a}The species currently assigned as *Vibrio*, including those of the related genera *Listonella* (*L.*) and *Photobacterium* (*P.*). *V. marinus* has been omitted based on poor sequence homologies on 5 S and 16 S rRNA sequences.
^{b}An incomplete list with examples of sites of isolation: S, marine sediments; W, water; F, fish; I, marine invertebrates; and —, other site.
^{c}HP, FP, and IP signify that these species have been shown to be pathogenic for humans, fish, and invertebrates, respectively.

ACKNOWLEDGMENTS. The pLOF Km vector was kindly provided by Dr. V. de Lorenzo. The antibodies raised against DnaK and GroEL were kindly provided by Drs. Graham S. Walker and David N. Nelson. Work in the authors' laboratory was made possible thanks to financial support by the Swedish Natural Science Research Council, the Nordic Ministry Council, the Swedish National Environmental Protection Agency, and the Swedish Forestry and Agricultural Science Research Council.

REFERENCES

Albertson, N. H., Nyström, T., and Kjelleberg, S., 1990a. Exoprotease activity of two marine bacteria during starvation, *Appl. Environ. Microbiol.* **56**:218–223.
Albertson, N. H., Nyström, T., and Kjelleberg, S., 1990b, Functional mRNA half-lives in the marine *Vibrio* sp. S14 during starvation and recovery, *J. Gen. Microbiol.* **136**:2195–2199.

Albertson, N. H., Nyström, T., and Kjelleberg, S., 1990c, Macromolecular synthesis during recovery of the marine *Vibrio* sp. S14 from starvation, *J. Gen. Microbiol.* **136**:2201–2207.

Albertson, N. H., Nyström, T., and Kjelleberg, S., 1990d, Starvation-induced modulations in binding protein-dependent glucose transport by the marine *Vibrio* sp. S14, *FEMS Microbiol. Lett.* **70**:205–210.

Almirón, M., Link, A. J., Furlong, D., and Kolter, R., 1992, A novel DNA-binding protein with regulatory and protective roles in starved *Escherichia coli, Genes and Development* **6**:2646–2654.

Amy, P. S., and Morita, R. Y., 1983, Starvation-survival patterns of sixteen freshly isolated open-ocean bacteria, *Appl. Environ. Microbiol.* **45**:1109–1115.

Austin, B., and Austin, D. A., 1987, *Bacterial Fish Pathogens, Disease in Farmed and Wild Fish*, Horwood, Chichester.

Baker, P., Singleton, F. L., and Hood, M. A., 1983, Effects on nutrient deprivation on *Vibrio cholerae, Appl. Environ. Microbiol.* **46**:930–940.

Bauman, P., and Bauman, L., 1977, Biology of the marine enterobacteria: Genera *Benecka* and *Photobacterium, Annu. Rev. Microbiol.* **31**:39–61.

Belas, M. R., and Colwell, R. R., 1982, Adsorption kinetics of laterally and polarly flagellated *Vibrio, J. Bacteriol.* **151**:1568–1580.

Buck, J. D., Overstrom, N. A., Patton, G. W., Anderson, H. F., and Gorzelany, J. F., 1991, Bacteria associated with stranded cetaceans from the northeast USA and southwest Florida Gulf Coast, *Dis. Aquat. Org.* **10**:147–152.

Button, D. K., Schut, F., Quang, P., Ravonna, M., and Robertsson, B. R., 1993, Viability and isolation of marine bacteria by dilution culture: theory, procedures and initial results, *Appl. and Environ. Microbiol.*, in press.

Chowdhury, M. A. R., Yamanaka, H., Miyoshi, S., and Shinoda, S., 1990, Ecology and seasonal distribution of *Vibrio parahaemolyticus* in aquatic environment of a temperate region, *FEMS Microbiol. Ecol.* **74**:1–10.

Colwell, R. R., 1984, *Vibrios in the Environment*, Wiley, New York.

Corpe, W. A., 1970, Attachment of marine bacteria to solid surfaces, in: *Adhesion in Biological Systems* (R. S. Mandy, ed.), Academic Press, New York, pp. 73–87.

Davis, B. D., Luger, S. M., and Tai, P. C., 1986, Role of ribosome degradation in the death of starved *Escherichia coli* cells, *J. Bacteriol.* **166**:439–445.

Davis, C. L., 1985, Physiological and ecological studies of mannitol utilizing marine bacteria, Ph.D. thesis, University of Cape Town.

Davis, C. L., 1992, Production of laminarinase and alginase by marine bacteria after starvation, *FEMS Microbiol. Ecol.* **86**:349–356.

Dawson, M. P., Humphrey, B. A., and Fournier, R. O., 1981, Adhesion: A tactic in the survival strategy of marine *Vibrio* during starvation, *Curr. Microbiol.* **6**:195–199.

Deane, S. M., Robb, F. T., and Woods, D. R., 1987, Production and activation of an SDS-resistant alkaline serine exoprotease of *Vibrio alginolyticus, J. Gen. Microbiol.* **133**:391–398.

Delong, E. F., and Franks, D. G., 1992, Phylogeny of barophilic and psychrophilic deep-sea bacteria, *Abstracts of the 92nd General Meeting of the American Society for Microbiology*, New Orleans.

de Lorenzo, V., Herrero, M., Jakubzik, U., and Timmis, K. N., 1990, Mini-Tn5 transposon derivatives for insertion mutagenesis, promoter probing, and chromosomal insertion of cloned DNA in gram-negative eubacteria, *J. Bacteriol.* **172**:6568–6572.

Dow, C. S., Whittenbury, R., and Carr, N. G., 1983, The 'shut down' or 'growth precursor' cell— an adaptation for survival in a potentially hostile environment, in: *Microbes in Their Natural Environments* (J. H. Slater, R. Whittenbury, and J. W. T. Wimpenny, eds.), Cambridge University Press, London, pp. 187–247.

Flärdh, K., Cohen, P. S., and Kjelleberg, S., 1992, Ribosomes exist in large excess over the apparent demand for protein synthesis during starvation in marine *Vibrio* sp. strain CCUG 15956, *J. Bacteriol.* **174:**6780–6788.

Foster, S. J., and Johnstone, K., 1989, The trigger mechanism of bacterial spore germination, in: *Regulation of Procaryotic Development* (I. Smith, R. A. Slepecky, and P. Setlow, eds.), American Society for Microbiology, Washington, D.C., pp. 89–108.

Gerhardt, P., and Marquis, R. E., 1989, Spore thermoresistance mechanisms, in: *Regulation of Procaryotic Development* (I. Smith, R. A. Slepecky, and P. Setlow, eds.), American Society for Microbiology, Washington, D.C., pp. 43–64.

Guckert, J. B., Hood, M. A., and White, D. C., 1986, Phospholipid ester-linked fatty acid profile changes during nutrient deprivation of *Vibrio cholerae*: Increases in the *trans/cis* ratio and proportions of cyclopropyl fatty acids, *Appl. Environ. Microbiol.* **52:**794–801.

Hecker, M., and Völker, U., 1990, General stress proteins in *Bacillus subtilis*, *FEMS Microbiol. Ecol.* **74:**197–213.

Holmquist, L., Jouper-Jaan, Å., Weichart, D., Nelson, D. R., and Kjelleberg, S., 1993, The induction of stress proteins in three marine *Vibrio* during carbon starvation, *FEMS Microbiol. Ecol.*, in press

Holmquist, L., and Kjelleberg, S., 1993a, Changes in viability, respiratory activity and morphology of the marine *Vibrio* sp. strain S14 during starvation for individual nutrient and subsequent recovery, *FEMS Microbiol. Ecol.*, in press

Holmquist, L., and Kjelleberg, S., 1993b, The carbon starvation stimulon in the marine *Vibrio* sp. S14 (CCUG 15956) includes three periplasmic space protein responders, *J. Gen. Microbiol.* **39:**209–215.

Hood, M. A., Guckert, J. B., White, D. C., and Deck, F., 1986, Effect of nutrient deprivation on lipid, carbohydrate, DNA, RNA, and protein levels in *Vibrio cholera*, *Appl. Environ. Microbiol.* **52:**788–793.

Hulton, C. S. J., Seirafi, A., Hinton, J. C. D., Sidebotham, J. M., Waddell, L., Pavitt, G. D., Oven-Hughes, T., Spassky, A., Buc, H., and Higgins, C. F., 1990, Histone-like protein H1 (H-NS), DNA supercoiling and gene expression in bacteria, *Cell* **63:**631–642.

Humphrey, B., Kjelleberg, S., and Marshall, K. C., 1983, Responses of marine bacteria under starvation conditions at a solid–water interface, *Appl. Environ. Microbiol.* **45:**43–47.

Jenkins, D. E., Auger, E. A., and Matin, A., 1992, Role of RpoH, a heat shock regulator protein, in *Escherichia coli* carbon starvation protein synthesis and survival, *J. Bacteriol.* **173:**1992–1996.

Jouper Jaan, Å., Dahllöf, B., and Kjelleberg, S., 1986, Changes in protein composition of three bacterial isolates from marine waters during short periods of energy and nutrient deprivation, *Appl. Environ. Microbiol.* **52:**1419–1421.

Jouper-Jaan, Å., Goodman, A., and Kjelleberg, S., 1992, Bacteria starved for prolonged periods develop increased protection against lethal temperatures, *FEMS Microbiol. Ecol.* **101:**229–236.

Kaneko, T., and Colwell, R. R., 1973, Ecology of *Vibrio parahaemolyticus* in Chesapeake Bay, *J. Bacteriol.* **113:**24–32.

Kaplan, R., and Apirion, D., 1975, The fate of ribosomes in *Escherichia coli* cells starved for a carbon source, *J. Biol. Chem.* **250:**1854–1863.

Kjelleberg, S., and Hermansson, M., 1984, Starvation induced effects on bacterial surface characteristics, *Appl. Environ. Microbiol.* **48:**497–503.

Kjelleberg, S., Hermansson, M., Mårdén, P., and Jones, G. W., 1987, The transient phase between growth and non-growth of heterotrophic bacteria, with emphasis on the marine environment, *Annu. Rev. Microbiol.* **41:**25–49.

Kjelleberg, S., Nyström, T., Albertson, N., and Flärdh, K., 1990, Nutrient limitation: Global responses and prokaryotic development, *FEMS Microbiol. Ecol.* 74.

Kjelleberg, S., Flärdh, K., Nyström, T., and Moriarty, D. J. W., 1993, Growth limitation and starvation of bacteria, in: *Aquatic Microbiology: An Ecological Approach* (T. Ford, ed.), Blackwell, Oxford.

Lange, R., and Hengge-Aronis, R., 1991, Identification of a central regulator of stationary phase gene expression in *Escherichia coli, Mol. Biol.* **5:**49–59.

MacDonell, M. T., and Hood, M. A., 1982, Isolation and characterization of ultramicrobacteria from a Gulf Coast estuary, *Appl. Environ. Microbiol.* **43:**566–571.

MacDonell, M. T., and Hood, M. A., 1984, Ultramicrovibrios in Gulf Coast estuarine water: Isolation, characterization and incidence, in: *Vibrios in the Environment* (R. R. Colwell, ed.), Wiley, New York, pp. 551–562.

Malmcrona-Friberg, K., Tunlid, A., Mårdén, P., Kjelleberg, S., and Odham, G., 1986, Chemical changes in cell envelope and poly-β-hydroxybutyrate during short term starvation of a marine bacterial isolate, *Arch. Microbiol.* **144:**340–345.

Malmcrona-Friberg, K., Goodman, A., and Kjelleberg, S., 1990, Chemotactic responses of a marine *Vibrio* sp. strain S14 (CCUG 15956) to low-molecular-weight substances under starvation-survival conditions, *Appl. Environ. Microbiol.* **56:**3699–3704.

Mårdén, P., Tunlid, A., Malmcrona-Friberg, K., Odham, G., and Kjelleberg, S., 1985, Physiological and morphological changes during short term starvation of marine bacterial isolates, *Arch. Microbiol.* **142:**326–332.

Mårdén, P., Nyström, T., and Kjelleberg, S., 1987, Uptake of leucine by a marine gram-negative heterotrophic bacterium during exposure to starvation conditions, *FEMS Microbiol. Ecol.* **45:**233–241.

Matin, A., Auger, E. A., Blum, P. H., and Schultz, J. E., 1989, Genetic basis of starvation survival in nondifferentiating bacteria, *Annu. Rev. Microbiol.* **43:**293–316.

Morita, R. Y., 1975, Psychrophilic bacteria, *Bacteriol. Rev.* **39:**144–167.

Morita, R. Y., 1985, Starvation and miniaturisation of heterotrophs, with special emphasis on maintenance of the starved viable state, in: *Bacteria in Their Natural Environments* (M. M. Fletcher and G. D. Floodgate, eds.), Academic Press, New York, pp. 111–130.

Morita, R. Y., 1988, Bioavailability of energy and its relationship to growth and starvation survival in nature, *Can. J. Microbiol.* **34:**436–441.

Moyer, C. L., and Morita, R. Y., 1989, Effect of growth rate and starvation-survival on cellular DNA, RNA, and protein of a psychrophilic marine bacterium, *Appl. Environ. Microbiol.* **55:**2710–2716.

Murrell, W. G., 1988, Bacterial spores—Nature's ultimate survival package, in: *Microbiology in Action* (W. G. Murrell and I. R. Kennedy, eds.), Wiley, New York, pp. 311–346.

Neidhardt, F. C., Ingraham, J. I., and Schaechter, M., 1990, *Physiology of the Bacterial Cell. A Molecular Approach*, Sinauer, Sunderland, Mass.

Novitsky, J. A., and Morita, R. Y., 1978, Starvation induced barotolerance as a survival mechanism of a psychrophilic marine vibrio in the waters of the Antarctic Convergence, *Mar. Biol.* **49:**7–10.

Nyström, T., and Kjelleberg, S., 1989, Role of protein synthesis in the cell division and starvation induced resistance to autolysis of a marine *Vibrio* during the initial phase of starvation, *J. Gen. Microbiol.* **135:**1599–1606.

Nyström, T., Mårdén, P., and Kjelleberg, S., 1986, Relative changes in incorporation rates of leucine and methionine during starvation survival of two bacterial isolates from marine waters, *FEMS Microbiol. Ecol.* **38:**285–292.

Nyström, T., Albertson, N., and Kjelleberg, S., 1988, Synthesis of membrane and periplasmic proteins during starvation of a marine *Vibrio* sp., *J. Gen. Microbiol.* **134:**1645–1651.

Nyström, T., Albertsson, N., Flärdh, K., and Kjelleberg, S., 1990a, Physiological and molecular adaption to starvation and recovery from starvation by the marine *Vibrio* sp. S14, *FEMS Microbiol. Ecol.* **74:**129–140.

Nyström, T., Flärdh, K., and Kjelleberg, S., 1990b, Responses to multiple-nutrient starvation in a marine *Vibrio* sp. strain CCUG 15956, *J. Bacteriol.* **172:**7085–7097.

Nyström, T., Olsson, R. M., and Kjelleberg, S., 1992, Survival, stress resistance and alterations in protein expression in the marine *Vibrio* sp. S14 during starvation for different individual nutrients, *Appl. Environ. Microbiol.* **58:**55–65.

Ochman, H., Gerber, A. S., and Hartl, D. L., 1988, Genetic application of an inverse polymerase chain reaction, *Genetics* **120:**621–623.

Okamura, S., Maruyama, H. B., and Yanagita, T., 1973, Ribosome degradation and degradation products in starved *Escherichia coli*. VI. Prolonged culture during glucose starvation, *J. Biochem.* **73:**915–922.

Oliver, J. D., and Stringer, W. F., 1984, Lipid composition of a psychrophilic marine *Vibrio* sp. during starvation-induced morphogenesis, *Appl. Environ. Microbiol.* **47:**461–466.

Onarheim, A. M., 1990, Characterization and possible biological significance of an autochthonous flora in the intestinal mucus of seawater fish, in: *Microbiology in Poicilotherms* (R. Lesel, ed.), Elsevier, Amsterdam, pp. 197–201.

Östling, J., Goodman, A., and Kjelleberg, S., 1991, Behaviour of IncP-1 plasmids and a miniMu transposon in a marine *Vibrio* sp.: Isolation of starvation inducible *lac* operon fusions, *FEMS Microbiol. Ecol.* **86:**83–94.

Philips, T. A., Vaughn, V., Bloch, P. L., and Neidhardt, F. C., 1987, Gene protein index of *Escherichia coli* K-12, Edition 2, in: *Escherichia coli and Salmonella typhimurium, Cellular and Molecular Biology* (F. C. Neidhardt, J. O. Ingraham, K. Brooks Low, B. Magasanik, M. Schaechter, and H. E. Umbarger, eds.), American Society for Microbiology, Washington, D. C., pp. 919–966.

Ruby, E. G., and McFall-Ngai, M. J., 1992, A squid that glows in the night: Development of an animal–bacterial mutualism, *J. Bacteriol.* **174:**4865–4870.

Ruby, E. G., and Morin, J. G., 1979, Luminous enteric bacteria of marine fishes: A study of their distribution, densities and dispersion, *Appl. Environ. Microbiol.* **38:**406–411.

Schmidt, T. M., DeLong, E. F., and Pace, N. R., 1991, Analysis of a marine picoplankton community by 16S rRNA gene cloning and sequencing, *J. Bacteriol.* **173:**4371–4378.

Setlow, P., 1992, I will survive: Protecting and repairing spore DNA, *J. Bacteriol.* **174:**2737–2741.

Simon, M., and Azam, F., 1989, Protein content and protein synthesis rates of planktonic marine bacteria, *Mar. Ecol. Prog. Ser.* **51:**201–213.

Smigielski, A. J., Wallace, B. J., and Marshall, K. C., 1989, Changes in membrane functions during short-term starvation of *Vibrio fluvialis* NCTC 11328, *Arch. Microbiol.* **151:**336–347.

Swoboda, U. K., Dow, C. S., and Vitkovic, L., 1982, Nucleoids of *Caulobacter crescentus* CB15, *J. Gen. Microbiol.* **128:**279–289.

Tabor, P. S., Ohwada, K., and Colwell, R. R., 1981, Filterable marine bacteria found in the deep sea: Distribution, taxonomy and response to starvation, *Microbiol. Ecol.* **7:**67–83.

Venkateswaran, K., Kim, S. W., Nakano, H., Onbe, T., and Hashimito, H., 1989, The association of *Vibrio parahaemolyticus* serotypes with zooplankton and its relationship with bacterial indication of pollution, *Syst. Appl. Microbiol.* **11:**194–201.

Westerdahl, A., Olsson, J. C., Kjelleberg, S., and Conway, P. L., 1991, Isolation and characterization of turbot (*Scophtalmus maximus*)-associated bacteria with inhibitory effect against *Vibrio anguillarum*, *Appl. Environ. Microbiol.* **57:**2223–2228.

Yu, C., and Lee, A. M., 1987, The sugar-specific adhesion/deadhesion apparatus of the marine bacterium *Vibrio furnissi* in a sensorium that continuously monitors nutrient levels in the environment, *Biochem. Biophys. Res. Commun.* **149:**86–92.

6

Global Systems Approach to the Physiology of the Starved Cell

Thomas Nyström

1. INTRODUCTION

Microorganisms have limited capacities to control their environment or to physically escape environmental stress. Thus, they almost invariably respond to changes in their environment by changing themselves. This can be achieved either by modifications in genetic constitution or by phenotypic adaptation. Only the latter of these processes will be considered in this chapter.

Bacteria generally need only express part of their genome in order to become a structural and functional unit of an environment. Continual changes in environmental conditions can be accommodated by complex, but efficient mechanisms of altering the pattern of gene expression. Among the environmental parameters that commonly modulate the properties of bacterial cells in nature, the concentration of macronutrients is of particular importance. Natural environments are frequently depleted of some or several essential nutrients, and microorganisms are exposed routinely to macronutrient insufficiency (Tempest and Neijssel, 1981; Morita; Moriarty and Bell; van Elsas and van Overbeek, this volume). While the exhaustion of a building block requires the induction of particular operons, macronutrient limitations are more complex stresses involv-

Thomas Nyström • Department of Microbiology and Immunology, University of Michigan Medical School, Ann Arbor, Michigan 48109-0620. *Present address*: Department of General and Marine Microbiology, University of Göteborg, S-413 19 Göteborg, Sweden.
Starvation in Bacteria, edited by Staffan Kjelleberg. Plenum Press, New York, 1993.

ing "global control systems" (Gottesman and Neidhardt, 1983) requiring regulation above the operon level. Global control systems consist of multiple unlinked genes and operons coordinately controlled by a common regulatory signal or regulatory gene (Neidhardt, 1983). Several global control systems are known to be activated in response to a variety of stresses including temperature shifts (Neidhardt et al., 1984; Neidhardt, 1987), irradiation, chemical and physical assault, and nutrient starvation (Gottesman and Neidhardt, 1983). Examples of global regulatory networks include the SOS (lexA-controlled) response (Walker, 1984), the oxidation stress (oxyR-controlled) response (Christman et al., 1985), the heat-shock (rpoH-controlled) response (Neidhardt et al., 1984), and the stringent (relA/spoT-controlled) response (Gallant, 1979).

Growth arrest resulting from severe stress conditions are likely to activate several signals and unlinked global regulatory systems. To avoid confusion, Smith and Neidhardt (1983) therefore proposed the use of the term stimulon to refer to the entire set of genes responding to a given environmental stimulus. The large number of proteins responding to an environmental stimulus can be studied and identified by means of two-dimensional (2-D) polyacrylamide gel electrophoresis (O'Farrell, 1975), which allows monitoring the pattern of gene expression of the whole cell. A 2-D gel can resolve up to 2500 protein spots (VanBogelen and Neidhardt, 1990), which can be detected by either direct staining or autoradiography. Stress-inducible proteins thus can be recognized by examining autoradiograms of 2-D gels of cultures that have been pulse-labeled with a radioactive amino acid for brief periods before and shortly after the imposition of stress. An example of the power of such an analysis is depicted in Fig. 1, which demonstrates the global effects of overexpressing the RelA protein of the stringent response (Gallant, 1979) and the accompanying elevated levels of guanosine 3', 5'-bispyrophosphate (ppGpp). Note that only major alterations representing markedly induced proteins are highlighted with arrowheads in the figure.

The purpose of this chapter is to review some findings obtained by the global systems approach regarding the response of different prokaryotes to nutrient starvation. Emphasis will be on the responses of the marine Vibrio sp. S14 (CCUG 15956) and E. coli to nutrient starvation and to relate the findings obtained for these organisms to the response of other prokaryotes, including some differentiating bacteria such as spore-forming Bacillus spp. and fruiting Myxococcus spp. No attempt will be made to review comprehensively the wealth of existing information about the global systems approach for which purpose the reader is referred to articles by Gottesman and Neidhardt (1983), Gottesman (1984), Neidhardt et al. (1990), and VanBogelen and Neidhardt (1990).

2. *PROTEIN EXPRESSION DURING STARVATION-INDUCED GROWTH ARREST*

2.1. *Physiology of the Starved Cell*

The mechanism of adaptation to nutrient exhaustion in copiotropic (Poindexter, 1981) nondifferentiating bacteria is a process of cellular development that involves sequential changes in cell physiology (e.g., Thomas and Batt, 1969; Dow and Whittenbury, 1980; Dow *et al.*, 1983; Morita, 1985) and minor changes in ultrastructure (Morita, 1983, 1985; Kjelleberg *et al.*, 1987). The physiological alterations and survival characteristics of the starved bacteria depend on the organism and such factors as the growth phase at which starvation was initiated, growth rate of the cells, nutritional status of the medium, population density of the starved cells, biological history of the cells, and the nature of the starvation environment (Dawes, 1976).

A sporulating *Bacillus* sp. cell proceeds through a series of defined morphological stages of differentiation that culminate in the formation of a stress-resistant endospore during nutrient depletion. In contrast, the mechanism of adaptation of nondifferentiating copiotrophic bacteria evidently occurs in the absence of such sophisticated and easily detected morphological changes. However, some of the numerous physiological changes that take place in response to carbon or multiple-nutrient starvation (carbon, nitrogen, and phosphorus) of the marine *Vibrio* sp. S14 bear obvious similarities to the sporulation responses of *Bacillus* spp. (see Östling *et al.*, this volume). Carbon- and multiple-nutrient-starved *Vibrio* sp. S14 cells are also more resistant to stresses such as ampicillin-induced autolysis and ultrasound treatment (Nyström and Kjelleberg, 1989), heat treatment, near-UV and UV irradiation, and exposure to high concentrations of $CdCl_2$ (Nyström *et al.*, 1992). Similarly, carbon or nitrogen starvation of *E. coli* cells significantly diminishes their sensitivity to a variety of stresses (Jenkins *et al.*, 1988, 1990; Matin, 1990). In addition, the time-dependent development of stress resistance in both *Vibrio* sp. S14 (Nyström *et al.*, 1988, 1990a, 1992) and *E. coli* (Groat *et al.*, 1986) is accompanied by sequential alterations in gene expression.

2.2. *Sequential Expression of Starvation-Inducible Proteins*

An examination of the protein profile of *Vibrio* sp. S14 cells starved for carbon, nitrogen, and phosphorus reveals that the synthesis of starvation-inducible proteins is time-dependent and that the proteins can be grouped in different classes with respect to their time of appearance during starvation (Nyström *et al.*, 1990a). Similar to the carbon-starvation response of *E. coli*

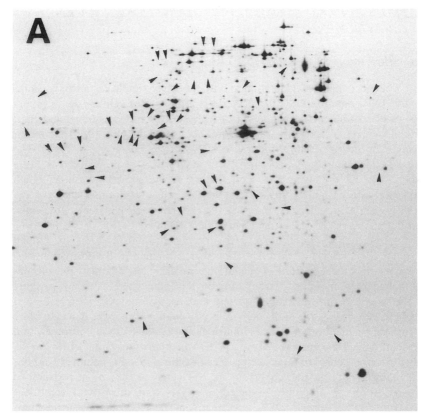

Figure 1. Two-dimensional polyacrylamide gel analysis of the effect of overproducing the RelA protein of the stringent response. In this *E. coli* construct, the *relA* gene is under control of a conditional P_{tac} promoter, such that transcription is controlled by IPTG (Schreiber *et al.*, 1991). (A) 2-D protein pattern of pulse-labeled steady-state growing, uninduced *E. coli* W3110 cells. (Figure continued on next page.)

(Groat *et al.*, 1986), most of these proteins are synthesized in a burst immediately at the onset of starvation. Inhibitors of protein synthesis added for a brief period at the onset of starvation dramatically decrease long-term survival, while later additions for extended periods have little or no effect on subsequent survival. Based on such results of experiments, it has been argued that the induction of some of the early starvation proteins is an absolute requirement for long-term survival of both *E. coli* (Reeve *et al.*, 1984; Matin, 1990) and *Vibrio*

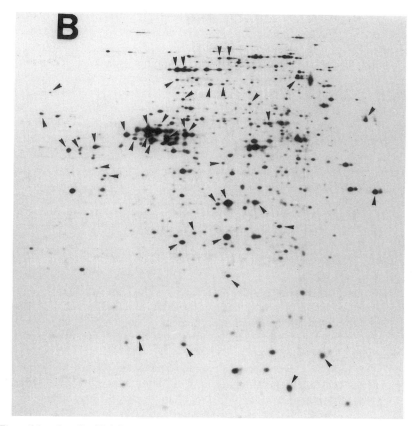

Figure 1 (continued). (B) 2-D protein pattern of cells 60 min after addition of 100 μM IPTG. The proteins whose expression is induced by this treatment are marked with arrowheads. (Nyström, unpublished data.)

sp. S14 (Nyström *et al.*, 1990a) during starvation. A similar finding has been reported for *Salmonella typhimurium* (Spector and Foster, this volume).

Sequence of Events

Determination of the regulation of the temporal classes of starvation proteins and their possible interrelationships is an important area in the field of starvation-survival studies regardless of whether one works with differentiating or nondifferentiating bacteria. Several pieces of information regarding the regulation of the ordered progression of gene expression during differentiation in *B. subtilis* exist, and a general picture is beginning to emerge. In contrast, the regulatory features behind the sequential alterations in gene expression in nondifferentiating bacteria are not well understood.

In an attempt to investigate whether protein synthesis at the onset of starvation in *Vibrio* sp. S14 was necessary for the orderly expression of proteins induced at later times, cells were incubated with chloramphenicol during the first hour of starvation, washed free from the inhibitor, and suspended in fresh starvation medium. Subsamples were subsequently removed at different times of starvation, pulse-labeled with radioactive methionine, and processed for 2-D gel electrophoresis analysis. Most (26 of 33) of the "early" starvation-induced proteins were immediately synthesized when the cells were washed free from the inhibitor (Nyström *et al.*, 1990a). However, five "early" starvation proteins were markedly delayed (more than 3 h) in expression while two failed to be expressed at any time subsequent to the chloramphenicol treatment. In addition, while most "late" starvation proteins were unaffected by the inhibition of protein synthesis during the first hour of starvation, the induction of six "late" starvation proteins, two of which normally were induced subsequent to 5 h of starvation, was totally abolished (Nyström *et al.*, 1990a). These observations indicate that further work may be warranted to elucidate whether sequential cascade pathways are operating during starvation in nondifferentiating bacteria. The differential effects of chloramphenicol on the temporal expression of starvation proteins also point to the fact that several different pathways run in parallel during starvation.

In addition to a hypothetical "cascade" pathway, akin to the proposed sigma factor "cascade" operating during sporulation in *B. subtilis* (Losick and Pero, 1981), different operons or global regulatory networks may be temporally induced as a consequence of sequential decreases in the pools of key metabolites below critical levels.

3. OVERLAP BETWEEN STARVATION STIMULONS

3.1. Unique, General, and Universal Starvation Proteins

Two-dimensional gel electrophoresis analysis of the proteins synthesized at the onset of starvation for different individual nutrients has revealed interesting patterns of overlap between starvation stimulons. While a minor fraction of one stimulon may overlap with another, the majority of the proteins induced by one specific stimulus are generally unique to that particular stimulon. An inspection of the "Gene–protein database of *Escherichia coli* K-12: Edition 3" (VanBogelen *et al.*, 1990), in which proteins of *E. coli* responsive to different stresses and starvation conditions are recorded, reveals that most proteins induced by one specific stimulus are unique to that stimulon. However, the starvation stimulons examined (carbon-, nitrogen-, phosphate-, and isoleucine-stimulon) share some 27 proteins (VanBogelen *et al.*, 1990). Twenty of these general starvation proteins are induced by two of the different starvation conditions, six are induced by three starvation conditions, while one (C13.5) appears to be a nonspecific responder to starvation-induced growth inhibition. Thirteen additional stress conditions that cause growth inhibition have subsequently been demonstrated to significantly induce the expression of the C13.5 protein (Nyström and Neidhardt, 1992), suggesting that C13.5 may be a universal responder to stress or stasis.

Similar to the results obtained for *E. coli* (Groat *et al.*, 1986; VanBogelen *et al.*, 1990), the carbon, nitrogen, and phosphorus starvation stimulons of *Vibrio* sp. S14 share some 18 genes (Nyström *et al.*, 1992), 9 of which are induced by all of the starvation conditions examined. Some of the general starvation proteins are also induced by stresses such as heat shock, $CdCl_2$, and nalidixic acid exposure (Nyström *et al.*, 1990a). Nalidixic acid primarily induces an SOS response, while $CdCl_2$ is a potent inducer of the oxidation stress response (VanBogelen *et al.*, 1987).

The overlap between starvation-inducible proteins and proteins induced by heat-shock, DNA-damaging, and oxidative agents may indicate that starvation causes the provocation of the heat-shock, oxidation stress, and SOS response. This does not appear to be the case, however, since only minor fractions of the heat-, $CdCl_2$-, and nalidixic acid-inducible stimulons are induced in response to starvation conditions (Nyström *et al.*, 1990a). This suggests that starvation may affect the expression of some members of these stress responses, but does not cause a full induction of each global regulatory network. This is consistent with the demonstration that gene members of global regulatory systems are independently responsive to individual controls under certain conditions (VanBogelen *et al.*, 1987).

All of the 18 proteins that are induced by two or all three of the starvation conditions examined in *Vibrio* sp. S14 belong to the "early" class of starvation proteins. Ten of these proteins are expressed persistently after induction through a 48-h period of starvation for carbon, nitrogen, and phosphorus, while the remaining eight are only transiently expressed (Nyström and Kjelleberg, unpublished). Notably, eight of the ten proteins persistently synthesized during prolonged periods of starvation belong to the group of proteins that are induced regardless of the identity of the depleted macronutrient.

3.2. Identity of Some General Starvation Proteins

While several starvation-specific proteins (i.e., proteins induced by the exhaustion of one specific nutrient) seem to be specifically involved with enhancing the cell's capacity of assimilation and uptake of the particular limiting factor, general and universal starvation proteins may have more general protective functions related to the growth arrest state. Fifteen of the twenty-six proteins reported to be induced by two or more starvation conditions (Van-Bogelen *et al.*, 1990) are products of identified genes or members of known global regulatory networks. These genes as well as their responsiveness to different stresses are recorded in Fig. 2. An inspection of Fig. 2 reveals that eight of these general starvation proteins are members of known global regulatory networks; six belong to the heat shock (*rpoH*-controlled) response and two to the oxidation stress (*oxyR*-controlled) response (VanBogelen *et al.*, 1990). Although heat shock proteins appear to possess several diverse functions in the cell, they seem to have a general role in protein processing (Neidhardt and VanBogelen, 1987). These functions include protein degradation, the prevention of the misfolding of other polypeptides, and allowing other proteins to reach the correct destination inside or outside the cell (Georgopoulos *et al.*, 1990). However, it is not known whether the induction of some members of the heat shock response during starvation is a result of an increased requirement for protein processing. Notably, none of the heat-shock proteins of *E. coli* are induced by nitrogen starvation at 37°C, and there is no overlap between the nitrogen starvation stimulon and the heat shock stimulon of *Vibrio* sp. S14 (Nyström *et al.*, 1990b).

It is noteworthy that protein H-NS can be found among the general starvation proteins (Fig. 2). Protein H-NS is an abundant protein of the bacterial nucleoid and may play an important role in the large-scale organization of the chromosome (Spassky *et al.*, 1984; Gualerzi *et al.*, 1986; Pon *et al.*, 1988). In addition, H-NS has been reported to stimulate the rate of spontaneous chromosomal deletions (Lejeune and Danchin, 1990), to alter thermoregulation of a pilus gene cluster (Göransson *et al.*, 1990) and the osmoregulation of *proU* (Higgins *et al.*, 1988), and to affect phase variation in fimbriae (Spears *et al.*, 1986) and the rates of mini-Mu transposition (Falconi *et al.*, 1988). The possible role of H-NS in modulating levels of gene expression and gene rearrangements during starvation remains to be elucidated.

3.3. Possible Regulation of General Starvation and Stress Proteins

3.3.1. Passive Control

Jensen and Pedersen (1990) have presented an intriguing model of "passive" growth rate control that may explain the direct growth rate-dependence of stringent promoters driving transcription of rRNAs and ribosomal protein mRNAs that may be applicable also to transcriptional control during starvation-induced growth arrest. Basically, the idea is that the pattern of transcription at

PROTEINS INDUCED BY TWO STARV. CONDITIONS			STARVATION CONDITIONS				OTHER STRESS [c] CONDITIONS						
Protein[a] design.	Gene name	Member of G.R.N[b].	NS	PS	CS	ILE	NA	42C	50C	10C	Cd	QN	HP
B15.0													
B66.0	*dnaK*	HTP											
C15.4	*mopB*	HTP											
C17.7													
D29.8													
D33.4	*htpH*	HTP											
E88.0													
F22.5	*eda*												
F37.9	*tyrA*												
F29.0													
F45.6	*gor*												
F84.1	*htpM*	HTP											
G15.4		OXY											
G38.2													
G42.1													
G48.0													
G52.0	*amn*												
G72.1													
G76.0													
H41.0													
PROTEINS INDUCED BY THREE STARVATION CONDITIONS													
B25.3	*grpE*	HTP											
B56.5	*mopA*	HTP											
F14.7	*hns*												
F50.6		OXY											
G57.1													
G60.1													
PROTEIN INDUCED BY FOUR STARVATION CONDITIONS													
C13.5													

Figure 2. Schematic diagram of the overlap among starvation and stress responses in *E. coli* W3110. The shaded boxes represent proteins that are induced by a given starvation or stress condition as determined by 2-D gel electrophoresis (Van Bogelen *et al.*, 1990). (a) Protein designations: Alphanumerics adapted from Van Bogelen *et al.* (1990). (b) G.R.N., global regulatory networks: HTP, heat-shock proteins (RpoH-controlled); OXY, oxidation stress proteins (OxyR-controlled). (c) Starvation and stress conditions: NS, nitrogen starvation; PS, phosphate starvation; CS, carbon starvation; ILE, isoleucine starvation; NA, nalidixic acid; 42C, shift in temperature from 28 to 42°C; 50C, shift in temperature from 28 to 50°C; 10C, shift in temperature from 37 to 10°C; Cd, cadmium chloride treatment; QN, treatment with the quinone ACDQ; HP, treatment with hydrogen peroxide.

different growth rates is dictated by competition between promoters for a limited supply of free RNA polymerase. The model centers around the proposition that the rate of elongation of RNA polymerase and ribosomes is variable (Pedersen, 1984) and will decrease with decreasing growth rates or at the cessation of growth as a result of nutrient limitation. Thus, more RNA polymerase molecules will become sequestered at elongation, less will be released per unit time leading to a fall in the availability of free RNA polymerase. As a consequence, the frequency of transcription initiation will decrease markedly at those promoters that are difficult to saturate. It is reasonable to propose that genes whose products contribute most to the rapid growth rate in rich media, i.e., genes encoding ribosomal proteins and rRNA, will possess such promoters with high maximal reaction velocity (rapid clearing of the promoter region). In addition, the model of Jensen and Pedersen (1990) implies that all promoters may in fact be inhibited during growth arrest (at least during conditions that induce stringency). However, promoters that bind RNA polymerase with a high affinity but initiate transcription at a low frequency will be less affected. Thus, general starvation or stress proteins may represent a group of proteins encoded by genes driven by easily saturated promoters that are less repressed (in comparison with genes driven by stringent promoters) during starvation-induced growth inhibition.

If in fact such passive control is the sole mechanism behind the regulation of general starvation proteins, then their rate of synthesis may be postulated to be inversely dependent on growth rate. In a study describing differential rates of synthesis of 140 individual proteins of *E. coli* as a function of growth rate (Pedersen *et al.*, 1978), the expression of 33 proteins was found to be inversely dependent on growth rate. Three of these 33 proteins, G76.0 and the products of the *htpM* and *hns* genes, belong to the group of general starvation and stress proteins. However, the expression of DnaK and GroEL (regarded as general starvation proteins in Fig. 2), the products of the *dnaK* and *mopA* genes, respectively, increase with an increased growth rate, and the expression of yet another general starvation protein, GroES (Fig. 2), the product of *mopB*, was found to be growth rate independent (Pedersen *et al.*, 1978). These results demonstrate that there is no common growth rate-dependent regulatory feature for all of the general starvation proteins.

From the point of economy, it is feasible that genes with "passive" regulation would be found among those with low levels of expression. This is consistent with the results of Pedersen *et al.* (1978), who found that 67% of the proteins with the lowest levels of expression in *E. coli* belong to the group of proteins the expression of which was inversely dependent on growth rate. This group contained only 20% of the proteins examined, indicating an overrepresentation within this group of genes with weak promoters (Pedersen *et al.*, 1978).

3.3.2. "Gearbox" Promoters and Growth Rate-Dependent Expression

Some of the genes found to be increasingly expressed when *E. coli* enters starvation-induced "stationary" phase have been shown to be regulated at the level of transcription initiation. A few of these genes are driven by so-called "gearbox" promoters. Gearboxes denote a distinct class of promoters characterized by showing an activity that is inversely dependent on growth rate (Aldea *et al.*, 1990; Connell *et al.*, 1987). Thus far, the cell division *ftsQAZ* gene cluster, the *bolA* morphogene (Aldea *et al.*, 1990), and the *mcbA* gene encoding Microcin B17 (Connell *et al.*, 1987) have been found to be driven by "gearboxes." The promoters of these genes share some interesting features including conserved and characteristic sequences around their -10 and -35 regions (Aldea *et al.*, 1990). The induction of expression of these genes during entrance into starvation-induced stationary phase has been interpreted as being a consequence of the activation of gearbox promoters caused by the gradual decrease in growth rate that occurs during growth rate transition (Aldea *et al.*, 1990). If this interpretation is correct, then gearbox promoters are probably activated regardless of the conditions causing the cessation of growth, and may be responsible for the induction of some general starvation and stress proteins.

The presence of the distinct and functionally essential -10 region of the gearbox promoter suggests that the regulation of this class of promoters may require a distinct sigma factor, or alternatively a modified form of RNA polymerase presumably activated in accordance with alterations in growth rate. The product of *katF*, a newly discovered global regulator of stationary-phase gene expression and a putative sigma factor (Mulvey and Loewen, 1989; Lange and Hengge-Aronis, 1991a), has been suggested to be the required factor regulating the expression of gearbox promoters. However, while the *katF* gene product seems to be required for *bolA* expression (Hengge-Aronis, this volume; Lange and Hengge-Aronis, 1991b), the *mcbA* gene was unaffected or even expressed at a somewhat higher level in a *katF* mutant strain (Bohannon *et al.*, 1991).

An additional mechanism responsible for the regulatory features of genes driven by gearbox promoters may be passive regulation as described in Section 3.3.1. In this context, it is interesting to note that several of the genes (*ftsA*, *ftsZ*, *ftsQ*, *bolA*; Aldea *et al.*, 1990) so far demonstrated to be gearbox dependent are involved in cell division and morphogenesis. Such proteins are usually found to have very low levels of expression and the products may be difficult to detect using standard protein electrophoresis analysis without the aid of specific antibodies. Gearbox promoters may thus be postulated to bind RNA polymerase with a high affinity but initiate transcription at a low frequency. In other words, gearbox promoters may be easily saturated (less affected by limitations in the availability of free RNA polymerase) and hence represent a class of weak promoters allowing only low levels of expression. This is consistent with the notion that base substitutions in the -10 region of the *bolA* gearbox promoter creating a TATAAT consensus sequence for sigma70-dependent "housekeeping" promoters resulted in a 20-fold increase in *bolA* expression (Aldea *et al.*, 1990). Moreover, the expression obtained from this mutated promoter was found to be constitutive with respect to growth rate, with an expression level similar to that of the fully induced wild-type *bolA* promoter (Aldea *et al.*, 1990).

3.3.3. *Perturbations in the Levels of Key Metabolites*

As pointed out by Sonenshein (1989), the search for the key metabolite that might monitor the availability of carbon, nitrogen, and phosphorus and function as a corepressor of sporulation has been the Holy Grail for *Bacillus* sporulation physiologists ever since the publication of the seminal paper on *B. subtilis* sporulation by Schaeffer and co-workers in 1965. The pertubation of the levels of this metabolite as a result of depletion for either external carbon, nitrogen, or phosphorus would activate the sporulation pathway unless other global regulatory responses provoked by the nutrient exhaustion allow the cell to "escape" starvation. It has been argued that GTP possesses the properties of such a hypothetical corepressor (Ochi *et al.*, 1982), and treatments that force the GTP pool to drop (e.g., decoyine additions) cause inhibition of balanced growth and initiation of sporulation (Freese *et al.*, 1979; Freese and Heinze, 1984). With the aid of 2-D gel electrophoresis, Hecker and Völker (1990) identified some 13 proteins of *B. subtilis* as general responders to starvation- and stress-induced growth arrest. However, an experimental decrease in the GTP content of the cells leading to sporulation did not trigger the induction of these general stress proteins (Hecker and Völker, 1990). In addition, all 13 general stress proteins were normally induced in *spoOA* mutants during starvation-induced growth arrest. Thus, the induction of general stress proteins by the starvation of *B. subtilis* is an early stationary-phase event rather than a sporulation event. It has been postulated that general stress proteins are dispensable for sporulation but may aid in structural stabilization and stress resistance of the "early" sporulating cell (Dowds *et al.*, 1987; Hecker and Völker, 1990).

An intact *relA* gene was found to be a prerequisite for maximal induction of 11 and 10 of the 13 general stress proteins during amino acid starvation and oxygen limitation, respectively (Hecker and Völker, 1990). While 5 of the general stress proteins required the *relA* gene for maximal expression during osmotic stress, all but one were induced in a *relA*-independent manner during heat stress (Hecker and Völker, 1990). Based on these results, the authors proposed that the induction of general stress proteins in *B. subtilis* may be either *relA*-dependent or -independent depending on the nature of the stress condition encountered by the cells. It is interesting to note, however, that the mechanism of both sporulation initiation and stress protein induction may be coupled to alterations in guanine metabolism in *B. subtilis*, at least during starvation-induced growth arrest.

Alteration in guanine metabolism during starvation for different individual nutrients does not seem to be the sole mechanism involved in the induction of general starvation proteins in *Vibrio* sp. S14. While the induction of general starvation proteins in *Vibrio* sp. S14 does in some cases coincide with the induction of stringency, these proteins are induced also during starvation conditions that alter the intracellular levels of neither ppGpp (Nyström *et al.*, 1990a) nor GTP/GDP (K. Flärdh, personal communication).

Perturbations of normal levels of the pyridine nucleotides NAD and NADP have been proposed to signal the induction of general starvation proteins in *Salmonella typhimurium* (Foster, 1983). It has been argued that alterations in the levels of the nucleotides NAD/NADP in response to glucose, ammonium, or phosphate starvation conditions should result in numerous stress signals which would lead to the induction of general starvation-inducible genes associated with maintenance and survival (Foster, 1983; Spector *et al.*, 1986, 1988), since glucose, ammonium, and phosphate are all essential for NAD biosynthesis and NAD(P) participates in a variety of catabolic and anabolic enzymatic reactions. By the use of the global systems approach, six proteins induced by nicotinate starvation were identified as being general responders to nutrient starvation (Spector *et al.*, 1986, 1988). The expression of these general starvation-inducible genes exhibits different mechanisms of regulation, including *relA*-dependent, *cya*-dependent, and *cya*-independent-*crp*-dependent mechanisms (Spector, 1990). Notably, four of the general starvation-inducible loci play key roles in the maintenance of viability during prolonged periods of simultaneous starvation for phosphate, carbon, and nitrogen (Spector, 1990). A detailed account of starvation-induced proteins in *Salmonella typhimurium* is presented by Spector and Foster (this volume).

4. ROLE OF GLOBAL REGULATORY NETWORKS IN STARVATION-SURVIVAL

4.1. Stringent Control

Inhibition of protein synthesis by chloramphenicol during the initial phase of carbon starvation results in a rapid loss of viability of both *E. coli* (Reeve *et al.*, 1984) and *Vibrio* sp. S14 (Nyström *et al.*, 1990a&b) cells. It is possible that this deleterious effect of chloramphenicol on starvation-survival is mediated through the relaxation of stringency rather than the inhibition of the synthesis of "starvation-survival-related" proteins, since the addition of chloramphenicol abolishes the induction of stringent control that normally occurs during different starvation conditions (Lund and Kjeldgaard, 1972; Nyström *et al.*, 1990a&b). Studies with an *E. coli relA* mutant indicated that the deleterious effect of chloramphenicol on survival during carbon starvation is not mediated through interference with the stringent response but is instead caused by the inhibition of protein synthesis *per se* (Reeve *et al.*, 1984). However, the *relA* mutant (*relA1*) employed in the study by Reeve *et al.* (1984) is unfortunately able to synthesize low levels of ppGpp during growth and accumulate normal levels of ppGpp during carbon-source starvation (Cashel and Rudd, 1987; Metzger *et al.*, 1989). An *E. coli* strain deleted for both *relA* and *spoT* (Metzger *et al.*, 1989) (genes encoding the only known ppGpp synthetases) generates quite different results, in that viability during carbon starvation was lost at the same rate as the isogenic parent treated with chloramphenicol (Fig. 3A; Nyström, unpublished). Moreover, induction of the RelA protein using the construct of Schreiber and co-workers (1991) for 1 h prior to carbon starvation

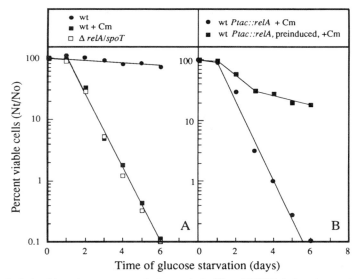

Figure 3. Role of the stringent response in the survival of *E. coli* W3110 during glucose starvation. (A) Viability of a Δ*relA/spoT* mutant (□), the isogenic wild-type (wt) parent (●), and the isogenic (wt) parent treated with chloramphenicol (■) during long-term glucose starvation. (B) Viability of *E. coli* W3110 P_{tac}::*relA* during glucose starvation in the presence of chloramphenicol. *E. coli* P_{tac}::*relA* pretreated with IPTG for 1 h (■), and with no pretreatments (●), prior to glucose starvation in the presence of chloramphenicol (Cm). (Nyström, unpublished data.)

markedly diminishes the effect of chloramphenicol on subsequent starvation-survival (Fig. 3B; Nyström, unpublished). It has previously been demonstrated that relaxed control *relB* mutants of *E. coli* exhibit a diminished ability to form colonies on solid medium when starved for nutrients, particularly glucose (Mosteller, 1978; Mosteller and Kwan, 1976). These results suggest an important role for the stringent response in starvation-survival (see also Spector and Foster, this volume).

There is no direct evidence for a correlation between the induction of stringency and starvation-survival of *Vibrio* sp. S14. However, in a study demonstrating that some prestarvation and prestress conditions greatly enhanced survival during subsequent C, N, and P starvation in the presence of chloramphenicol or rifampicin, it was found that the highest degree of survival was achieved using prestress conditions that provoked a stringent response (Nyström *et al.*, 1990a).

4.2. The Heat-Shock Response

A mutation in the gene encoding the major heat shock protein DnaK of *E. coli* results in pleiotropic effects with regard to the cell's response to nutrient starvation (Spence *et al.*, 1990). The *dnaK* mutant is highly susceptible to killing by starvation for carbon and analysis of proteins induced during starvation for different individual nutrients showed that the *dnaK* mutant is defective for the expression of some carbon starvation proteins as well as general starvation proteins (Spence *et al.*, 1990). Moreover, the *dnaK* mutant is markedly impaired in its ability to utilize alternative carbon sources (Spence *et al.*, 1990). A deletion in the *rpoH* gene encoding the heat shock-specific sigma factor results in a similar increased sensitivity to carbon starvation conditions (Jenkins *et al.*, 1991). However, elucidating the effect of an *rpoH* mutation on starvation-survival is obscured by the fact that the $\Delta rpoH$ mutant is also defective in bulk protein synthesis during starvation (Jenkins *et al.*, 1990), which itself affects the cell's capacity to withstand starvation conditions.

In a variety of organisms including *Saccharomyces cerevisiae* (Kurtz *et al.*, 1986), *Drosophila melanogaster* (Zimmerman *et al.*, 1983), and mice (Kothary *et al.*, 1987), some heat shock proteins are expressed during specific stages of cellular differentiation. In addition, myxospore formation in *Myxococcus xanthus* is accelerated by heat shock administered prior to the induction of development (Killeen and Nelson, 1988) and it has been proposed that the response to starvation-induced fruiting body formation in *M. xanthus* shares common genes with the heat-shock response (Nelson and Killeen, 1986). It has been hypothesized by Kurtz and co-workers (1986) that the heat-shock response may be an ancient response to stress from which many organisms have evolved developmental pathways. The induction of some heat shock proteins by non-sporulating prokaryotes such as *E. coli* and *Vibrio* spp. may thus represent rudiments of this ancient development pathway (Spence *et al.*, 1990; Östling *et al.*, this volume).

5. *PROTEIN EXPRESSION DURING RECOVERY FROM STARVATION*

Two important aspects of starvation-survival are that the copiotrophic, nondifferentiating organisms must preserve their genomes during prolonged periods of energy and nutrient exhaustion, and also retain the ability to recover during sporadic inputs of nutrients (see Morita, this volume). While the nondifferentiating bacteria lack the genetic security of spore formation during growth arrest, they maintain the ability to rapidly metabolize substrates by retaining a functional proton flow, electron transport chain, and active membrane-bound ATPases (Kjelleberg *et al.*, 1983) during prolonged periods of starvation. Even long-term (200 h)-starved cells of *Vibrio* sp. S14 are not dormant but respond immediately to nutrient addition (Mårdén *et al.*, 1987; Albertson *et al.*, 1990a, b; Östling *et al.*, this volume).

Substrate addition to long-term energy- and nutrient-starved *Vibrio* sp. S14 cells results in pronounced alterations in gene expression (Albertson *et al.*, 1990c). Some 23 proteins are sequentially repressed during the maturation phase preceding regrowth (Albertson *et al.*, 1990c). Most of these belong to different temporal classes of starvation-inducible proteins (see Section 2.2.). In addition, 21 proteins exhibit novel or markedly increased expression as a consequence of substrate addition. These proteins are induced in a sequential manner during maturation and can be divided into at least three temporal groups (Albertson *et al.*, 1990c). Most interestingly, 11 of these proteins seem to be maturation-specific in that they are exclusively synthesized during maturation and are repressed as soon as regrowth occurs. A question of interest concerning these maturation-specific proteins relates to their possible functional similarity with products of the *ger* genes (genes involved in spore germination) of *B. subtilis* (Foster and Johnstone, 1989). The role of some maturation (mat) proteins may be to degrade starvation-inducible proteins whose functions may become superfluous or even inhibitory to the vegetative cells. Such functions may include structural stabilization of the nonreplicating nucleoid and the cell wall (Nyström and Kjelleberg, 1989). In contrast to *ger* genes in *B. subtilis* which are expressed during sporulation since the initial phase of germination takes place in the absence of *de novo* protein synthesis (Foster and Johnstone, 1989), the mat proteins of *Vibrio* sp. S14 are expressed exclusively during the maturation phase. Thus, the *ger* genes are classified in accordance with their function rather than the timing of their synthesis. Future analysis will be needed to determine the function of mat proteins and whether mutations in loci encoding these proteins will indeed impair the nondifferentiating bacterial cell's ability to exit starvation.

6. PERSPECTIVES

A significant barrier to progress in understanding the mechanisms of adaptation of bacteria such as *E. coli* and *Vibrio* spp. to nutrient stress and growth arrest may be the absence of easily detectable morphological changes. Clones of developmental genes cannot be directly selected for with respect to distinctive morphological phenotypes, and mutants that are blocked at discrete developmental stages are likewise not available. The effects for example of *Mud-lacZ* insertional disruption of starvation-inducible loci, therefore, are often characterized according to their effects on the cell's: (1) survival during starvation; (2) ability to recover from starvation; or (3) induce cross-protection to other stresses. The global systems approach, however, making use of a nearly complete picture of the cell's complement of proteins and patterns of gene expression, permits analysis of complex cellular responses and regulatory networks. This approach will complement the analysis of individual genes and their involvement in the starvation-survival characteristics of nondifferentiating bacteria. In addition, the time-dependent alterations in the "2-D protein profiles" during starvation of nondifferentiating microbes may compensate for the lack of easily detected morphological stages. Mutations in starvation-inducible genes can thus be analyzed as to their effect on the ordered progression of the "2-D protein profiles" during starvation.

The global systems approach generates massive amounts of data and as described in this chapter, hundreds of proteins have been shown to exhibit markedly altered expression in response to starvation in nondifferentiating bacteria. While a few of the early starvation-inducible proteins are products of known genes (VanBogelen *et al.*, 1990), essentially all of the genes activated in a sequential and ordered fashion after the initial burst of synthesis of early proteins are of unknown identity. Clearly, elucidating the properties of these proteins (as well as proteins being sequentially repressed during starvation) and their genes will be of considerable importance to our understanding of the physiology of starved nondifferentiating bacteria. Toward the goal of tracing these starvation-inducible proteins to their structural genes, the technique of "reverse genetics" in combination with the global systems approach may be an invaluable tool. Automated protein-sequencing instrumentations and methodologies have evolved considerably in recent years (LeGendre and Matsudaira, 1989), and it is possible to obtain sequence data with less than 10 pmol protein (Hunkapiller *et al.*, 1986). Thus, protein samples can be recovered from 2-D gels by electroelution or electroblotting onto a solid support and sequenced directly without further purification. The amino acid sequence information obtained is subsequently used for construction of degenerate oligonucleotide probes which make possible cloning the corresponding gene by screening genomic libraries followed by mutant constructions. Analysis of these clones and mutants will increase our understanding of the physiological adaptation of nondifferentiating bacteria to starvation. In addition, comparison between the regulatory programs provoked in response to growth arrest in nondifferentiating bacteria and the sporulation pathway for example of *Bacillus* spp., can be performed on a molecular basis that will eventually answer questions of whether the ability to sporulate was lost by nondifferentiating bacteria or acquired by sporulators after evolutionary divergence (Sonenshein, 1989).

ACKNOWLEDGMENTS. I thank N. H. Albertson and G. W. Jones for helpful and critical comments on the manuscript. The author's work on global systems analysis has been supported by grants from the Swedish Natural Science Research Council and by Public Health Service grant GM-17892 from the National Institutes of Health to Frederick C. Neidhardt.

REFERENCES

Albertson, N. H., Nyström, T., and Kjelleberg, S., 1990a, Exoprotease activity of two marine bacteria during starvation, *Appl. Environ. Microbiol.* **56:**218–223.

Albertson, N. H., Nyström, T., and Kjelleberg, S., 1990b, Starvation-induced modulations in binding protein-dependent glucose transport by the marine *Vibrio* sp. S14, *FEMS Microbiol. Lett.* **70:**205–210.

Albertson, N. H., Nyström, T., and Kjelleberg, S., 1990c, Macromolecular synthesis during recovery of the marine *Vibrio* sp. S14 from starvation, *J. Gen. Microbiol.* **136:**2201–2207.

Aldea, M., Garrido, T., Pla, J., and Vincente, M., 1990, Division genes in *Escherichia coli* are expressed coordinately to cell septum requirements by gearbox promoters, *EMBO J.* **11:**3787–3794.

Bohannon, D. E., Connell, N., Keener, J., Tormo, A., Espinosa-Urgel, M., Zambrano, M. M., and Kolter, R., 1991, Stationary-phase-inducible "gearbox" promoters: Differential effects of *katF* mutations and role of σ^{70}, *J. Bacteriol.* **173:**4482–4492.

Cashel, M., and Rudd, K. E., 1987, The stringent response, in: *Escherichia coli and Salmonella typhimurium, Cellular and Molecular Biology,* Vol. 2 (F. C. Neidhardt, J. L. Ingraham, K. B. Low, B. Magasanik, M. Schaechter, and H. E. Umbarger, eds.), American Society for Microbiology, Washington, D.C., pp. 1410–1438.

Christman, M. F., Morgan, R. W., Jacobson, F. S., and Ames, B. N., 1985, Positive control of a regulon for defense against oxidative stress and some heat shock proteins in *Salmonella typhimurium, Cell* **41:**753–762.

Connell, N., Han, Z., Moreno, F., and Kolter, R., 1987, An *E. coli* promoter induced by the cessation of growth, *Mol. Microbiol.* **1:**195–201.

Dawes, E. A., 1976, Endogenous metabolism and the survival of starved procaryotes, in: *The Survival of Vegetative Microbes* (T. R. G. Gray and R. R. Postgate, eds.), Cambridge University Press, London, pp. 19–53.

Dow, C. S., and Whittenbury, R., 1980, Prokaryotic form and function, in: *Contemporary Microbial Ecology* (D. C. Ellwood, J. N. Hedger, M. J. Latham, J. M. Lynch, and J. H. Slater, eds.), Academic Press, New York, pp. 391–417.

Dow, C. S., Whittenbury, R., and Carr, N. G., 1983, The 'shut down' or 'growth precursor' cell— An adaptation for survival in a potentially hostile environment, in: *Microbes in Their Natural Environments*, Vol. 34 (J. H. Slater, R. Whittenbury, and J. W. T. Wimpenny, eds.), Cambridge University Press, London, pp. 187–247.

Dowds, B. C. A., Murphy, P., McConnell, D. J., and Devine, K. M., 1987, Relationship among oxidative stress, growth cycle, and sporulation in *Bacillus subtilis*, *J. Bacteriol.* **169**:5771–5775.

Falconi, M., Gualtieri, M. T., La teana, A., Losso, M. A., and Pon, C. L., 1988, Proteins from the prokaryotic nucleoid: Primary and quaternary structure of the 15kD *Escherichia coli* DNA-binding protein H-NS, *Mol. Microbiol.* **2**:323–329.

Foster, J. W., 1983, Identification and characterisation of a *relA*-dependent starvation-inducible locus (*sin*) in *Salmonella typhimurium*, *J. Bacteriol.* **156**:424–428.

Foster, S. J., and Johnstone, K., 1989, The trigger mechanism of bacterial spore germination, in: *Regulation of Procaryotic Development* (I. Smith, R. A. Slepecky, and P. Setlow, eds.), American Society for Microbiology, Washington, D.C., pp. 89–108.

Freese, E., and Heinze, J., 1984, Metabolic and genetic control of bacterial sporulation, in: *The Bacterial Spore*, Vol. 2 (A. Hurst, G. Gould, and J. Dring, eds.), Academic Press, New York, pp. 101–172.

Freese, E., Heinze, J., and Galliers, E. M., 1979, Partial purine deprivation causes sporulation of *Bacillus subtilis* in the presence of excess ammonium, glucose and phosphate, *J. Gen. Microbiol.* **115**:193–205.

Gallant, J. A., 1979, Stringent control in *Escherichia coli*, *Annu. Rev. Genet.* **13**:393–415.

Georgopoulos, C., Ang, D., Liberek, K., and Zylicz, M., 1990, Properties of the *Escherichia coli* heat shock proteins and their role in bacteriophage lambda growth, in: *Stress Proteins in Biology and Medicine* (R. I. Morimoto, A. Tissieres, and C. Georgopoulos, eds.), Cold Spring Harbor Laboratory, Cold Spring Harbor, N.Y., pp. 191–221.

Göransson, M., Sunden, B., Nilsson, P., Dagberg, B., Forsman, K., Emanuelsson, K., and Uhlin, B. E., 1990, Transcriptional silencing and thermoregulation of gene expression in *Escherichia coli*, *Nature* **344**:682–685.

Gottesman, S., 1984, Bacterial regulation: Global regulatory networks, *Annu. Rev. Genet.* **18**:415–441.

Gottesman, S., and Neidhardt, F. C., 1983, Global control systems, in: *Gene Function in Procaryotes* (J. Beckwith, J. Davies, and J. A. Gallant, eds.), Cold Spring Harbor Laboratory, Cold Spring Harbor, N.Y., pp. 163–184.

Groat, R. G., Schultz, J. E., Zychlinsky, E., Bockman, A., and Matin, A., 1986, Starvation proteins in *Escherichia coli*: Kinetics of synthesis and role in starvation survival, *J. Bacteriol.* **168**:486–493.

Gualerzi, C. O., Losso, M. A., Lammi, M., Friedrich, K., Pawlik, R. T., Canonaco, M. A., Gianfranceschi, G., Pingoud, A., and Pon, C. L., 1986, Proteins from the prokaryotic nucleoid. Structural and functional characterization of the *Escherichia coli* DNA-binding protein NS (HU) and H-NS, in: *Bacterial Chromatin* (C. O. Gualerzi and C. L. Pon, eds.), Springer-Verlag, Berlin, pp. 101–134.

Hecker, M., and Völker, U., 1990, General stress proteins in *Bacillus subtilis*, *FEMS Microbiol. Ecol.* **74**:197–214.

Higgins, C. F., Dorman, C. J., Stirling, D. A., Waddell, L., Booth, I. R., May G., and Bremer, E., 1988, A physiological role for DNA supercoiling in the osmotic regulation of gene expression in *S. typhimurium* and *E. coli*, *Cell* **52**:569–584.

Hunkapiller, M. W., Grundlund-Moyer, K., and Whiteley, N. W., 1986, Analysis of phenylthiohydantoin amino acids by HPLC, in: *Methods of Protein Microcharacterization* (J. E. Shively, ed.), Humana Press, Clifton, N.J., pp. 315–327.

Jenkins, D. E., Schultz, J. E., and Matin, A., 1988, Starvation-induced cross protection against heat or H_2O_2 challenge in *Escherichia coli*, *J. Bacteriol.* **170:**3910–3914.

Jenkins, D. E., Chaisson, S. A., and Matin, A., 1990, Starvation induced cross protection against osmotic challenge in *Escherichia coli*, *J. Bacteriol.* **172:**2779–2781.

Jenkins, D. E., Auger, E. A., and Matin, A., 1991, Role of RpoH, a heat shock regulator protein, in *Escherichia coli* carbon starvation protein synthesis and survival, *J. Bacteriol.* **173:**1992–1996.

Jensen, K. F., and Pedersen, S., 1990, Metabolic growth rate control in *Escherichia coli* may be a consequence of subsaturation of the macromolecular biosynthetic apparatus with substrates and catalytic components, *Microbiol. Rev.* **54:**89–100.

Killeen, K. P., and Nelson, D. R., 1988, Acceleration of starvation- and glycerol-induced myxospore formation by prior heat shock in *Myxococcus xanthus*, *J. Bacteriol.* **170:**5200–5207.

Kjelleberg, S., Humphrey, B. A., and Marshall, K. C., 1983, Initial phase of starvation and activity of bacteria at surfaces, *Appl. Environ. Microbiol.* **46:**1166–1172.

Kjelleberg, S., Hermansson, M., Mårdén, P., and Jones, G. W., 1987, The transient phase between growth and nongrowth of heterotrophic bacteria, with special emphasis on the marine environment, *Annu. Rev. Microbiol.* **41:**25–49.

Kothary, R. K., Perry, M. D., Clapoff, S., Maltby, U., Moran, L. A., and Rossant, J., 1987, Heat shock gene expression in mouse embryogenesis, *ISCU Short Rep.* **7:**85.

Kurtz, S., Rossi, J., Petko, L., and Lindquist, S., 1986, An ancient developmental induction: Heat shock proteins induced in sporulation and oogenesis, *Science* **231:**1154–1157.

Lange, R., and Hengge-Aronis, R., 1991a, Identification of a central regulator of stationary phase gene expression in *Escherichia coli*, *Mol. Microbiol.* **5:**49–59.

Lange, R., and Hengge-Aronis, R., 1991b, Growth phase regulated expression of *bolA* and morphology of stationary-phase *Escherichia coli* cells are controlled by the novel sigma factor σ^s, *J. Bacteriol.* **173:**4474–4481.

LeGendre, N., and Matsudaira, P. T., 1989, Purification of proteins and peptides by SDS-PAGE, in: *A Practical Guide to Protein and Peptide Purification for Microsequencing* (P. T. Matsudaira, ed.), Academic Press, New York, pp. 49–69.

Lejeune, P., and Danchin, A., 1990, Mutations in the *bglY* gene increase the frequency of spontaneous deletions in *Escherichia coli* K-12, *Proc. Natl. Acad. Sci. USA* **87:**360–363.

Losick, R., and Pero, J., 1981, Cascades of sigma factors, *Cell* **25:**582–584.

Lund, E., and Kjeldgaard, N. O., 1972, Metabolism of guanosine tetraphosphate in *Escherichia coli*, *Eur. J. Biochem.* **28:**316–326.

Mårdén, P., Nyström, T., and Kjelleberg, S., 1987, Uptake of leucine by a gram-negative heterotrophic bacterium during exposure to starvation conditions, *FEMS Microbiol. Ecol.* **45:**233–241.

Matin, A., 1990, Molecular analysis of the starvation stress in *Escherichia coli*, *FEMS Microbiol. Ecol.* **74:**185–196.

Metzger, S., Sarubbi, E., Glaser, G., and Cashel, M., 1989, Protein sequence encoded by the *relA* and the *spoT* genes of *Escherichia coli* are interrelated, *J. Biol. Chem.* **264:**9122–9125.

Morita, R. Y., 1982, Starvation-survival of heterotrophs in the marine environment, in: *Advances in Microbial Ecology*, Vol. 6 (K. C. Marshall, ed.), Plenum Press, New York, pp. 171–198.

Morita, R. Y., 1985, Starvation and miniaturisation of heterotrophs, with special emphasis on maintenance of the starved viable state, in: *Bacteria in Their Natural Environments* (M. M. Fletcher and G. D. Floodgate, eds.), Academic Press, New York, pp. 111–130.

Mosteller, R. D., 1978, Evidence that glucose starvation-sensitive mutants are altered in the *relB* locus, *J. Bacteriol.* **133**:1034–1037.

Mosteller, R. D., and Kwan, S. F., 1976, Isolation of relaxed-control mutants of *Escherichia coli* K-12 which are sensitive to glucose starvation, *Biochem. Biophys. Res. Commun.* **69**: 325–332.

Mulvey, M. R., and Loewen, P. C., 1989, Nucleotide sequence of *katF* of *Escherichia coli* suggests KatF protein is a novel sigma transcription factor, *Nucleic Acids Res.* **17**:9979–9991.

Neidhardt, F. C., 1983, Multigene regulation and the growth of the bacterial cell, in: *The Molecular Biology of Bacterial Growth* (M. Schaechter, F. C. Neidhardt, J. L. Ingraham, and N. O. Kjeldgaard, eds.), Jones & Bartlett, Boston, pp. 218–223.

Neidhardt, F. C., 1987, Multigene systems and regulons, in: *Escherichia coli and Salmonella typhimurium, Cellular and Molecular Biology*, Vol 2 (F. C. Neidhardt, J. L. Ingraham, K. B. Low, B. Magasanik, M. Schaechter, and H. E. Umbarger, eds.), American Society for Microbiology, Washington, D.C., pp. 1313–1317.

Neidhardt, F. C., and VanBogelen, R. A., 1987, Heat shock response, in: *Escherichia coli and Salmonella typhimurium, Cellular and Molecular Biology*, Vol. 2 (F. C. Neidhardt, J. L. Ingraham, K. B. Low, B. Magasanik, M. Schaechter, and H. E. Umbarger, eds.), American Society for Microbiology, Washington, D. C., pp. 1334–1345.

Neidhardt, F. C., VanBogelen, R. A., and Vaughn, V., 1984, The genetics and regulation of heat shock proteins, *Annu. Rev. Genet.* **18**:295–329.

Neidhardt, F. C., Ingraham, J. L., and Schaechter, M., 1990, *Physiology of the Bacterial Cell*, Sinauer, Sunderland, Mass., pp. 351–388.

Nelson, D. R., and Killeen, K. P., 1986, Heat shock proteins of vegetative and fruiting *Myxococcus xanthus* cells, *J. Bacteriol.* **168**:1100–1106.

Nyström, T., and Kjelleberg, S., 1989, Role of protein synthesis in the cell division and starvation induced resistance to autolysis of a marine vibrio during the initial phase of starvation, *J. Gen. Microbiol.* **135**:1599–1606.

Nyström, T., Albertson, N. H., and Kjelleberg, S., 1988, Synthesis of membrane and periplasmic proteins during starvation of a marine *Vibrio* sp., *J. Gen. Microbiol.* **134**:1645–1651.

Nyström, T., Flärdh, K., and Kjelleberg, S., 1990a, Responses to multiple-nutrient starvation in the marine *Vibrio* sp. strain CCUG 15956, *J. Bacteriol.* **172**:7085–7097.

Nyström, T., Albertson, N. H., Flärdh, K., and Kjelleberg, S., 1990b, Physiological and molecular adaptation to starvation and recovery from starvation by the marine *Vibrio* sp. S14, *FEMS Microbiol. Ecol.* **74**:129–140.

Nyström, T., Olsson, R. M., and Kjelleberg, S., 1992, Survival, stress resistance, and alteration in protein expression in the marine *Vibrio* sp. strain S14 during starvation for different individual nutrients, *Appl. Environ. Microbiol.* **58**:55–65.

Ochi, K., Kandala, J. C., and Freese, E., 1982, Evidence that *Bacillus subtilis* sporulation induced by the stringent response is caused by a decrease in GTP or GDP, *J. Bacteriol.* **151**:1062–1065.

O'Farrell, P. H., 1975, High resolution two-dimensional electrophoresis of proteins, *J. Biol. Chem.* **250**:4007–4021.

Pedersen, S., 1984, *Escherichia coli* ribosomes translate *in vivo* with variable rate, *EMBO J.* **3**:2895–2898.

Pedersen, S., Bloch, P. L., Reeh, S., and Neidhardt, F. C., 1978, Patterns of protein synthesis in *Escherichia coli*: A catalog of the amount of 140 individual proteins at different growth rates, *Cell* **14**:179–190.

Poindexter, J. S., 1981, The caulobacters: Ubiquitous unusual bacteria, *Microbiol. Rev.* **45**: 123–179.

Pon, C. L., Calogero, R. A., and Gualerzi, C. O., 1988, Identification, cloning, nucleotide sequence and chromosomal map location of *hns*, the structural gene for *Escherichia coli* DNA-binding protein H-NS, *Mol. Gen. Genet.* **212:**199–212.

Reeve, C. A., Bockman, A. T., and Matin, A., 1984, Role of protein synthesis in the survival of carbon-starved *Escherichia coli* K-12, *J. Bacteriol.* **160:**1041–1046.

Schaeffer, P., Millet, J., and Aubert, J.-P., 1965, Catabolite repression of bacterial sporulation, *Proc. Natl. Acad. Sci. USA* **54:**704–711.

Schreiber, G., Metzger, S., Aizenman, E., Roza, S., Cashel, M., and Glaser, G., 1991, Pleiotropic effects of overexpression of the *relA* gene in *Escherichia coli*, *J. Biol. Chem.* **266:**3760–3767.

Smith, M. W., and Neidhardt, F. C., 1983, Proteins induced by anaerobiosis in *Escherichia coli*, *J. Bacteriol.* **154:**344–350.

Sonenshein, A. L., 1989, Metabolic regulation of sporulation and other stationary-phase phenomena, in: *Regulation of Procaryotic Development* (I. Smith, R. A. Slepecky, and P. Setlow, eds.), American Society for Microbiology, Washington, D. C., pp. 109–130.

Spassky, A., Rimsky, S., Garreau, H., and Buc, H., 1984, H1a, an *E. coli* DNA-binding protein which accumulates in stationary phase, strongly compacts DNA in vitro, *Nucleic Acids Res.* **12:**5321–5340.

Spears, P. A., Schauer, D., and Orndorff, P. E., 1986, Metastable regulation of type 1 piliation in *Escherichia coli* and isolation and characterization of a phenotypically stable mutant, *J. Bacteriol.* **168:**179–185.

Spector, M. P., 1990, Gene expression in response to multiple nutrient starvation conditions in *Salmonella typhimurium*, *FEMS Microbiol. Ecol.* **74:**175–184.

Spector, M. P., Aliabadi, Z., Gonzalez, T., and Foster, J. W., 1986, Global control in *Salmonella typhimurium: Two-dimensional electrophoretic analysis of starvation-, anaerobiosis-, and heat shock-inducible proteins, *J. Bacteriol.* **168:**420–424.

Spector, M. P., Park, Y. K., Tirgari, S., Gonzalez, T., and Foster, J. W., 1988, Identification and characterization of starvation-regulated genetic loci in *Salmonella typhimurium* by using Mu d-directed *lacZ* operon fusions, *J. Bacteriol.* **170:**345–351.

Spence, J., Cegielska, A., and Georgopoulos, C., 1990, Role of *Escherichia coli* heat shock proteins DnaK and HtpG (C62.5) in response to nutritional deprivation, *J. Bacteriol.* **172:** 7157–7166.

Tempest, D. W., and Neijssel, O. M., 1981, Metabolic compromises involved in the growth of microorganisms in nutrient-limited (chemostat) environments, in: *Basic Life Sciences*, Vol. 18 (A. Hollaender, ed.), Plenum Press, New York, pp. 335–356.

Thomas, T. D., and Batt, R. D., 1969, Survival of *Streptococcus lactis* in starvation conditions, *J. Gen. Microbiol.* **50:**367–382.

VanBogelen, R. A., and Neidhardt, F. C., 1990, Global systems approach to bacterial physiology: Protein responders to stress and starvation, *FEMS Microbiol. Ecol.* **74:**121–128.

VanBogelen, R. A., Kelly, P. M., and Neidhardt, F. C., 1987, Differential induction of heat shock, SOS, and oxidation stress regulons and accumulation of nucleotides in *Escherichia coli*, *J. Bacteriol.* **169:**26–32.

VanBogelen, R. A., Hutton, M. E., and Neidhardt, F. C., 1990, Gene–protein database of *Escherichia coli* K-12: Edition 3, *Electrophoresis* **11:**1131–1166.

Walker, G. C., 1984, Mutagenesis and inducible responses to deoxyribonucleic acid damage in *Escherichia coli*, *Microbiol. Rev.* **48:**60–93.

Zimmerman, J. L., Petri, W., and Meselson, M., 1983, Accumulation of a specific subset of *D. melanogaster* heat shock mRNAs in normal development without heat shock, *Cell* **32:**1161–1170.

7

Approaches to the Study of Survival and Death in Stationary-Phase Escherichia coli

Deborah A. Siegele, Marta Almirón, and Roberto Kolter

1. INTRODUCTION

When pure cultures of the bacterium *Escherichia coli* are grown in standard laboratory media, the organisms grow exponentially until conditions no longer support growth and the cells enter stationary phase. During growth the change in the number of viable cells is predominantly the result of cell division. There is little or no detectable cell death during exponential growth as judged by the close correlation between viable counts and total cells observed microscopically. Once growth is arrested, some cells survive while others die. The large questions to be addressed are: Why do some cells die during starvation and what is the cause of their death? Why do some cells survive during starvation and what functions are required for their survival? While these questions remain largely unanswered, the approaches that have been made possible in recent years by the application of molecular genetics have come to complement the predominantly physiological work that had been carried out earlier (reviewed in Dawes, 1989; Matin *et al.*, 1989; Matin, 1991; Siegele and Kolter, 1992). Here we present descriptions of the main approaches our labora-

Deborah A. Siegele, Marta Almirón, and Roberto Kolter • Department of Microbiology and Molecular Genetics, Harvard Medical School, Boston, Massachusetts 02115. *Present address of D.S.*: Department of Biology, Texas A&M University, College Station, Texas 77843.
Starvation in Bacteria, edited by Staffan Kjelleberg. Plenum Press, New York, 1993.

tory uses to address these questions. We describe the methodologies we use to incubate *E. coli* in stationary phase and the methodologies used to define, distinguish, and separate viable and nonviable cells in our experiments. We also describe genetic approaches we have taken to study cell physiology during stationary phase. This presentation should serve as an indication that much remains to be explored; the prospects of gaining new insights in the understanding of microbial death and survival make the work particularly exciting.

2. TERMINOLOGY

Medium composition and the circumstances that lead to the cessation of growth play a critical role in the physiology of cells in stationary phase. In this chapter *stationary phase* will be used as a very broad and general term to define "the time after the bacteria have grown to saturation in liquid growth medium" and refers exclusively to batch cultures. *Starvation* will mean "the absence, in the incubation medium, of a specified essential nutrient" and will always be used in conjunction with the particular nutrient that is lacking. Two different methods that lead to a condition of starvation are commonly used. One method is to centrifuge or filter growing cells and resuspend them in medium lacking the nutrient under consideration; this has been referred to as *starvation by resuspension*. The second method is to starve the cells by purposely providing a low enough amount of the nutrient under consideration so as to ensure that growth is arrested as a result of the exhaustion of that nutrient; this has been referred to as *starvation by exhaustion*. All of the results presented here are from starvation by exhaustion experiments.

3. CULTURE CONDITIONS

The physical setup for our incubations is as follows: Three milliliters of liquid medium is placed in 18×150-mm glass test tubes with plastic caps. After inoculation, the tubes are incubated at 37°C in a roller drum at a 5° angle and aerated by rotating at 60 rpm. The samples are kept rotating at all times, except for short periods of sampling. Sampling is done by removing 10-μl aliquots which can then be diluted for determination of viable counts or for preparation of samples for microscopy. Loss of water by evaporation is not a major problem during incubations of 1 to 2 weeks, because our warm room is humidified. However, water lost by evaporation can be replaced with sterile deionized water.

The experiments described use four basic media: LB, M63, MOPS and Davis (Miller, 1972; Neidhardt *et al.*, 1974; Davis *et al.*, 1986). Because of variations from the published methods and for the convenience of the reader we reproduce the recipes in Appendix 8.1.

4. METHODS TO DISTINGUISH VIABLE AND NONVIABLE CELLS

To study the survival of bacteria in stationary phase an assay for cell viability is needed. Some of the methods used in our laboratory to assess the viability of cells and to separate viable and nonviable cells will be described here (for details of the procedure see Appendixes 8.1 and 8.2). Additional methods of assessing cell viability have been reviewed by Roszak and Colwell (1987) and Oliver (this volume).

4.1. By Plating

Perhaps the most obvious assay of viability is the ability of cells to grow and form colonies on fresh medium. Colony-forming units (CFU) are routinely determined by making serial dilutions and plating aliquots of the dilutions. However, measurements of CFU may not always accurately reflect the number of viable cells present in a culture. There are numerous reports of starved cells that can take up and metabolize nutrients, but do not resume growth under standard plating conditions (Roszak and Colwell, 1987). Such cells have been shown to be able to resume growth under special conditions (Roszak *et al.*, 1983; Nilsson *et al.*, 1991). Therefore, it is useful to consider viability as an operational rather than as an absolute definition.

4.2. By Microscopy

Acridine orange (AO) staining of methanol-fixed cells, coupled with fluorescence microscopy, can serve as an independent method to determine the number of viable cells in a culture. When AO intercalates into nucleic acid, it fluoresces if excited with UV light. At pH 4, AO bound to RNA fluoresces orange, while AO bound to DNA fluoresces green (Kasten, 1967). The rationale for using AO staining as a viability assay is that a living *E. coli* cell usually contains more RNA than DNA and thus stains orange. A cell in which much of the RNA has been degraded, but still retains its DNA, will stain green. Degradation of ribosomal RNA has been correlated with loss of viability assayed by CFU under some conditions of starvation (Davis *et al.*, 1986). By this criterion, green fluorescing cells are considered nonviable. Clearly a cell that has degraded all of its ribosomes will be unable to resume growth and can be considered dead.

However, there are limitations to this assay. Some mechanisms of cell death might not result in loss of RNA; such nonviable cells would not be detected by AO staining. In addition, it is not known how much RNA is needed for a cell to retain viability. Thus, cells that give rise to only a faint orange fluorescence or fluoresce green when stained with AO may be viable and might be detected in a CFU assay. Despite these theoretical limitations, in practice AO staining is a useful procedure to identify different populations of cells in stationary-phase cultures. We have found that the correlation between estimates of the fraction of viable cells in stationary-phase cultures assessed by staining with AO versus measurements of CFU depends on the particular starvation condition. In some starvation conditions there is good agreement between the results of the two assays, while in other conditions there is not. The conditions giving rise to these results will be described in a later section.

4.3. By Centrifugation

During exponential phase, *E. coli* cultures are remarkably homogeneous. In contrast, stationary-phase cultures may contain mixtures of viable and nonviable cells. It would be useful to have a method to separate viable and nonviable cells so that the biochemical properties of the two cell types could be determined. Depending on the mechanisms leading to cell death, viable and nonviable cells may have different compositions, which would result in their banding differently in density gradients. Renografin-76 equilibrium density gradients can be used to detect changes in cell density (Cahn and Fox, 1968). We were able to use such gradients to separate viable and nonviable cells from stationary-phase LB cultures, as will be described in Section 5. Details for gradient preparation are described in Appendix 8.3.

5. STATIONARY-PHASE SURVIVAL STUDIES WITH ZK126

We have used these three methods (CFU determination, AO staining, and Renografin density gradients) to study our standard laboratory "wild type" strain, ZK126 (*E. coli* K-12 W3110 *tna2* Δ*lacU169*). From results presented below we conclude that the fraction of the population that survives in stationary phase depends on the particular starvation conditions. The conditions tested include what nutrient has been exhausted, whether the cells are in minimal or rich medium, and the buffer used for minimal medium (see Mason and Egli, this volume). We also observe that CFU or AO staining assays for cell viability give comparable results under some, but not all conditions.

When cells enter stationary phase as a result of glucose exhaustion in M63 medium (0.2% glucose), the number of CFU/ml does not vary appreciably

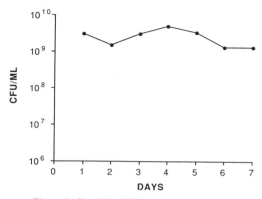

Figure 1. Survival of ZK126 in M63 glucose.

during the first 7 days in stationary phase (Fig. 1). During this time, all of the cells fluoresce orange when stained with AO. When M63 glucose-exhausted cultures are examined after 20 days, the correlation between CFU/ml and AO staining is not always seen; the cells continue to fluoresce orange but there is a drop in the CFU/ml. We do not understand the reasons for this occasional drop in CFU/ml after prolonged starvation. During the first 7 days in stationary phase, cells from M63 glucose cultures band as a single band in Renografin-76 density gradients consistent with there being a pure population of viable cells (Fig. 2). However, the cell density observed depends on the age of the culture. Stationary-phase cells from an overnight culture are more dense than exponential-phase cells. A further increase in cell density is observed after 3 days in stationary phase. No further changes in cell density were observed

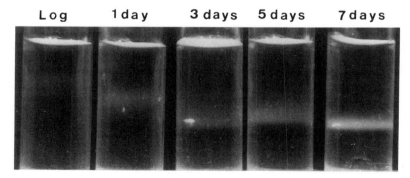

Figure 2. Renografin gradients of ZK126 cultures in M63 glucose.

between 3 and 7 days. These changes in cell density indicate that changes in cell composition occur during starvation in glucose-exhausted cultures.

We studied the viability of ZK126 in conditions where cells entered stationary phase by exhaustion of phosphate or ammonium from the medium. MOPS buffered medium was used for these experiments. The results are shown in Fig. 3. In summary, glucose exhaustion and phosphate exhaustion in MOPS medium did not lead to a significant decrease in the fraction of viable cells assayed by CFU over 10 days. However, ammonium exhaustion did lead to a slow decrease in CFU, with only 14% of the CFU/ml present in an overnight culture still able to plate after 10 days. The fraction of cells fluorescing orange after AO staining was determined on day 11. In the glucose- and phosphate-exhausted cultures, there was good correlation between the estimates of the fraction of viable cells assayed by either CFU or AO staining. Samples from the glucose- and phosphate-exhausted cultures contained 93 to 98% orange fluorescing cells. In contrast, this correlation was not seen for ammonium-exhausted cultures. Under these starvation conditions, 98% of the cells fluoresced orange after AO staining, while only 14% of the initial CFU/ml remained after 10 days.

The finding that ZK126 survived phosphate exhaustion in MOPS buffered medium (Fig. 3) was in contrast to the results found by Davis and colleagues for strain D10 under different conditions of phosphate starvation (Davis *et al.*, 1986). The medium used by Davis *et al.* is unusual in that there was no added

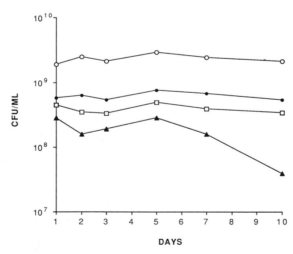

Figure 3. Survival of ZK126 in MOPS buffered media. Growth was arrested by exhaustion of (○) 0.2% glucose, (●) 0.05% glucose, (□) 6.6×10^{-5} M phosphate, or (▲) 2.4×10^{-3} M ammonia.

Mg^{2+}. To determine whether the differences in viability observed were the result of differences in the media used, we carried out phosphate exhaustion experiments in the medium used by Davis *et al.*, both with and without added Mg^{2+} (Fig. 4). In the absence of added Mg^{2+}, we observed a rapid loss of viability assayed by CFU/ml very similar to that observed by Davis *et al.* Addition of Mg^{2+} led to a delay of the onset, and slower kinetics, of cell death, but still more cell death was observed during 10 days than was seen during phosphate exhaustion in MOPS medium (Fig. 3). A major difference in the composition of the MOPS and Davis media used for phosphate starvation is the presence of Tris-HCl in the latter. These results suggest that the marked differences seen by us in comparison with Davis *et al.* are caused by differences in the medium used rather than by strain differences.

We also studied the viability of ZK126 after growth to saturation in LB. We believe that in our experiments, cells enter stationary phase after exponential growth in LB upon carbon source starvation, because when spent medium from a 1-day-old LB culture is used as growth medium for a fresh inoculation, it does not support growth unless a carbon source is added. A note of caution here is important. To test the ability of "conditioned" or "spent" medium to support growth, we routinely remove the cells by centrifugation followed by filtration through 0.2-μm-pore filters (Acrodisc, Gelman Sciences). We found that the filters must be extensively washed with sterile water prior to use, because if only small volumes, such as a few milliliters, are filtered they can release enough carbon source to support the growth of about 10^8 cells/ml.

Figure 5 shows a typical viability curve for ZK126 after exponential growth in LB. Cultures generally saturate at about $5-10 \times 10^9$ CFU/ml. The

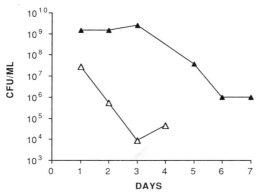

Figure 4. Survival of ZK126 in Davis medium (Davis *et al.*, 1986). Growth was arrested by exhaustion of phosphate. △, medium without added Mg^{2+}; ▲, medium with 1 mM $MgSO_4$.

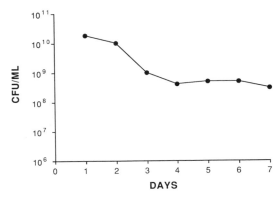

Figure 5. Survival of ZK126 in LB.

counts remain more or less constant for the first day or two, but in the subsequent 2 to 4 days there is a drop of one or two orders of magnitude in the CFU/ml. The number of CFU/ml then stabilizes and further decreases occur very slowly. The results seen when following the AO staining properties of these cells are consistent with the colony counts. All of the cells from exponentially growing or overnight cultures fluoresce orange after staining with AO (seen in Fig. 6A,B as bright gray). Stationary-phase cells from an overnight culture are smaller and appear more round than the rod-shaped cells seen during exponential phase. After 4 days in stationary phase, the number of orange cells decreases paralleling the decrease in CFU/ml. As the number of orange cells decreases, green cells appear (Fig. 6C; orange—bright gray, green—dull gray). Renografin gradients from LB cultures after 8 days reveal two major populations (Fig. 7). A faint band near the top of the gradient contains $< 5\%$ of the optical density, but $> 99\%$ of the CFU found in the gradient; most of the cells in this fraction fluoresce orange with AO. A band of greater density, which contains most of the optical density in the gradient, contains $< 0.1\%$ of the CFU of the gradient; nearly all of the cells in this fraction fluoresce green.

The results presented in the preceding paragraphs serve to illustrate that the fraction of cells surviving in stationary phase is affected by the starvation conditions. Other effects on survival caused by variation in the starvation conditions have been noted. Matin and co-workers observed significant differences in the fraction of *E. coli* cells surviving glucose starvation depending on whether starvation was the result of exhaustion or resuspension (Reeve *et al.*, 1984a; Groat *et al.*, 1986). Dependence of survival on starvation conditions has

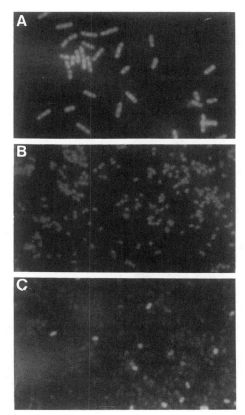

Figure 6. Acridine orange (AO) staining of ZK126 cells from LB cultures. Orange fluorescing cells appear bright gray, green fluorescing cells appear dull gray. (A) Exponential-phase cells; (B) 1-day-old cells; (C) 4-day-old cells.

also been seen in the marine *Vibrio* sp. S14 (Nyström *et al.*, 1992; Östling *et al.*, this volume).

6. SURVIVAL GENES

A number of stationary-phase or starvation-induced genes are regulated by the sigma factor, RpoS (Lange and Hengge-Aronis, 1991; McCann *et al.*, 1991). Inactivation of this regulatory molecule leads to decreased viability in stationary phase after carbon starvation and, in addition, RpoS activity is needed for starved cells to develop increased resistance to environmental assaults such as high temperature or oxidative damage (Lange and Hengge-Aronis, 1991;

Figure 7. Renografin gradient of an 8-day-old ZK126 culture in LB.

McCann *et al.*, 1991; Hengge-Aronis, this volume). Which members of the *rpoS* regulon are needed for starvation survival remains to be determined.

The rate of protein turnover in *E. coli* increases during starvation for carbon, nitrogen, or amino acids (Mandelstam, 1960; Goldberg and St. John, 1976). This may provide a source of energy as well as of amino acids needed for synthesis of starvation-induced proteins. Matin and co-workers have shown that peptidase mutants of *E. coli* K12 or *Salmonella typhimurium* LT-2 with decreased protein turnover have decreased viability during prolonged glucose starvation (Reeve *et al.*, 1984b). Mutants strains lacking multiple peptidases had the lowest levels of protein degradation and died more rapidly during glucose starvation than a strain lacking a single peptidase activity. The peptidase mutants also had decreased rates of protein synthesis during starvation, indicating that new protein synthesis is required for maintaining cell viability during starvation. These findings show that protein turnover plays an important role in survival during carbon starvation.

Increased levels of the nucleotide ppGpp have an important role in maintaining viability during starvation for amino acids and probably play a similar role after starvation for glucose or phosphate (Mach *et al.*, 1989). Intracellular levels of ppGpp increase after starvation for either amino acids or carbon and this correlates with a number of physiological changes which collectively are termed the stringent response (Cashel and Rudd, 1987). *relA* mutants do not accumulate ppGpp after amino acid starvation and also have decreased viability under these conditions (Mach *et al.*, 1989). However, after glucose or phosphate starvation, *relA* mutants have the same viability as the isogenic $relA^+$ parent (Mach *et al.*, 1989). The accumulation of ppGpp in response to carbon starvation is not dependent on *relA* (Metzger *et al.*, 1989), and Mach *et al.* detected no difference in the ppGpp concentration in $relA^+$ and $relA^-$ strains after either glucose or phosphate starvation. How do increases in the intracellular concentration of ppGpp lead to enhanced stationary-phase survival? Increased rates of protein degradation could play a role as could some of the other changes in gene expression, metabolism, and cell composition that result from induction of the stringent response. The transcriptional changes that occur during the stringent response may result from interaction of ppGpp with RNA polymerase holoenzyme containing sigma70 (Cashel and Rudd, 1987). Changes in ppGpp concentration might also affect the expression or activity of the sigma factor RpoS, whose function is needed for maximal survival. Other aspects of the role of the stringent response during starvation of *E. coli* and *S. typhimurium* are addressed by Nyström, and Spector and Foster (this volume).

Spence *et al.* (1990) investigated the role of the highly conserved heat shock proteins DnaK and HtpG during starvation. Mutants lacking HtpG had no differences in the rate of survival during starvation for either carbon, nitrogen, and phosphate, nor were any differences in the pattern of starvation-induced proteins observed under any of these conditions. In contrast, strains containing the *dnaK103* mutation had greatly reduced viability during carbon starvation. Loss of DnaK also led to decreased viability during nitrogen or phosphate starvation, but the effect was much less severe than after carbon starvation. The *dnaK103* mutation also had pleiotropic effects on the pattern of proteins synthesized under each of these starvation conditions. This correlation between decreased viability and alteration of the pattern of starvation-induced proteins is consistent with the observation that new protein synthesis at early starvation times is required for full viability (Groat *et al.*, 1986). Exactly how the *dnaK103* mutation leads to decreased starvation survival is not known. *In vitro* purified DnaK binds to unfolded peptides and it appears to function as a chaperone *in vivo*. DnaK is involved directly or indirectly in multiple cellular processes and *dnaK* mutations result in pleiotropic phenotypes including decreased proteolysis (Straus *et al.*, 1988). Spence *et al.* have proposed that one way DnaK may promote stationary-phase survival is through its role in proteolysis. DnaK may be needed either for protein turnover in general, or it could be involved in degradation of specific proteins that may be deleterious if present during starvation.

We designed a genetic screen to identify other genes that are dispensable during logarithmic growth, but essential during stationary phase (Tormo *et al.*, 1990). In these screens, random insertion mutants are isolated and an aliquot of each culture is frozen before the cells enter stationary phase. The remainder of each culture is incubated under specific stationary-phase conditions and for specific times. The number of CFU/ml in each culture is determined by directly plating 1–10 µl of the culture. Mutants that show few or no CFU remaining are chosen. Because of the practical constraints in dealing with thousands of mutants in such a screen, we do not generate serial dilutions but rather plate the cultures directly. Thus, only mutations that lead to a decrease of several orders of magnitude in the CFU/ml are detected in these screens. We expected that this screen could yield two classes of mutants: (1) mutants that die in stationary phase and (2) mutants that remain viable, but are unable to form colonies because they cannot exit from stationary phase.

From screens of this type we have obtained four insertions in previously unknown genes, designated *surA–D*, that resulted in what we term a survival defective phenotype (Sur⁻). We have used this designation as a general term to describe the phenotype, not because we know that the role of the gene products is exclusively in the cell's survival. These four new genetic loci were identified after screening nearly 20,000 insertion mutants. Since no duplicates were obtained, it is most likely that there are many more loci where insertions will lead to a Sur⁻ phenotype. Besides their genetic location, little is known about the physiological function of *surC* and *surD* genes. In contrast, the *surA* and *surB* loci have been studied in some detail.

The *surA* gene is located in a complex operon at minute 1 in the *E. coli* chromosome. Mutants with an insertion in *surA* die 4 to 5 days after entering stationary phase in LB (Tormo *et al.*, 1990). The fraction of cells that fluoresce orange after AO staining decreases along with the decrease in the number of CFU/ml. Since the lack of *surA* leads to cell death, it appears that *surA* is essential for maintaining viability during stationary phase in LB. Even though the insertion mutation has a polar effect in the downstream gene, *pdxA*, mutations in this gene as well as in the others that are part of the same operon (*ksgA*, *apaG* and *apaH*) do not show the survival deficient phenotype. The relation among these genes is still unclear. Suppressors of the *surA1* mutation can arise during stationary phase and are caused by excisions of the transposable element from the *surA* gene. The frequency of appearance of these suppressors per bacterium-hour increases with prolonged incubation in stationary phase (Tormo *et al.*, 1990). This time-dependent increase in the frequency of excision during starvation has been observed for other transposable elements (Shapiro, 1984; Hall, 1988; Mittler and Lenski, 1990).

The *surB1* mutation was obtained in a screen for loss of plating ability after glucose exhaustion in M63 (Siegele and Kolter, 1993). The behavior of cells with *surB* mutations is very different from that of *surA* mutants. Cells with the *surB1* mutation grow exponentially at 37°C, but lose the ability to form colonies at 37°C 2 to 3 h after the onset of stationary phase. However, the inability to form colonies after entering stationary phase is a temperature-sensitive phenotype. The mutant cells can regrow at 30°C, indicating that they remain viable in stationary phase. These results suggest that the product of *surB* is required for cells to exit from stationary phase at 37°C.

7. REVERSE GENETICS TO STUDY STATIONARY-PHASE GENE EXPRESSION

Dramatic changes in gene expression occur as cells enter stationary phase (Groat *et al.*, 1986; Spector *et al.*, 1986; Connell *et al.*, 1987; Bohannon *et al.*, 1991; Lange and Hengge-Aronis, 1991; Spector and Cubitt, 1992). Genes induced as cells enter stationary phase are likely to include functions required for survival. Many of the properties of stationary-phase cells, such as their increased resistance to high temperature or oxidative agents, are the result of expression of the stationary phase-induced *rpoS* regulon (Lange and Hengge-Aronis, 1991; McCann *et al.*, 1991). There may also be other starvation-induced functions that are not regulated by *rpoS*. Identifying the genes encoding these stationary phase-induced proteins will be needed to understand their function and regulation.

One method to identify genes expressed during stationary phase is to use fusions to a reporter gene such as *lacZ*. This method has been used to identify stationary-phase-specific genes in a number of organisms (Schultz *et al.*, 1988; Spector *et al.*, 1988; Lange and Hengge-Aronis, 1991; Östling *et al.*, 1991; Nyström *et al.*, 1992). Another method is to use the techniques of "reverse genetics," which involves identifying genes starting from their protein products. To isolate proteins specific for stationary phase, a method is needed to identify those that are synthesized after the end of exponential growth, because most of the proteins present in a stationary-phase cell were made during exponential phase. One such method is to pulse-label cultures that have been starved for a few hours with radioactive amino acids, then resolve the proteins by two-dimensional gel electrophoresis. This methodology has been used to catalog many of the proteins induced at the onset of stationary phase (Groat *et al.*, 1986; Lange and Hengge-Aronis, 1991; Nyström, this volume).

To characterize the pattern of proteins synthesized during prolonged starvation, we have labeled cultures that have been starved for several days using pulses of up to 10 h. We observed that the pattern of gene expression during a period of 1 week was in continuous change. Three-day-old cells expressed a pattern of proteins than differed from those expressed by 1-day-old cells or by cells just entering stationary phase.

Taking advantage of this finding and using the "reverse genetics" approach mentioned above, we have identified several genes that are expressed in 3-day-old cultures after glucose exhaustion. One of these genes encodes a novel DNA binding protein, named Dps, with regulatory and protective roles in starved cells (Almirón *et al.*, 1992). Mutants lacking Dps show a greatly altered pattern of gene expression during stationary phase suggesting that Dps has a global regulatory role. In addition, these mutants are unable to develop the characteristic starvation-induced resistance to oxidative damage despite the fact that they express normal levels of the stationary-phase-specific catalase HPII.

8. APPENDIXES

8.1. Culture Media

The medium is made with water deionized by reverse osmosis. We routinely include 1 μg/ml thiamine in minimal medium because of the *thiA* marker in many of our strains.

LB medium contains, per liter:

10 g tryptone
5 g yeast extract
5 g NaCl
For plates, 20 g agar is added per liter

M63 medium contains, per liter:

3 g KH_2PO_4
7 g K_2HPO_4
2 g $(NH_4)_2SO_4$

$MgSO_4$ is added sterilely to a final concentration of 1 mM after the medium is autoclaved. We do not include any additional iron in our preparation of M63. Unless stated otherwise, sugars are added to a final concentration of 0.2%. For plates 15 g agar is added per liter.

MOPS medium contains, per liter:

40 ml MOPS (1.0 M, pH 7.4)
4 ml Tricine (1.0 M, pH 7.4)
1 ml $FeSO_4$ (0.01 M)
1 ml K_2SO_4 (0.276 M)
1 ml $CaCl_2$ (0.5 mM)

1 ml $MgCl_2$ (0.528 M)
10 ml NaCl (5 M)
1 ml micronutrients solution
5 ml NH_4Cl (1.9 M)
10 ml K_2HPO_4 (0.132 M)
10 ml glucose (20%)

The micronutrients solution contains 3 μM $(NH_4)_6(MoO_7)_{24}$, 400 μM H_3BO_3, 30 μM $CoCl_2$, 10 μM $CuSO_4$, 80 μM $MnCl_2$, and 10 μM $ZnSO_4$.

The concentrations of ammonium, phosphate, and glucose used depend on the experiment. For ammonia starvation by exhaustion, 1.25 ml NH_4Cl (1.9 M) was added per liter. For phosphate starvation by exhaustion, 0.5 ml K_2HPO_4 (0.132 M) was added per liter. To allow cells to enter stationary phase by glucose exhaustion at the same cell density reached after ammonia or phosphate exhaustion, the concentration of glucose was reduced to 0.05%.

Davis glucose exhaustion medium contains, per liter:

1 g $(NH_4)_2SO_4$
3 g KH_2PO_4
7 g K_2HPO_4
0.5 g sodium citrate
5 mg $CaCl_2$
100 mg $FeSO_4$
2 g glucose

For phosphate exhaustion experiments the medium used was as described above, except that 11 mg KH_2PO_4 and 25 mg K_2HPO_4 were used. Trizma Base (12.1 g) was added for buffering and the pH adjusted to 7.6 with HCl.

8.2. AO Staining

The procedure we employ to stain and visualize cells is basically that recommended by Difco Laboratories and is as follows:

1. Place a 10-μl sample directly from the culture onto a glass microscope slide. Let it air-dry.
2. To fix the sample, cover the slide with 100% methanol and let it air-dry.
3. Dispense a few drops of SpotTest™ acridine orange stain (Difco #3561-26-5) on the fixed sample.
4. Wait 2 min.
5. Wash with water for a few seconds. Allow it to dry.
6. Cover sample with one drop of immersion oil and examine with 100× objective. We use filter 487709 on a Zeiss fluorescence microscope.

8.3. Renografin-76 Equilibrium Density Gradients

The density gradients are made of "Renografin-76+" (Diatrizoate Meglumine and Diatrizoate Sodium Injection USP) purchased from Squibb Diagnostics, Catalog # NDC0003082155.

1. The following working solutions are prepared:

   Solution A: 1.125 g/ml (17.4 ml Renografin + 32.6 ml water)
   Solution B: 1.225 g/ml (31.25 ml Renografin + 18.75 ml water)

2. Six solutions of different densities are made as follows:

   Solution 1: 5 ml of A
   Solution 2: 4 ml A + 1 ml B
   Solution 3: 3 ml A + 2 ml B
   Solution 4: 2 ml A + 3 ml B
   Solution 5: 1 ml A + 4 ml B
   Solution 6: 5 ml B

3. The gradients are made in 13×51-mm polycarbonate centrifuge tubes for the Beckman TLA100.3 rotor. Step gradients, total volume 2.4 ml, can be made with any desired combination of Solutions 1–6. We start with 0.4-ml fractions of each solution and then adjust the gradients to give the greatest resolution between bands. This procedure is analytical; the size of the gradients can be enlarged for preparative purposes.

4. Cells to be centrifuged in these gradients are first pelleted and rinsed twice in sterile saline solution (0.9% NaCl) and resuspended in 0.1 ml sterile saline prior to loading. Generally the analytical gradients described above can be layered with 10^7–10^8 cells without loss of resolution.

5. Gradients are spun at 65,000 rpm in the TLA100.3 rotor in a Beckman TL100 tabletop ultracentrifuge at 4°C for 1 h. Bands can be collected from the top, using a syringe and needle. After collection, the cells in the sample are pelleted, rinsed, and resuspended in sterile saline and used for whatever analysis is desired.

8.4. Notes about Reverse Genetics

Partial amino acid sequence can be obtained from individual labeled proteins after the gel is transferred to a solid support (Tempst *et al.*, 1990). To determine which labeled proteins are present at high enough levels for sequencing, the protein blot can be stained with Ponceau S and compared with an autoradiograph from the same gel. As long as a protein spot can be visualized by Ponceau S staining, there is enough protein to obtain a partial amino acid sequence (Tempst *et al.*, 1990).

The partial amino acid sequence can then be used to identify the gene encoding the protein. The amino acid sequence can be used to scan the sequence data bases for matches with regions that have already been sequenced. In the event that the sequence does not match any known *E. coli* sequence, radiolabeled degenerate oligonucleotides based on the amino acid sequence (Sambrook *et al.*, 1989) can be used to locate the gene by hybridization to libraries of the *E. coli* genome, such as the Kohara ordered library of the *E. coli* chromosome (Kohara *et al.*, 1987). When the gene has been identified and cloned, mutations in the chromosomal locus can be made by allele replacement (Winans *et al.*, 1985). Once a mutation is generated, the mutant can be tested for stationary-phase phenotypes.

ACKNOWLEDGMENTS. This chapter was conceived from a suggestion of Dr. Bernard Davis, who has always served as inspiration in guiding our work in this field. The authors are grateful for the helpful discussions with the rest of the members of the laboratory, in particular María Mercedes Zambrano, Dian Bohannon, Sara Lazar, and Gjalt Huisman. The work presented in this chapter was supported by a grant from the National Science Foundation to R.K. (MCB-9207323). D.S. was the recipient of a USPHS postdoctoral fellowship (GM13781). R.K. was the recipient of an American Cancer Society Faculty Research Award (FRA 350).

REFERENCES

Almirón, M., Link, A. J., Furlong, D., and Kolter, R., 1992, A novel DNA binding protein with regulatory and protective roles in starved *E. coli*, *Genes and Dev.* **6:**2646–2654.

Bohannon, D. E., Connell, N., Keener, J., Tormo, A., Espinosa-Urgel, M., Zambrano, M. M., and Kolter, R., 1991, Stationary-phase-inducible "gearbox" promoters: Differential effects of *katF* mutations and the role of σ^{70}, *J. Bacteriol.* **173:**4482–4492.

Cahn, F. H., and Fox, M. S., 1968, Fractionation of transformable bacteria from component cultures of *Bacillus subtilis* on renografin gradients, *J. Bacteriol.* **95:**867–875.

Cashel, M., and Rudd, K. E., 1987, The stringent response, in: *Escherichia coli and Salmonella typhimurium, Cellular and Molecular Biology*, Vol. 2 (F. C. Neidhardt, J. L. Ingraham, K. B. Low, B. Magasanik, M. Schaechter, and H. E. Umbarger, eds.), American Society for Microbiology, Washington, D.C., pp. 1410–1438.

Connell, N., Han, Z., Moreno, F., and Kolter, R., 1987, An *E. coli* promoter induced by the cessation of growth, *Mol. Microbiol.* **1:**195–201.

Davis, B. D., Luger, S. J., and Tai, P. C., 1986, Role of ribosome degradation in the death of starved *E. coli* cells, *J. Bacteriol.* **166:**439–445.

Dawes, E. A., 1989, Growth and survival of bacteria, in: *Bacteria in Nature*, Vol. 3 (J. S. Poindexter and E. R. Ledbetter, eds.), Plenum Press, New York, pp. 67–187.

Goldberg, A. L., and St. John, A. C., 1976, Intracellular protein degradation in mammals and bacterial cells: Part 2, *Annu. Rev. Biochem.* **45**:747–803.

Groat, R. G., Schultz, J. E., Zychlinsky, E., Bockman, A., and Matin, A., 1986, Starvation proteins in *Escherichia coli*: Kinetics of synthesis and role in starvation survival, *J. Bacteriol.* **168**:486–493.

Hall, B. G., 1988, Adaptive evolution that requires multiple spontaneous mutations. I. Mutations involving an insertion sequence, *Genetics* **120**:887–897.

Kasten, F. H., 1967, Cytochemical studies with acridine orange and the influence of dye contaminants in the staining of nucleic acids, *Int. Rev. Cytol.* **21**:141–202.

Kohara, Y., Akiyama, K., and Isono, K., 1987, The physical map of the whole *E. coli* chromosome: Application of a new strategy for rapid analysis and sorting of a large genomic library, *Cell* **50**:495–508.

Lange, R., and Hengge-Aronis, R., 1991, Identification of a central regulator of stationary phase gene expression in *Escherichia coli*, *Mol. Microbiol.* **5**:49–59.

McCann, M. P., Kidwell, J. P., and Matin, A., 1991, The putative σ factor KatF has a central role in development of starvation-mediated general resistance in *Escherichia coli*, *J. Bacteriol.* **173**: 4188–4194.

Mach, H., Hecker, M., Hill, I., Schroeter, A., and Mach, F., 1989, Physiologische Bedeutung der "stringent control" bei *Escherichia coli* unter extremen Hungerbedingungen, *Z. Naturforsch.* **44c**:838–844.

Mandelstam, J., 1960, The intracellular turnover of protein and nucleic acids and its role in biochemical differentiation, *Bacteriol. Rev.* **24**:289–308.

Matin, A., 1991, The molecular basis of carbon-starvation-induced general resistance in *Escherichia coli*, *Mol. Microbiol.* **5**:3–10.

Matin, A., Auger, E. A., Blum, P. H., and Schultz, J. E., 1989, Genetic basis of starvation survival in nondifferentiating bacteria, *Annu. Rev. Microbiol.* **43**:293–316.

Metzger, S., Schreiber, G., Aizenman, E., Cashel, M., and Glaser, G., 1989, Characterization of the *relA1* mutation and a comparison of *relA1* with new *relA* null alleles in *Escherichia coli*, *J. Biol. Chem.* **264**:21146–21152.

Miller, J. H., 1972, *Experiments in Molecular Genetics*, Cold Spring Harbor Laboratory Press, Cold Spring Harbor, N.Y.

Mittler, J. E., and Lenski, R. E., 1990, New data on excisions of Mu from *E. coli* MCS2 cast doubt on directed mutation hypothesis, *Nature* **344**:173–175.

Neidhardt, F. C., Bloch, P. L., and Smith, D. F., 1974, Culture medium for enterobacteria, *J. Bacteriol.* **119**:736–747.

Nilsson, L., Oliver, J. D., and Kjelleberg, S., 1991, Resuscitation of *Vibrio vulnificus* from the viable but nonculturable state, *J Bacteriol.* **173**:5054–5059.

Nyström, T., Olsson, R. M., and Kjelleberg, S., 1992, Survival, stress resistance and alterations in protein expression in the marine *Vibrio sp.* S14 during starvation for different individual nutrients, *Appl. Environ. Microbiol.* **58**:55–65.

Östling, J., Goodman, A. E., and Kjelleberg, S., 1991, Behaviour of IncP-1 plasmids and a miniMu transposon in a marine *Vibrio sp.*: Isolation of starvation inducible *lac* operon fusions, *FEMS Microbiol. Ecol.* **86**:83–94.

Reeve, C. A., Amy, P. S., and Matin, A., 1984a, Role of protein synthesis in the survival of carbon-starved *Escherichia coli* K-12, *J. Bacteriol.* **160**:1041–1046.

Reeve, C. A., Bockman, A. T., and Matin, A., 1984b, Role of protein degradation in the survival of carbon-starved *Escherichia coli* and *Salmonella typhimurium*, *J. Bacteriol.* **157**:758–763.

Roszak, D. B., and Colwell, R. R., 1987, Survival strategies of bacteria in the natural environment, *Microbiol. Rev.* **51**:365–379.

Roszak, D. B., Grimes, D. J., and Colwell, R. R., 1983, Viable but nonrecoverable stage of *Salmonella enteritidis* in aquatic systems, *Can. J. Microbiol.* **30**:334–337.

Sambrook, J., Fritsch, E. F., and Maniatis, T., 1989, *Molecular Cloning: A Laboratory Manual*, Cold Spring Harbor Laboratory Press, Cold Spring Harbor, N.Y.

Schultz, J. E., Latter, G. I., and Matin, A., 1988, Differential regulation by cyclic AMP of starvation protein synthesis in *Escherichia coli*, *J. Bacteriol.* **170**:3903–3909.

Shapiro, J. A., 1984, Observations on the formation of clones containing *araB–lacZ* cistron fusions, *Mol. Gen. Genet.* **194**:79–90.

Siegele, D. A., and Kolter, R., 1992a, Life after log, *J. Bacteriol.* **174**:345–348.

Siegele, D. A., and Kolter, R., 1993, Isolation and characterization of an *Escherichia coli* mutant unable to resume growth after starvation, *Genes and Dev.*, in press.

Spector, M. P., and Cubitt, C. L., 1992, Starvation-inducible loci of *Salmonella typhimurium*: Regulation and roles in starvation-survival, *Mol. Microbiol.* **6**:1467–1476.

Spector, M. P., Aliabadi, Z., Gonzalez, T., and Foster, J. W., 1986, Global control in *Salmonella typhimurium*: Two-dimensional electrophoretic analysis of starvation-, anaerobiosis-, and heat shock-inducible proteins, *J. Bacteriol.* **168**:420–424.

Spector, M. P., Park, Y. K., Tirgari, S., Gonzalez, T., and Foster, J. W., 1988, Identification and characterization of starvation-regulated genetic loci in *Salmonella typhimurium* by using Mud-directed *lacZ* operon fusions, *J. Bacteriol.* **170**:345–351.

Spence, J., Cegielska, A., and Georgopoulos, C., 1990, Role of Escherichia coli heat shock proteins Dnak and HtpG (C62.5) in response to nutritional deprivation, *J. Bacteriol.* **172**:7157–7166.

Straus, D. B., Walter, W. A., and Gross, C., 1988, *Escherichia coli* heat shock gene mutants are defective in proteolysis, *Genes Dev.* **2**:1851–1858.

Tempst, P., Link, A. J., Riviere, L. R., Fleming, M., and Elicone, C., 1990, Internal sequence analysis of proteins separated on polyacrylamide gels at the submicrogram level: Improved methods, applications and gene cloning strategies, *Electrophoresis* **11**:537–553.

Tormo, A., Almirón, M., and Kolter, R., 1990, *surA*, an *Escherichia coli* gene essential for survival in stationary phase, *J. Bacteriol.* **172**:4339–4347.

Winans, S. C., Elledge, S. J., Krueger, J. H., and Walker, G. C., 1985, Site-directed insertion and deletion mutagenesis with cloned fragments in *Escherichia coli*, *J. Bacteriol.* **161**:1219–1221.

8

The Role of rpoS in Early Stationary-Phase Gene Regulation in Escherichia coli K12

Regine Hengge-Aronis

1. INTRODUCTION

For several decades the enterobacteriaceum *Escherichia coli* has been a favorite experimental organism of prokaryotic geneticists and molecular biologists. A wealth of knowledge including many of the most fundamental concepts of modern molecular biology has been obtained using *E. coli* cells that were usually grown rapidly under laboratory conditions. More recently, there has been a growing interest in global regulatory phenomena elicited by changing environmental conditions. Unfortunately, the natural environments of *E. coli* do not have much in common with a laboratory culture. Both the mammalian intestine as the primary habitat as well as soil, sediments, and sea- or freshwater as potential secondary habitats are frequently characterized by nutrient limitation and thus sustain only sporadic growth (Morita, 1988, this volume; Roszak and Colwell, 1987; Moriarty and Bell, this volume; van Elsas and van Overbeek, this volume). This situation much better resembles stationary phase under laboratory conditions.

Limitation of a single nutrient does not necessarily result in a transition

Regine Hengge-Aronis • Department of Biology, University of Konstanz, 7750 Konstanz, Germany.

Starvation in Bacteria, edited by Staffan Kjelleberg. Plenum Press, New York, 1993.

into stationary phase but usually induces alternative nutrient scavenging systems. Well-characterized examples are the cAMP/CAP system (Botsford and Harman, 1992; Ullman and Danchin, 1983), the *ntr* (Kustu *et al.*, 1989; Magasanik and Neidhardt, 1987) and the *pho* regulons (Torriani-Gorini *et al.*, 1987). Once these systems are induced, growth of the cells can continue. Total starvation for any essential nutrient, however, induces the entry into stationary phase. As described in this review and several other chapters in this volume, this transition is a highly ordered process which involves dramatic changes in cellular physiology and even in morphology. During this process, at least the transition phase or early stationary phase and the late stationary phase have to be distinguished. This review focuses on the molecular events during early stationary phase, since little is known about later events.

Using two-dimensional gel electrophoresis, it has been shown that 30–50 proteins are induced with distinct kinetics by carbon starvation (Groat *et al.*, 1986). Only recently systematic genetic and molecular studies have been undertaken with the aim of elucidating both the regulation of stationary phase-induced genes as well as the functions of their gene products within the larger context of survival under highly unfavorable environmental conditions.

2. IDENTIFICATION OF rpoS AS A CENTRAL REGULATORY GENE FOR EARLY STATIONARY-PHASE GENE EXPRESSION

The *rpoS* gene was discovered several times independently. Initially a mutation (*nur*) conferring near UV sensitivity was described (Tuveson, 1981). Then a regulatory gene (*katF*) for the expression of the *katE*-encoded catalase HPII (Loewen and Triggs, 1984) and of exonuclease III (*xthA*) (Sak *et al.*, 1989) was identified, and *nur* and *katF* were shown to be allelic (Sammartano *et al.*, 1986). Mutations in *katF* (Sammartano *et al.*, 1986) or *xthA* (Demple *et al.*, 1983) render the cells sensitive against hydrogen peroxide. A gene (*appR*) mapping in the same chromosomal region as *katF* (59 min) was then shown to be required for high-level expression of acid phosphatase (the *appA* gene product) (Touati *et al.*, 1986). Finally, a systematic search for stationary phase (or carbon starvation)-induced genes was undertaken using the *lacZ* fusion technique (Lange and Hengge-Aronis, 1991b). This approach yielded a series of *csi::lacZ* fusions, which in rich medium exhibited induction during transition into stationary phase that was either transient or continued well into the nongrowing state (Weichart *et al.*, 1993). All fusion strains were tested for a potential reduction in long-term starvation survival and for alterations in the two-dimensional gel pattern of total protein synthesis in comparison with the isogenic wild-type strain. The most drastically reduced starvation survival as well as pleiotropic changes in the 2-D gel pattern were found for the *csi-2::lacZ* fusion, indicating that the *csi-2* gene probably encoded a regulatory protein. Moreover, the *csi-2::lacZ* fusion strain was defective for a variety of stationary-phase functions, such as glycogen synthesis and resistance against various stress conditions (Lange and Hengge-Aronis, 1991b). *csi-2* was mapped at 58.9 min (2880 bp on the physical map, Kohara *et al.*, 1987) of the *E. coli* chromosome (Lange and Hengge-Aronis, 1991b), i.e., in the region where *katF* (*nur*) and *appR* had been localized. With the idea in mind that all these genes might be allelic, the *csi-2::lacZ* strain was shown to be H_2O_2-sensitive and defective in *appA* expression. Cloning of *csi-2* yielded plasmids carrying the *katF* open reading frame, and *csi-2::lacZ* as well as a *csi-2::Tn10* insertion could be localized within *katF* in the chromosome (Lange and Hengge-Aronis, 1991b). Moreover, it was shown that *katF* plasmids could complement *appR* mutations (Touati *et al.*, 1991). Thus, *csi-2*, *katF*, and *appR* are in fact the same gene.

The *katF* sequence (Mulvey and Loewen, 1989) turned out to share extensive homology with *rpoD* which encodes σ^{70}, the vegetative sigma subunit of RNA polymerase in *E. coli*. This indicated the intriguing possibility that its gene product acted as an alternative sigma factor. From the results described in detail below, it has become clear that this novel sigma factor is a central regulator of stationary-phase gene expression. According to this function, the novel designations "*rpoS*" and "σ^S" were introduced for the gene and its product, respectively (where "S" stands for *s*tarvation or *s*tationary phase) (Hengge-Aronis, 1993).

rpoS not only controls other growth phase-regulated genes, but is itself stationary phase-induced (approximately five- to tenfold in rich medium). It is, however, not autoregulated since a *rpoS::lacZ* fusion integrated into the chromosome on a lambda phage exhibits identical kinetics of induction in *rpoS*$^+$ and *rpoS::Tn10* backgrounds. A potential σ^F recognition site upstream of the *rpoS* open reading frame also does not play a role in its expression which is similar in *flbB*$^+$ and *flbB* strains (R. Lange and R. Hengge-Aronis, unpublished results). Expression of *rpoS::lacZ* fusions is inversely correlated with growth rate, and cannot be induced by abrupt starvation for carbon or nitrogen, but by any kind of nutrient limitation that results in a gradual reduction of the growth rate (Lange and Hengge-Aronis, 1991b; Mulvey *et al.*, 1990). An unidentified substance in spent medium from a stationary-phase culture as well as some aromatic acids can induce *rpoS* (Mulvey *et al.*, 1990; Schellhorn and Stones, 1992), whereas addition of cAMP reduces *rpoS* expression at least in an adenylate cyclase mutant (Lange and Hengge-Aronis, 1991b). Taken together, not much is known about the regulatory mechanism as well as the primary signal responsible for *rpoS* induction. Elucidation of these mechanisms, however, might also shed light on the regulation of other non-*rpoS*-regulated growth phase-controlled genes.

3. THE ROLE OF rpoS IN STATIONARY-PHASE STRESS RESISTANCE

Although the enteric bacterium *E. coli* cannot compete with free-living starvation specialists such as marine *Vibrio* species (see Morita, this volume; Östling *et al.*, this volume), it can survive prolonged starvation relatively well. The plating efficiency of an *E. coli* culture starved for carbon or nitrogen sources decreases with a half-life of 5 to 10 days (Lange and Hengge-Aronis, 1991; Reeve *et al.*, 1984). A culture starved for phosphate exhibits faster die-off kinetics (N. Henneberg and R. Hengge-Aronis, unpublished results) possibly because of an excessive loss of ribosomes (Davis *et al.*, 1986). As already mentioned, *rpoS* mutants exhibit drastically reduced starvation survival in all media tested (Lange and Hengge-Aronis, 1991b; McCann *et al.*, 1991; Mulvey *et al.*, 1990). Plating efficiencies may, however, overestimate death rates within a culture, since also in *E. coli* there could be a viable, but nonculturable state as was found for other bacterial species (Roszak and Colwell, 1987; Oliver, this volume). Therefore, the low plating efficiency of starved *rpoS* mutants may be the result either of cellular death or of a faster transition into a nonculturable state.

Besides surviving starvation stress *per se*, stationary-phase *E. coli* cells also develop a remarkable resistance against additional stress factors, such as heat shocks ($> 50°C$) and hydrogen peroxide (Jenkins *et al.*, 1988) or high salt concentrations (Jenkins *et al.*, 1990). Resistance in this context is defined not as the ability to grow under stress conditions, but as high plating efficiency of a culture following high dose stress exposure. A certain resistance against these stress conditions can also be induced in growing cells by exposure to sublethal doses of the respective stress agent. These adaptive resistances are related to the induction of specific stimulons or regulons designed to cope with a specific stress condition (see Nyström, this volume). Interestingly, the adaptive systems usually confer a lower degree of protection than the stationary-phase resistance systems (Jenkins *et al.*, 1988). This suggests that these systems only partially overlap or are even totally different. Moreover, stationary-phase stress resistance develops in cells that have never been exposed to any of the specific stress conditions. Therefore, the regulatory mechanisms underlying the induction of adaptive systems and of stationary-phase systems must be different. Stationary-phase stress tolerance develops within 4 h after the onset of starvation, requires protein synthesis (Jenkins *et al.*, 1988), and as described in detail below, requires *rpoS* in all cases studied so far (Hengge-Aronis, 1993).

3.1. Thermotolerance

Although *E. coli* cannot grow at temperatures above 46°C (Gross *et al.*, 1990), it is able to maintain its cellular integrity and thus survive for a short time also at 50°C or more. Increased heat shock resistance is found when a growing culture is preadapted for at least 30 min to an elevated but not yet lethal temperature (42°C) (Yamamori and Yura, 1982) or in stationary phase (Jenkins *et al.*, 1988).

Originally, induction of the heat shock system was implicated in adaptive thermotolerance (Yamamori and Yura, 1982), but it was subsequently shown that strong induction of the heat shock proteins in a strain which carried the heat shock sigma factor (σ^{32}) gene *rpoH* under the control of the IPTG-inducible P_{tac} promoter was not sufficient to produce adaptive thermotolerance (vanBogelen *et al.*, 1987). It was then suggested that the alternate sigma factor σ^E is involved, which stimulates transcription of *rpoH* and perhaps other heat-inducible but non-σ^{32}-dependent genes at very high temperature (Erickson and Gross, 1989).

On the other hand, *rpoS* mutants are deficient in stationary-phase thermo-tolerance (Lange and Hengge-Aronis, 1991b; McCann *et al.*, 1991), but can develop adaptive thermotolerance normally when grown at 42°C before being challenged with the lethal temperature (Hengge-Aronis *et al.*, 1991). This indicates that at least the regulation, and maybe also the structural genes involved in these two types of thermotolerance are different.

Besides its regulatory dependence on *rpoS*, little is known about the actual mechanism of thermoprotection in stationary-phase cells. It has been shown that an *E. coli* strain which is unable to synthesize the disaccharide trehalose [O-α-D-glucosyl(1-1)-α-D-glucoside] is impaired in stationary-phase thermo-tolerance, but not in adaptive thermotolerance. The genes responsible for trehalose synthesis (*otsA*, *otsB*) are induced during transition into stationary phase in an *rpoS*-dependent manner (Hengge-Aronis *et al.*, 1991) (see Section 4.3). Heat shock interferes with the integrity of membranes (Tsuchido *et al.*, 1985, 1989) and causes the denaturation of proteins. Trehalose binds to the polar head groups of phospholipids (Lee *et al.*, 1986) and has a membrane protective effect (Crowe *et al.*, 1984). *In vitro* experiments have shown that it prevents membrane fusion (Womersley *et al.*, 1986) and stabilizes enzyme function (Crowe *et al.*, 1988). Sugars and polyols in general also increase the thermal stability of proteins (Back *et al.*, 1979). Thus, trehalose is obviously an appropriate substance to counteract the deleterious effects of heat shocks. Interestingly, trehalose is not only a thermoprotectant, but a general stress protectant (van Laere, 1989) which is found in prokaryotic as well as in eukaryotic cells. Its role in osmoprotection of *E. coli* is well established

(Csonka, 1989), and in various eukaryotic organisms, including animals, fungi, and lower plants, it plays an important role in the survival of dehydration.

A strain with a defect in trehalose synthesis is clearly impaired in stationary-phase thermotolerance, but *rpoS* mutants are even more thermosensitive under all conditions tested. This indicates that trehalose synthesis is not the only factor involved in *rpoS*-dependent thermotolerance. Magnesium ions might play an additional role, since thermosensitivity of a trehalose-free strain increases further when the Mg^{2+} concentration of the medium is reduced below 1 mM (Hengge-Aronis *et al.*, 1991). Mg^{2+} was reported to stabilize ribosomes (Hurst and Hughes, 1978), but its precise role in thermoprotection has not been studied in detail.

The expression of the heat shock proteins DnaK and GroEL slightly increases in carbon-starved cells (Groat *et al.*, 1986). DnaK and GroEL, however, apparently do not contribute to stationary phase-induced thermoprotection since *rpoH* mutants, which have strongly reduced amounts of DnaK and GroEL, still exhibited an increase in thermotolerance but not in DnaK and GroEL expression during carbon starvation (Jenkins *et al.*, 1991). Moreover, *rpoS* is not required for increased *rpoH* transcription and heat shock protein expression either in stationary phase or at very high temperature (R. Hengge-Aronis, unpublished results). Yet it seems reasonable to assume the participation of chaperone proteins in thermoprotection. *rpoS* might therefore regulate the expression of some unknown stationary phase-specific chaperone protein(s).

3.2. Oxidative Stress Resistance

The mammalian intestine as the primary habitat of *E. coli* provides an anaerobic environment. However, *E. coli* also grows well under aerobic conditions where it depends on aerobic respiration and reactive oxygen species are ubiquitous. In addition, at least pathogenic *E. coli* strains, which leave the intestine, have to cope with high concentrations of reactive oxygen species (which for instance are released by macrophages). Reactive oxygen in the form of superoxide radical ($O_2^{\cdot-}$), hydrogen peroxide, and hydroxyl radical ($OH^{\cdot}$) is also produced during the sequential one-electron reduction of O_2, which is catalyzed by respiratory chain enzymes and various dehydrogenases. $O_2^{\cdot-}$ is generated by redox-cycling agents such as paraquat (methyl viologen) or plumbagin, and is dismutated to H_2O_2 by superoxide dismutases (SOD). A major source of $OH^{\cdot}$ radicals is the reaction of H_2O_2 with reduced iron (Fenton reaction) (Farr and Kogoma, 1991).

Reactive oxygen species cause severe damage to all macromolecular cellular components, including DNA, RNA, membranes, and proteins, by a variety of reactions. Especially $OH^{\cdot}$ radicals can react with almost any biomolecule (Farr and Kogoma, 1991). In their defense against oxidative stress, *E. coli* cells use several strategies in parallel. One is the synthesis of enzymes like superoxide dismutase or catalase that directly destroy reactive oxygen species and thus prevent oxidative damage. In addition, there are various repair systems which eliminate cellular damage, once it has occurred.

Similarly as described above for thermotolerance, increased resistance against oxidative stress can be observed either in cells pretreated with sublethal adaptive doses of superoxide radicals (Farr *et al.*, 1985) or hydrogen peroxide (Demple *et al.*, 1983) or in stationary-phase cells (Jenkins *et al.*, 1988).

The adaptive systems induced by $O_2^{\cdot-}$ and H_2O_2 are the *soxRS* and *oxyR* regulons, respectively (for recent reviews see Demple and Amabile-Cuevas, 1991; Farr and Kogoma, 1991; Storz *et al.*, 1990b). *soxR* and *soxS* (Wu and Weiss, 1991) probably encode a sensor/regulator system which acts sequentially (Wu and Weiss, 1992) in the induction of MnSOD (*sodA*), endonuclease IV (*nfo*), glucose-6-phosphate dehydrogenase (*zwf*) and several other superoxide-inducible (*soi*) genes (Walkup and Kogoma, 1989). *oxyR* encodes a H_2O_2-activated regulator, which in its oxidized form directly activates (Storz *et al.*, 1990a) the transcription of catalase HPI (*katG*), alkyl hydroperoxide reductase (*ahpC*, *ahpF*), glutathione reductase (*gor*), and approximately half a dozen other proteins of unidentified functions (Storz *et al.*, 1990b).

Stationary-phase oxidative stress resistance requires *rpoS*. This was found for resistance against H_2O_2 (Lange and Hengge-Aronis, 1991b; McCann *et al.*, 1991; Sammartano *et al.*, 1986) and against the $O_2^{\cdot-}$ generator paraquat (N. Henneberg and R. Hengge-Aronis, unpublished results). H_2O_2 resistance in starved cells is dependent on the *rpoS*-controlled repair enzyme exonuclease III (*xthA*) (Demple *et al.*, 1983) and a novel DNA-binding protein (*dps*) (Almirón *et al.*, 1992). Also, catalase HPII (encoded by *katE*) belongs to the *rpoS* regulon (Loewen and Triggs, 1984). Since H_2O_2 rapidly diffuses into and within the cells, the presence of catalase does not render single cells H_2O_2 resistant. However, it increases the rate of survival of dense populations in the presence of exogenous H_2O_2 (Ma and Eaton, 1992). Interestingly, there appears to be no overlap between adaptive and stationary-phase H_2O_2 resistance mechanisms. Thus, pairs of structurally different enzymes with similar functions have evolved in *E. coli* rather than a complex differential regulation of singular enzymes.

3.3. Other Stationary-Phase Stress Resistances

Besides being more resistant against heat shock and oxidative stress, stationary-phase *E. coli* cells exhibit enhanced tolerance against many other stress conditions. In all cases studied so far, *rpoS* mutants are more sensitive than the isogenic *rpoS*$^+$ parental strains.

Resistance against near-UV irradiation (300–400 nm) was shown to require *rpoS* and *xthA* (Tuveson, 1981; Sak *et al.*, 1989). Since H_2O_2 is a photoproduct of near-UV irradiation (Eisenstark, 1989), near-UV resistance is probably related to stationary-phase oxidative stress resistance.

A stationary *E. coli* wild-type culture survives exposure to an extremely high NaCl concentration (2.5 M) for many hours surprisingly well (Jenkins *et al.*, 1990). *rpoS* mutants are not only much more sensitive against this extreme osmolarity in stationary phase (McCann *et al.*, 1991), but also grow less well exponentially in minimal medium of increased osmotic strength (> 0.3 M NaCl added; R. Hengge-Aronis, unpublished results). Wild-type cells grown under these conditions accumulate trehalose as an osmoprotectant (Giaever *et al.*, 1988). Osmotic exponential-phase induction of the trehalose-synthesizing enzymes (encoded by *otsA* and *otsB*) also requires *rpoS*, indicating a basal but essential level of *rpoS* also in growing cells (Hengge-Aronis *et al.*, 1991). Since trehalose acts as an osmoprotectant in growing cells, and the trehalose-synthesizing enzymes are also induced in stationary phase (Hengge-Aronis *et al.*, 1991), this sugar may be also involved in stationary-phase protection against extreme osmolarities. In addition, stationary-phase *E. coli* cells are spherical rather than rod-shaped and smaller than growing cells. These morphological changes also are regulated by *rpoS* (Lange and Hengge-Aronis, 1991a). Moreover, there is evidence for changes in the amount and composition of membrane proteins [e.g., the lipoprotein encoded by *osmB* (Hengge-Aronis *et al.*, 1991; Jung *et al.*, 1990)], phospholipids (Souzu, 1982), lipopolysaccharide (Ivanov and Fomchenkov, 1989), and maybe peptidoglycan (Aldea *et al.*, 1989; Buchanan and Sowell, 1982). All of these factors could increase the physical stability of the cell envelope and thus contribute to high osmolarity resistance.

Recently, increased *rpoS*-dependent resistance in stationary-phase *E. coli* cells against high doses of the DNA-alkylating mutagen *N*-methyl-*N'*-nitro-*N*-nitrosoguanidine (MNNG), of ethanol, acetone, toluene, and the detergent deoxycholate was found (N. Henneberg and R. Hengge-Aronis, unpublished results). The *rpoS*-regulated genes involved in these resistances are unknown. A specific set of proteins is induced in growing cells exposed to MNNG (the *ada*-regulon) (Lindahl *et al.*, 1988). A potential overlap between this adaptive system and the *rpoS* regulon has not been studied in detail.

3.4. Implications for the Functions of the Structural Genes Involved in Multiple Stationary-Phase Stress Resistance

The various deleterious treatments for which increased resistance in stationary phase was observed, have very broad effects on cell physiology: Different DNA-damaging substances were tested as well as agents which perturb membranes or denature proteins. In all cases, stationary-phase cells carrying the wild-type *rpoS* gene developed resistance. It seems unlikely that there is a separate system for each specific stress condition. These resistance systems are not induced by exposure to a particular stress situation, but under conditions of beginning nutrient limitation—just in case there might be a future need for them, when rapid *de novo* protein synthesis will not be possible. Simply for economical reasons there are probably a few *rpoS*-regulated resistance mechanisms with rather broad specificities. These might include:

1. Some enzymes that directly inactivate poisonous chemical substances (e.g., catalase HPII).
2. A broad-specificity DNA repair system which includes exonuclease III (*xthA*). DNA might also be in a better protected form, since its linking number (Balke and Gralla, 1987) and compactness (Spassky *et al.*, 1984) are probably different in nongrowing cells. Recently, a novel stationary phase-specific DNA-binding protein (Dps) with protective and regulatory functions has been shown to be under *rpoS* control (Almiron *et al.*, 1992). The expression of the histonelike protein H-NS (H1a), which is encoded by *osmZ* (May *et al.*, 1990), strongly increases during entry into stationary phase (Dersch *et al.*, 1993). Transcription of *osmZ*, however, does not depend on *rpoS*.
3. Membrane-protecting mechanisms: Trehalose, whose synthesis is under *rpoS* control, acts as a membrane protectant under various stress conditions. Moreover, there is evidence for changes in the amount and composition of cell envelope constituents (see above). An altered outer and inner membrane permeability as well as an enhanced physical stability of the cell envelope could be the consequence of these changes.
4. Protein-protecting systems: Under starvation conditions a delicate balance between protection of essential proteins and degradation of dispensable proteins must be established—such that at least a certain precursor pool for *de novo* protein synthesis is maintained. This suggests the existence of stationary phase-specific chaperone proteins essential for survival.

A majority of *rpoS*-controlled genes remain to be discovered. It can be expected that many of their gene products play a role in multiple stress protection in stationary phase.

4. *rpoS*-REGULATED GENES AND THEIR CELLULAR FUNCTIONS

On two-dimensional O'Farrell gels, the absence or reduced expression of 18 (Lange and Hengge-Aronis, 1991b) or up to 32 protein spots (McCann *et al.*,

1991) was observed for *rpoS* mutants. In these studies also some additional protein spots were found for the mutants, indicating that some genes might be under negative, maybe indirect control of *rpoS*. An increasing number of *rpoS*-dependent genes have been identified recently. These genes or operons are dispersed all over the chromosome (Fig. 1) and their gene products have a broad spectrum of unrelated functions. This diversity reflects the global nature of the physiological and morphological changes in early stationary phase and underlines the central importance of *rpoS*.

4.1. The Morphogene bolA and Stationary-Phase Cell Morphology

Size and shape of *E. coli* cells are dependent on the growth rate of the culture. Nutritional upshifts result in increased cell size (Kubitschek, 1990), whereas starved cells are smaller and spherical rather than rod-shaped. Size reduction is a continuous process along the growth curve (Lange and Hengge-Aronis, 1991a) and results from a rate of cell division which is higher than the rate of cell mass increase during entry into stationary phase (Maaloe and Kjeldgaard, 1966). In contrast to wild-type *E. coli* cells, *rpoS* mutants remain rod-shaped in stationary phase (Lange and Hengge-Aronis, 1991a).

The *bolA* gene is apparently involved in these morphological changes since its artificial overexpression causes spherical morphology of growing cells (Aldea *et al.*, 1988). Its major promoter (*bolA*p1) is growth phase-regulated, and is responsible for an approximately tenfold induction of *bolA* during entry into stationary phase (Aldea *et al.*, 1989). This induction requires the *rpoS* gene product (Bohannon *et al.*, 1991; Lange and Hengge-Aronis, 1991a). Unlike most other *rpoS*-dependent genes, induction of *bolA* is balanced with cell size

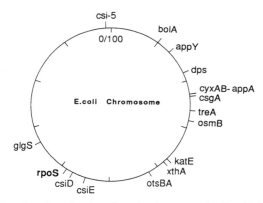

Figure 1. Location of *rpoS* and *rpoS*-regulated genes on the *E. coli* chromosome.

reduction in such a way that its gene product is synthesized at a constant amount per cell. A similar mode of expression was shown for the cell division operon *ftsQAZ* (Aldea *et al.*, 1990). Growth phase-controlled promoters were found upstream of *ftsQAZ* (Aldea *et al.*, 1990) and within the coding sequence of the operon (Dewar *et al.*, 1989). A potential regulation by *rpoS* is under study.

The observation that *bolA* requires an intact *ftsZ* gene for exerting its function (Aldea *et al.*, 1988) points to a role of the *bolA* gene product in cell division. Its sequence, which encodes a 13.5-kDa protein, exhibits a helix–turn–helix motif characteristic for a DNA-binding domain (Aldea *et al.*, 1989). In a *bolA* mutant the synthesis of penicillin-binding protein 6 (PBP6) is not induced in stationary phase (Aldea *et al.*, 1989; Buchanan and Sowell, 1982). PBP6 is a carboxy peptidase involved in providing precursors for septum peptidoglycan synthesis (Begg *et al.*, 1990). However, PBP6-deficient mutants could produce round cells when *bolA* was overexpressed (Aldea *et al.*, 1989), indicating that additional processes in cell division are influenced by *bolA* and therefore also by *rpoS*.

4.2. *xthA, katE, dps and Hydrogen Peroxide Resistance*

Exonuclease III (*xthA*) is the major AP endonuclease of *E. coli*. It removes nucleoside 5'-monophosphates near apurinic and apyrimidinic sites in damaged DNA (a detailed list of its catalytic activities is given in Saporito *et al.*, 1988). It is an essential DNA repair enzyme involved in eliminating damage that has occurred under conditions of H_2O_2 stress or any other stress that causes similar DNA lesions (e.g., near-UV irradiation). Its expression is entirely dependent on *rpoS*, since in *rpoS* mutants no exonuclease III activity is found even in the presence of a plasmid carrying *xthA* (Sak *et al.*, 1989). However, *xthA* is not exclusively expressed in stationary phase, since its mRNA could be isolated from exponentially growing cells (Saporito *et al.*, 1988).

Catalase HPII is one of two structurally different catalases in *E. coli* (Loewen and Triggs, 1984). The *katE*-encoded catalase HPII is induced approximately 30-fold during entry into stationary phase in an *rpoS*-dependent manner (Mulvey *et al.*, 1990; Schellhorn and Hassan, 1988), whereas catalase HPI (the *katG* gene product) belongs to the oxidative stress-induced *oxyR* regulon (Loewen *et al.*, 1985; Tartaglia *et al.*, 1989). *katE* is not under aerobic/anaerobic control (Mulvey *et al.*, 1990) nor is its level of expression influenced by the presence of hydrogen peroxide (Loewen *et al.*, 1985; Schellhorn and Hassan, 1988). Interestingly, the cellular location of the two catalases is different. Whereas catalase HPI is a periplasmic enzyme probably involved in the defense against exogenous oxidative stress, catalase HPII is located in the cytoplasm (Heimberger and Eisenstark, 1988). This indicates that at least under starvation conditions, significant concentrations of H_2O_2 can be produced endogenously. It has been suggested that the expression of catalase does not protect single cells against H_2O_2, which rapidly diffuses into the cell. However, in dense cultures or in colonies, H_2O_2 is efficiently eliminated by catalase, which thus is highly advantageous to the survival of the population (Ma and Eaton, 1992).

The promoter regions of *xthA* and *katE* do not exhibit similarity to the σ^{70} promoter consensus (Saporito *et al.*, 1988; von Ossowski *et al.*, 1991), but share significant homology in their -35 and -10 regions. This may indicate that σ^S does not directly recognize these promoters (see 5.2).

A third component in H_2O_2 resistance is the histone-like protein encoded by *dps*. Dps forms highly structured complexes with DNA *in vitro*. A *dps* mutant is extremely sensitive against H_2O_2, indicating that DNA might be physically protected when complexed to Dps. In addition, *dps* also influences global gene expression in the stationary phase (Almirón *et al.*, 1992).

4.3. The role of otsA, otsB, and treA in Trehalose Metabolism, Osmoprotection, and Thermotolerance

E. coli can use trehalose as a carbon source. Depending on the osmolarity of the medium, different mechanisms of uptake and metabolism are involved. At low osmolarity, trehalose is transported into the cell via a phosphotransferase system [$EII^{Tre}(treB)/EIII^{Glc}$], and the resulting trehalose-6-phosphate is probably immediately hydrolyzed to free trehalose. Degradation of trehalose is catalyzed by amylotrehalase (*treC*) (Boos *et al.*, 1990). Induction of *treB* and *treC* requires low osmolarity and the presence of exogenous trehalose with trehalose-6-phosphate as an internal inducer (Klein *et al.*, 1991). At high osmolarity, *E. coli* uses a periplasmic trehalase (*treA*) that releases glucose which then enters the cytoplasm via the glucose phosphotransferase system (Boos *et al.*, 1987; Gutierrez *et al.*, 1989). Under these conditions the cell also synthesizes and accumulates trehalose as an osmoprotectant (Csonka, 1989). Trehalose synthesis requires *otsA* and *otsB*, which are induced at high osmolarity and encode trehalose-6-phosphate synthase and trehalose-6-phosphatase, respectively (Giaever *et al.*, 1988).

The osmotically regulated genes *otsA*, *otsB*, and *treA* are also induced during transition into stationary phase (approximately tenfold). This increase in expression is entirely dependent on *rpoS* in the case of *otsA* and *otsB*, whereas *treA* expression is only partially reduced in a *rpoS* mutant. Also osmotic control of *otsA* and *otsB* in growing cells is abolished in an *rpoS* mutant (Hengge-Aronis *et al.*, 1991; Kaasen *et al.*, 1992), indicating that probably the same promoter is used for this dual regulation. There is evidence that the two genes constitute an operon (*otsBA*), which has been cloned (Kaasen *et al.*, 1992), but its sequence and the location of promoters have not been reported. For *treA* the osmotically regulated promoter exhibits little similarity to σ^{70}-dependent promoters, although evidence was presented that σ^{70} was involved in *treA* expression (Repoila and Gutierrez, 1991). At present it is unclear whether the same promoter is also used for stationary-phase induction. If so, *rpoS* might indirectly regulate it by controlling the expression of an unknown secondary regulator which could act together with σ^{70} to stimulate *treA* transcription, or a σ^{70}-regulated activator protein might be required together with σ^S to stimulate *treA* transcription.

As mentioned earlier, trehalose has membrane- and protein-protective properties (see Section 3.1) and has the additional advantage of not interfering with metabolic reactions even when present at very high intracellular concentrations (Csonka, 1989). These properties make trehalose an ideal stress protectant (van Laere, 1989). In minimal medium where the cells have to produce stress protective compounds endogenously, trehalose acts both as an osmoprotectant (Giaever *et al.*, 1988) and as a stationary phase-specific thermoprotectant (Hengge-Aronis *et al.*, 1991) (see Section 3.1).

4.4. glgS and Glycogen Synthesis

rpoS mutants are unable to synthesize glycogen (Lange and Hengge-Aronis, 1991b). The *glgCAP* operon which encodes ADP-glucose pyro-phosphorylase, glycogen synthase, and glycogen phosphorylase (Preiss and Romeo, 1989) is induced during transition into stationary phase (Okita *et al.*, 1981). However, this induction does not require *rpoS*, indicating that there must be other *rpoS*-controlled genes essential for glycogen synthesis (Hengge-Aronis and Fischer, 1992).

Such a gene was identified recently and was termed *glgS* (Hengge-Aronis and Fischer, 1992). Glycogen synthesis (but not the expression of *glgC* and *glgA*) is strongly stimulated by *glgS* overproduction, and is reduced by a chromosomal deletion covering part of *glgS*. *glgS* constitutes a monocistronic operon, which codes for a hydrophilic and highly charged protein of 7.886 kDa. Experiments with *glgS::lacZ* gene fusions demonstrated that *glgS* exhibits strong stationary-phase induction (50-fold in LB medium). Its expression is reduced by introducing either *rpoS::Tn10* or Δcya mutations and is abolished in an *rpoS cya* double mutant. At least two mRNA start sites have been mapped in the region upstream of *glgS*. Transcription from one of these promoters requires *rpoS*, transcription from the other is dependent on cAMP. Since starvation conditions, which do not induce *rpoS* transcription, result in the formation and accumulation of cAMP and vice versa, the combined activity of these two promoters ensures a rather strong stationary-phase induction of *glgS* under most kinds of starvation conditions (Hengge-Aronis and Fischer, 1992).

The level of *glgS* expression is one of several factors that determine the actual amount of glycogen accumulated in a stationary-phase cell. As mentioned above, the level of the glycogen synthetic enzymes GlgC and GlgA is growth phase-regulated (Okita *et al.*, 1981). In vitro expression of *glgC* could be increased by adding cAMP/CAP (Romeo and Preiss, 1989). Moreover, GlgC is strongly stimulated by fructose-1,6-bisphosphate which acts as an allosteric activator (Preiss and Romeo, 1989). Substantial glycogen accumulation requires an excess of glucose in the medium which has to be limited for other nutrients, i.e., conditions that result in an increase in the cellular level of fructose-1,6-bisphosphate. The actual concentration of glycogen accumulated in a cell appears to be determined mainly by two factors: this allosteric activation of ADP-glucose pyrophosphorylase, and the expression of *rpoS*-controlled genes such as *glgS* (and maybe additional unknown genes).

The precise function of GlgS remains to be determined. Preliminary results (R. Hengge-Aronis, unpublished data) indicate that it is involved in glycogen priming, a process which has not been studied in *E. coli*, but which is highly complex in eukaryotic cells (Calder, 1991; Smythe and Cohen, 1991).

4.5. *csi-5 (osmY): A Periplasmic Protein of Unknown Function*

The *csi*-5 gene was identified as a stationary phase-induced *lacZ* fusion that was not expressed in an *rpoS*::Tn*10* background (Lange and Hengge-Aronis, 1991b). As several other *rpoS*-regulated genes, *csi*-5 exhibits dual regulation, being induced in stationary phase as well as by osmotic upshift in growing cells. In both cases induction requires *rpoS* (Hengge-Aronis *et al.*, 1993).

The sequence of *csi*-5 (N. Henneberg and R. Hengge-Aronis, unpublished results) turned out to be identical to that of *osmY*, recently described as a hyperosmotically inducible gene (Yim and Villarejo, 1992). The gene product is a periplasmic protein of 21.090 kDa and 18.150 kDa for the precursor and the mature protein, respectively. An internal region of Csi-5 (OsmY) exhibits homology to the N-terminal domain of HU-like proteins which is involved in the formation of dimers (Tanaka *et al.*, 1984). This homology includes identical (34%) and conserved amino acids as well as the potential to form similar secondary structures. Since HU-β and IHF-α have only 21% identical amino acids in this domain, but nevertheless are thought to form similar 3-D structures and dimers (Bonnefoy and Rouviere-Yaniv, 1991), it seems likely that also the *csi*-5 (*osmY*) gene product is active as a dimer.

The sequence of the gene did not provide any clue to its function. Survival of long-term starvation and resistance to many different stress conditions are not affected in a *csi*-5::*lacZ* mutant (N. Henneberg and R. Hengge-Aronis, unpublished results). The only phenotype of an *osmY*::Tn*phoA* mutation is a slightly increased sensitivity to hyperosmotic stress when the cells are grown on a complex medium in the presence of NaCl (Yim and Villarejo, 1992). Together with the periplasmic location of the gene product, this finding could indicate a role in the uptake of some stress protective compound present in complex medium.

4.6. The Lipoprotein Encoded by osmB

The *osmB* gene was identified as one of a series of osmotically inducible chromosomal Tn*phoA* fusions (Gutierrez *et al.*, 1987). Its gene product is a small lipoprotein (4.23 kDa after processing) which is localized in the outer membrane (Jung *et al.*, 1989).

Besides being osmotically regulated, *osmB* exhibits even stronger induction in stationary phase (Jung *et al.*, 1990) which is abolished in an *rpoS* mutant (Hengge-Aronis *et al.*, 1991). The same promoter (*osmB*p2) is used for osmotic and for growth phase-dependent control (Jung *et al.*, 1990). Flanking the putative -35 region of *osmB*p2 an inverted repeat is found. A deletion of the upstream 6 bp of this repeat results in a loss of osmotic regulation but does not affect stationary-phase induction (Jung *et al.*, 1990). Although this implicates this inverted repeat in osmoregulation, it is a unique structure not found in other osmotically regulated genes.

The function of the OsmB protein has remained unclear although there is evidence that it is involved in cell surface alterations which take place in stationary phase. Whereas nongrowing wild-type cells form aggregates, such a clustering cannot be observed in mutants deficient in *osmB* (Jung *et al.*, 1990) or in *rpoS* (R. Hengge-Aronis, unpublished results). Despite the osmotic induction of *osmB*, which might indicate an osmoprotective function, an *osmB* mutant even grows better in medium of increased osmolarity than the isogenic $osmB^+$ strain (Jung *et al.*, 1989).

4.7. appY: A Secondary Regulator Involved in the Control of Acid Phosphatase (appA) and a Third Cytochrome Oxidase (cyxAB)

appA is the structural gene for a periplasmic acid phosphatase with an exotic pH optimum of 2.5 and unclear physiological function (Touati *et al.*, 1987). *appA* exhibits very complex regulation. Its expression is stimulated by entry into stationary phase, by anaerobiosis, by inorganic phosphate starvation (Touati *et al.*, 1987), or by overproduction of the AppY protein (Atlung *et al.*, 1989). Reduced *appA* expression is found in naturally occurring *rpoS* (*appR*) mutants (Touati *et al.*, 1986) or in *rpoS* null mutants (Lange and Hengge-Aronis, 1991b; Touati *et al.*, 1991). Recently it was found that *appA* is the third gene in an operon together with *cyxA* (*appC*) and *cyxB* (*appB*) which encode the subunits of one of three cytochrome oxidases in *E. coli*. The major operon promoter (*cyx*Ap) responds to all environmental conditions mentioned above and is subject to regulation by *rpoS* and *appY* (Dassa *et al.*, 1992; T. Atlung, personal communication). An internal promoter immediately upstream of *appA* appears to be constitutive (Dassa *et al.*, 1992) or only weakly regulated by growth phase and phosphate starvation (T. Atlung, personal communication).

appY was isolated by screening a plasmid-borne *E. coli* gene library for clones overproducing acid phosphatase. Its sequence revealed a helix–turn–helix motif indicative of a DNA-binding domain (Atlung *et al.*, 1989). A pleiotropic regulatory function of the AppY protein is also suggested by the finding that AppY overproduction affects the expression of several cellular proteins, and that AppY is identical with protein M5 (Atlung *et al.*, 1989), which was implicated in the control of enzymes of polysaccharide biosynthesis (Gayda *et al.*, 1979). Using *appY*::*lacZ* fusions it was recently found that also *appY* expression is reduced in *rpoS* mutants. Thus, the AppY protein seems to be a secondary regulator which is itself under the control of *rpoS*. It is therefore possible that regulation of the *cyxAB-appA* operon by *rpoS* is only indirect and involves *appY* (T. Atlung, personal communication).

4.8. mcc-Directed Synthesis of Microcin C7

Microcin C7 (MccC7) is a peptide antibiotic that inhibits protein synthesis in *E. coli* cells (Garcia-Bustos *et al.*, 1985) and is produced by *E. coli* strains carrying a plasmid (pMccC7) with the *mcc* genes (Novoa *et al.*, 1986). The *mcc* region can be divided into four subregions (α, β, τ, and δ) and *lacZ* fusions to all of them have been constructed. Fusions in α and β are strongly induced during transition into stationary phase whereas the expression of τ and δ appears to be constitutive. Growth phase regulation of α and β is abolished in naturally occurring *rpoS* (*appR*) mutants, and mutations isolated by screening for a loss of MccC7 production were mapped in *rpoS* (Diaz-Guerra *et al.*, 1989). Interestingly, microcin B17, which is another peptide antibiotic with similar growth phase regulation, requires OmpR for full expression (Connell *et al.*, 1987; Hernandez-Chico *et al.*, 1986) but is not under the control of *rpoS* (Bohannon *et al.*, 1991; Lange and Hengge-Aronis, 1991a). The production of microcins seems to be a way of competing with other microcin-sensitive strains under conditions of nutrient limitation.

4.9. Genes Involved in Pathogenesis (csgA, spv)

Certain strains of *E. coli* produce surface fibers termed curli that are involved in the binding of fibronectin and laminin and thus in the adhesion to eukaryotic tissue (Olsén *et al.*, 1989). *csgA*, the curli subunit gene, is under the control of *rpoS*, and *rpoS* mutants are unable to bind to fibronectin. Unlike other *rpoS*-dependent genes, *csgA* is best expressed in starved cells at low temperature and low osmolarity conditions, i.e., probably in an extraintestinal environment (Olsén *et al.*, 1993).

Another line of evidence indicates that *rpoS* is involved in pathogenesis. *rpoS* was shown to exist also in *Salmonella* where it controls the expression of virulence genes (*spv*). *Salmonella rpoS* mutants exhibit reduced stress survival and significantly reduced virulence in mice (Fang *et al.*, 1992). When cloned in *E. coli*, *spvB* requires *rpoS* for expression (Norel *et al.*, 1992). Taken together, these data suggest that both the physiological role and the promoter specificity of σ^S are the same in the two species.

4.10 rpoS-dependent Genes of Unknown Function (csiD, csiE, pex)

csiD and *csiE* were identified as stationary phase-inducible *lacZ* fusions. Expression of both is significantly reduced by mutations in *rpoS*. Besides this requirement for σ^S, *csiD* and *csiE* are also dependent on cAMP/CRP. So far, no functions could be assigned to these two genes (Weichart *et al.*, 1993).

On two-dimensional O'Farrell gels, a group of proteins was identified which was induced by starvation regardless of which kind of nutrient was absent (Groat *et al.*, 1986). These proteins were termed Pex proteins, and it was suggested that they played a role in the development of starvation-induced stress resistance (Matin, 1991). Several Pex proteins are not expressed in an *rpoS* mutant (McCann *et al.*, 1991). The identification of *pex* genes, however, has not been reported, and the function of the various Pex proteins is unknown.

It is possible that some of the Pex proteins correspond to the products of the *rpoS*-controlled genes described above. A recent report stated that *pexA* and *otsB* are identical, but no original data were shown (Strøm and Kaasen, 1993).

5. THE MECHANISM OF GENE REGULATION BY rpoS

5.1. rpoS Encodes a σ^{70}-Like Sigma Factor (σ^S)

Bacterial sigma factors are reversibly bound subunits of RNA polymerase that exhibit several activities which ultimately result in the initiation of transcription. These functions include (1) RNA polymerase core binding, (2) recognition of specific promoter sequences, and (3) formation of the "open" complex, i.e, DNA melting. Sigma factor binding to RNA polymerase core also reduces nonspecific DNA binding of the core enzyme (Helmann and Chamberlin, 1988).

A given bacterial species usually possesses a major "housekeeping" sigma factor for vegetative functions (e.g., σ^{70} encoded by *rpoD* in *E. coli*), but in addition produces alternative sigma factors for the transcription of specific sets of genes that have to be activated under certain environmental conditions, and which are characterized by alternative promoter sequences. Thus, *E. coli* has alternative sigma factors for the transcription of heat shock genes (Gross *et al.*, 1990), of genes activated under conditions of nitrogen limitation (Kustu *et al.*, 1989), and of genes involved in flagellar synthesis and chemotaxis (Arnosti and Chamberlin, 1989). In *B. subtilis* a functional cascade of five sigma factors is involved in spore formation (Stragier and Losick, 1990), and also in other bacterial species alternative sigma factors have been found to govern developmental processes (Apelian and Inouye, 1990; Buttner, 1989). σ^{70}-related factors share homology throughout their entire sequences, with four regions being most conserved (Helmann and Chamberlin, 1988). Among these, regions 2.4 and 4.2 have been implicated in promoter binding (Siegele *et al.*, 1989; Waldburger *et al.*, 1990).

The amino acid sequence derived from the *rpoS* gene (Mulvey and Loewen, 1989) displays strong homology to σ^{70} (Fig. 2). As in the case of *B. subtilis* genes, for which considerably less homology allowed the successful prediction of their sigma factor function (Stragier, 1991), this finding suggested that *rpoS* is the structural gene for a novel sigma factor. As described earlier in this chapter, the phenotypic analysis of *rpoS* mutants as well as the regulatory characteristics of *rpoS*-controlled genes demonstrate that the *rpoS* gene product controls genes involved in the physiological and morphological changes that take place in early stationary phase (hence the designation "σ^S"). The *rpoS* gene product indeed acts as a sigma subunit of RNA polymerase in an *in vitro* transcription system consisting of reconstituted purified components. It was found that several σ^{70}-dependent promoters were also recognized by σ^S, whereas only one promoter (*fic*) was specific for σ^S (Tanaka *et al.*, 1993). Unfortunately, none of the *rpoS*-dependent genes identified genetically was tested in these *in vitro* transcription experiments.

Based on sequence homology, σ^{70}-like sigma factors have recently been classified into three groups (Lonetto *et al.*, 1992). These are (1) the vegetative sigma factors essential for exponential growth (such as σ^{70}), (2) nonessential sigma factors with strongest homology to σ^{70}, especially in the RpoD-box (Tanaka *et al.*, 1988), in a 14mer in the 2.4 region and in a 20mer in the 4.2 region (Fig. 2), and (3) alternative sigma factors with less overall homology to σ^{70}, in which these specific regions are not conserved. σ^S belongs to the second group and is the closest relative of σ^{70} in *E. coli*.

5.2. What Is the Consensus Promoter Recognized by σ^S?

Sigma factors that recognize similar or identical -35 and -10 promoter regions, usually also display very good homology or even identity in their 2.4 and 4.2 domains. This is best exemplified by the two vegetative sigma factors

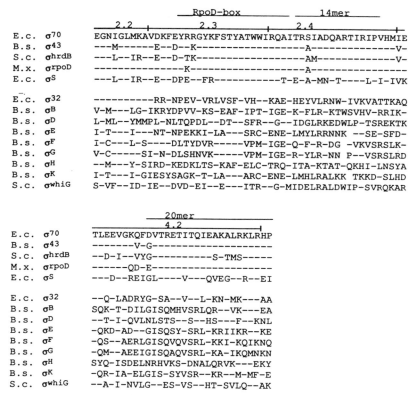

		RpoD-box	14mer
		2.2	2.3	2.4

```
E.c.  σ70     EGNIGLMKAVDKFEYRRGYKFSTYATWWIRQAITRSIADQARTIRIPVHMIE
B.s.  σ43     ---M------E--D--K-------------------A---------------V-
S.c.  σhrdB   ---L--IR--E--D-TK------------------AM--------------V-
M.x.  σrpoD   --------------K--------------------A----------------
E.c.  σS      ---L--IR--E--DPE--FR-----------T-E-A-MN-T----L-I-IVK

E.c.  σ32     ----------RR-NPEV-VRLVSF-VH--KAE-HEYVLRNW-IVKVATTKAQ
B.s.  σB      V-M---LG-IKRYDPVV-KS-EAF-IPT-IGE-K-FLR-KTWSVHV-RRIK-
B.s.  σD      L-ML--YMMPL-NLTQPDL--DT--SFR--G--IDGLRKEDWLP-TSREKTK
B.s.  σE      I-T---I---NT-NPEKKI-LA---SRC-ENE-LMYLRRNNK --SE-SFD-
B.s.  σF      I-C---L-S----DLTYDVR-----VPM-IGE-Q-F-R-DG -VKVSRSLK-
B.s.  σG      V-C-----SI-N-DLSHNVK-----VPM-IGE-R-YLR-NN P--VSRSLRD
B.s.  σH      --M---Y-SIRD-KEDKLTS-KAF-ELC-TRQ-ITA-KTAT-QKHI-LNSYA
B.s.  σK      I-T---I-GIESYSAGK-T-LA---ARC-ENE-LMHLRALKK TKKD-SLHD
S.c.  σwhiG   S-VF--ID-IE--DVD-EI--E---ITR--G-MIDELRALDWIP-SVRQKAR
```

		20mer
		4.2

```
E.c.  σ70     TLEEVGKQFDVTRETITQIEAKALRKLRHP
B.s.  σ43     -------V-G--------------------
S.c.  σhrdB   --D-I--VYG----------S-TMS-----
M.x.  σrpoD   ------QD-E--------------------
E.c.  σS      ---D--REIGL----V---QVEG--R--EI

E.c.  σ32     --Q-LADRYG-SA--V--L-KN-MK---AA
B.s.  σB      SQK-T-DILGISQMHVSRLQR--VK---EA
B.s.  σD      --T-I-QVLNLSTS--S--HS---F--KNL
B.s.  σE      -QKD-AD--GISQSY-SRL-KRIIKR--KE
B.s.  σF      -QS--AERLGISQVQVSRL-KKI-KQIKNQ
B.s.  σG      -QM--AEEIGISQAQVSRL-KA-IKQMNKN
B.s.  σH      SYQ-ISDELNRHVKS-DNALQRVK---EKY
B.s.  σK      -QR-IA-ELGIS-SYVSR--KR--M-MF-E
S.c.  σwhiG   --A-I-NVLG--ES-VS--HT-SVLQ--AK
```

Figure 2. Homology of various vegetative and alternative sigma factors to *E. coli* σ^{70}. Only the highly conserved regions involved in promoter recognition are shown. Amino acids different from the σ^{70} sequence are indicated; identical amino acids are represented by a dash. The assignment of specific regions is according to Lonetto *et al.* (1992). E.c., *Escherichia coli*; B.S., *Bacillus subtilis*; S.c., *Streptomyces coelicolor*; M.x., *Myxococcus xanthus*. Sequences are taken from the following references: σ^{E}, σ^{F}, σ^{G}, σ^{K} (Stragier *et al.*, 1989); σ^{S} (Mulvey and Loewen, 1989); σ^{hrdB} (Buttner, 1989); σ^{whiG} (Chater *et al.*, 1989); other sequences (Helmann and Chamberlin, 1988).

σ^{70} (*E. coli*) and σ^{43} (*B. subtilis*) (Helmann and Chamberlin, 1988). From this it might be inferred that σ^{S} as a closely σ^{70}-related sigma factor recognizes sequence determinants similar to the σ^{70} consensus (−35: TTGACA, −10: TATAAT). In fact, well-studied σ^{70} promoters such as *lacUV*p and *trp*p were recognized by σ^{S} *in vitro*. However, other σ^{70}-dependent promoters (mostly stringently controlled) were not transcribed by a σ^{S}-containing RNA polymerase holoenzyme. A comparison of these promoters does not reveal any sequence specific for σ^{S} recognition (Tanaka *et al.*, 1993).

Recently, promoter sequences of various *rpoS*-controlled genes have become available. Regulation by σ^S of these promoters might be direct or indirect, i.e., involve secondary regulators. These promoters include *katE*p, *xthA*p, *bolA*p1, *glgS*p2, *cyxA*p (*appC*p), *osmB*p2, *treA*p and *csi*-5p (*osm*Yp) (Fig. 3). Concerning homology in their −35 and −10 regions, these promoters fall in at least two classes (Fig. 3). Class I promoters (*katE*p, *xthA*p and *bolA*p1) strongly deviate from the σ^{70} consensus both in their −35 and −10 regions. Class II promoters exhibit homology to the −10(σ^{70}) consensus, but are very heterogenous in their −35 regions. Within this class, the promoters for which direct recognition by σ^S has been shown *in vitro* have an increased occurrence

```
                      -35                                    -10          +1
                  A---G--GTTAAGC                           CGTCCAG


katEp    AACTGTAGTTTAGC  CGATTTAGCCCCTG TA  CGTCCCG CTTT   G
xthAp    AACAGGCGGTAAGC  AACGCGAAATTCTGCTAC CATCCAC GC     A
bolAp1   ATTTGTTGTTAAGC TGCAATGGAAACGGTAAAAG CGGCGAG TATTT A
                                             ****  **  **


                      -35                                    -10          +1
                       ?                                   TATACT


cyxAp    TCGTAGAGTTTCA  GGATAAAGAGGGAGATCTAC CATTAT CGGGTT A
osmBp2   CGGAATAATTTCA  CCAGACTTATTCTTAGC    TATTAT AGTTAT A
glgSp2   ATATATTTACGCA  CGTTATGTTTAAAGGCAC   TACACT GATTGGGA
treAp    TTTCGATCATGCA  GCTAGTGCGATCCTGAAC   TAAGGT TTT    C
csi-5p   TGATATCCCGAGC  GGTTTCAAAATTGTGATC   TATATT TAA    C
ficp*      GCTCTCCGGC   GTAACCCGATTTGCCGCT   TATACT TGTGGCA
dnaQp2*    AAATTTCTACC  TGTTTAAGCATCTCTGG    TAGACT TCCTGTA
lacUV5p*   AGGCTTTACA   CTTTATGCTTCCGGCTCG   TATAAT GTGTGGA
rnaIp*     GTTCTTGAAG   TAGTGGCCCGACTACGGC   TACACT AGAAGGA
trpp*      AGCTGTTGACA  ATTAATCATCGAACTAG    TTAACT AGTACGC
```

Figure 3. Sequence homology in the −35 and −10 regions of *rpoS*-regulated genes. The potential consensus sequences are shown at the top of the promoter sequences that exhibit homology to each other. Nucleotides identical to the consensus are underlined. Asterisks indicate the "gearbox" motif in *bolA*p1. mRNA start sites are at +1. Promoters labeled by an asterisk are recognized by σ^S-containing RNA polymerase in vitro. Sequences are taken from the following sources: *katE* (von Ossowski *et al.*, 1991), *xthA* (Saporito *et al.*, 1988), *bolA* (Aldea *et al.*, 1989), *glgS* (Hengge-Aronis and Fischer, 1992), *cyxA* (identical to *appC*) (Dassa *et al.*, 1992), *osmB* (Jung *et al.*, 1990), *treA* (Repoila and Gutierrez, 1991), *csi*-5 (Barth and Hengge-Aronis, unpublished results). All other promoter sequences are as in Tanaka *et al.* (1993).

of C instead of A in the fifth position of the -10 region (Fig. 3). From the σ^S sequence as well as from the *in vitro* data, it seems highly unlikely that σ^S recognizes class I promoters directly. Since homology among these promoters concentrates in the -35 and -10 regions, one might speculate that an unknown secondary sigma factor is involved in the transcription of these promoters and that the structural gene of this unknown sigma factor is under *rpoS* control.

Some growth phase-regulated promoters, such as *bolA*p1 and *mcbA*p, contain the so-called "gearbox" sequence (Fig. 3), which has been implicated in stationary phase induction (Aldea *et al.*, 1989, 1990) as a potential recognition site for σ^S (Vicente *et al.*, 1991). However, the *mcbA* gearbox promoter (which directs the expression of the microcin B17 genes) is not under the control of σ^S (Bohannon *et al.*, 1991; Lange and Hengge-Aronis, 1991a), and other σ^S-regulated promoters do not contain gearboxes (Fig. 3). Thus, the gearbox cannot be the σ^S consensus sequence and the term "gearbox promoters" for stationary phase-inducible promoters is misleading. A potential gearbox binding protein and its role in regulation have yet to be identified.

5.3. Regulatory Subfamilies of rpoS-Controlled Genes

As an increasing number of *rpoS*-dependent genes have been identified, it has become apparent that there are subsets of genes whose expression responds to specific conditions in addition to being σ^S-controlled. This finding possibly points to the existence of specific regulatory proteins that act together with σ^S to control transcription.

otsBA, *treA*, and *osmB* were first identified as osmotically regulated genes (Boos *et al.*, 1987; Giaever *et al.*, 1988; Gutierrez *et al.*, 1987). Therefore, also other *rpoS*-dependent genes and phenotypes were systematically tested for a potential induction after osmotic medium upshift (Hengge-Aronis *et al.*, 1993). It was found that *rpoS*-dependent thermotolerance as well as H_2O_2 resistance could be induced by an increase in osmolarity. Also, *lacZ* fusions to *bolA*, *csi*-5 (*osmY*), and *glgS* were induced after the addition of 0.3 M NaCl to the growth medium. However, the extent of stationary-phase and osmotic induction did not correlate for a given gene. For instance, *glgS* shows at least 30- to 50-fold stationary-phase induction, but only a 2-fold osmotic induction (just as *rpoS* itself), whereas for the *ots* genes or *csi*-5 (*osmY*) osmotic and growth phase-dependent induction were approximately equal. On two-dimensional gels at least 18 proteins could be detected that exhibited *rpoS*-dependent osmotic induction. However, not all *rpoS*-dependent proteins previously identified are also osmotically induced (Hengge-Aronis *et al.*, 1993), and at least one gene (*csgA*) requires low osmolarity conditions for σ^S-dependent induction (Olsén *et al.*, 1993). The transcription of *rpoS* was only stimulated twofold by osmotic upshift, and this increase was too late to account for the rapid osmoinduction of *otsBA*, *csi*-5 and other *rpoS*-dependent genes (Hengge-Aronis *et al.*, 1993). Moreover, osmotic induction of *csi*-5 could also be observed in an *rpoS*-deficient strain, although at strongly reduced absolute levels (Hengge-Aronis *et al.*, manuscript in preparation). Taken together, these data indicate that only a subset of *rpoS*-dependent genes is hyperosmotically inducible and that σ^S alone is not sufficient for this regulation. In the case of the *osmB*p2 promoter, it was shown that a region which is part of an inverted repeat flanking the -35 region is required for osmotic but not for growth phase control (Jung *et al.*, 1990). This could indicate the requirement of some osmotic regulator protein in addition to σ^S.

The *cyxAB-appA* operon is induced in stationary phase as well as by a shift to anaerobic conditions. Also, the latter type of control requires *rpoS* (Dassa *et al.*, 1992). This operon could be a member of a subset of *rpoS*-regulated genes which is also controlled by the availability of oxygen.

Additional global regulators have been found to participate in the regulation of *rpoS*-dependent genes. *glgS*, *csiD*, *csiE* and *mcc* are under positive control of both σ^S and cAMP/CRP (Hengge-Aronis and Fischer, 1992; Marschall and Hengge-Aronis, unpublished results; Moreno *et al.*, 1992; Weichart *et al.*, 1993). *csi*-5(*osmY*) is negatively controlled by Lrp, a global regulator of not yet fully defined function (Hengge-Aronis *et al.*, manuscript in preparation). *csgA* and *mcc* belong to a group of σ^S-controlled genes that are also subject to repression by the histone-like protein H-NS (Olsén *et al.*, 1993; Moreno *et al.*, 1992).

These cases of a differential control within the *rpoS* regulon indicate that the role of σ^S in gene regulation is more complex than that of several other alternative σ^{70}-like sigma factors. For instance, the cellular concentration of σ^{32} alone appears to determine the level of expression of the heat shock genes. σ^S, on the other hand, seems to function more like σ^{70}, such that transcription initiation by σ^S is often also controlled by additional regulatory proteins. σ^S may be a second vegetative sigma subunit of RNA polymerase, possibly the predominant one in stationary phase cells, with a promoter specificity very similar to that of σ^{70}. As a consequence, σ^S would be able to recognize theoretically (and *in vitro*) many ordinary σ^{70}-dependent promoters, although *in vivo* these promoters may be inactivated during transition into stationary phase due to the activity of other global regulators (e.g., H-NS). Why a second vegetative sigma factor is necessary in stationary phase, remains open to speculation.

5.4. Multiple Promoters in the Control Region of rpoS-Dependent Genes

Most *rpoS*-controlled genes, for which transcripts have been mapped, are not transcribed from a single directly or indirectly σ^S-dependent promoter alone, but also possess additional promoters. In the cases of *bol*Ap2 (Aldea *et al.*, 1989), *app*Ap (Dassa *et al.*, 1992), and *osmB*p1 (Jung *et al.*, 1990), these are apparently weak promoters, whereas *glgS*p1 requires cAMP/CAP and exhibits similar activity as the σ^S-controlled *glgS*p2 during transition into stationary phase in rich medium (Hengge-Aronis and Fischer, 1992). Most likely these promoters are recognized by σ^{70}. Therefore, expression of these genes is not absolutely restricted to conditions where the cells produce active σ^S, and they could belong to other stimulons or regulons besides being members of the *rpoS* regulon. An interesting exception is provided by *katE* and *xthA*, which both appear to be transcribed from single *rpoS* dependent promoters (Saporito *et al.*, 1988; von Ossowski *et al.*, 1991). However, enzymes with similar activities as the *katE* and the *xthA* gene products exist that are induced under other environmental conditions (e.g., oxidative stress). This functional redundancy may render catalase HPII and exonuclease III dispensable under conditions where there is no or little active σ^S in the cell.

6. PERSPECTIVES

Although much has been learned about the physiology of stationary-phase *E. coli* cells, the investigation of the molecular details has merely begun. *rpoS* is probably just the first of a series or of a network of regulatory genes that play a role in stationary-phase gene expression.

Many directions for future research can be pointed out. It is clear that the majority of *rpoS*-controlled genes and their cellular functions have yet to be identified. The interaction of σ^S with promoter sequences will be another important topic, as well as the identification of additional regulatory proteins that positively or negatively act at σ^S-recognized promoters. Since σ^S is active in early stationary phase, it might be an early regulator in a temporal regulatory cascade or network. Three secondary regulators, BolA, AppY, and Dps which are themselves controlled by σ^S, have already been identified, and at least *dps* and *appY* have pleiotropic regulatory functions (Almirón *et al.*, 1992; Atlung *et al.*, 1989). Also, there is first albeit rather vague evidence for a secondary sigma factor (see Section 5.2). Furthermore, "cross-regulation" models that involve σ^S as well as σ^{70} (or other sigma factors) are readily conceivable. For instance, σ^S might be required for the expression of regulatory proteins that act on σ^{70}-dependent promoters (or vice versa). There are also genes that do not need *rpoS* for stationary-phase induction. *rpoS* itself, which is not autoregulated, belongs to this class of genes. Therefore, there must be other growth phase-associated regulatory mechanisms that act independently of *rpoS*. Finally, the initial intracellular signal that triggers the stationary-phase response and the mechanism of transduction of this signal are totally unknown.

Taken together, stationary phase in *E. coli* is still a largely unexplored but fascinating field. Even our present limited knowledge, however, indicates that also in *E. coli*, growth phase-associated changes in cellular physiology are much more complex than previously assumed and might be the equivalent of differentiation processes observed in other bacterial species.

ACKNOWLEDGMENTS. I thank Winfried Boos, in whose laboratory our work was done, for his continuous interest and many helpful discussion. I am indebted to my students Roland Lange, Dieter Weichart, Nicola Henneberg, Jürgen Böhringer, Christoph Marschall, Mechthild Barth and Margit Kreimer, who did many of the experiments mentioned in this review, as well as to Daniela Fischer for expert technical assistance and for all her help in general. I am also grateful to several colleagues who have communicated results prior to publication.

The work from this laboratory described herein was supported by the Deutsche Forschungsgemeinschaft (SFB 156).

REFERENCES

Aldea, M., Hernández-Chico, C., de la Campa, A. G., Kushner, S. R., and Vicente, M., 1988, Identification, cloning and expression of *bolA*, an *ftsZ*-dependent morphogene of *Escherichia coli*, *J. Bacteriol.* **170:**5169–5176.

Aldea, M., Garrido, T., Hernández-Chico, C., Vicente, M., and Kushner, S. R., 1989, Induction of a growth-phase-dependent promoter triggers transcription of *bolA*, an *Escherichia coli* morphogene, *EMBO J.* **8:**3923–3931.

Aldea, M., Garrido, T., Pla J., and Vicente, M., 1990, Division genes in *Escherichia coli* are expressed coordinately to cell septum requirements by gearbox promoters, *EMBO J.* **9:**3787–3794.

Almirón, M. Link, A., Furlong, D., and Kolter, R., 1992, A novel DNA binding protein with regulatory and protective roles in starved *Escherichia coli*, *Genes Dev.* **6:**2646–2654.

Apelian, D., and Inouye, S., 1990, Development-specific σ-factor essential for late-state differentiation of *Myxococcus xanthus*, *Genes Dev.* **4:**1396–1403.

Arnosti, D. N., and Chamberlin, M. J., 1989, Secondary σ factor controls transcription of flagellar and chemotaxis genes in *Escherichia coli*, *Proc. Natl. Acad. Sci. USA* **86:**830–834.

Atlung, T., Nielsen, A., and Hansen, F. G., 1989, Isolation, characterization, and nucleotide sequence of *appY*, a regulatory gene for growth phase-dependent gene expression in *Escherichia coli*, *J. Bacteriol.* **171:**1683–1691.

Back, J. F., Oakenfull, D., and Smith, M. B., 1979, Increased thermal stability of proteins in the presence of sugars and polyols, *Biochemistry* **18:**5191–5196.

Balke, V. L., and Gralla, J. D., 1987, Changes in the linking number of supercoiled DNA accompany growth transitions in *Escherichia coli*, *J. Bacteriol.* **169:**4499–4506.

Begg, K. J., Takasuga, A., Edwards, D. H., Dewar, S. J., Spratt, B. G., Adachi, H, Ohta, T., Matsuzawa, H., and Donachie, W. D., 1990, The balance between different peptidoglycan precursors determines whether *Escherichia coli* cells will elongate or divide, *J. Bacteriol.* **172:**6697–6703.

Bohannon, D. E., Connell, N., Keener, J., Tormo, A., Espinosa-Urgel, M., Zambrano, M. M., and Kolter, R., 1991, Stationary-phase-inducible "gearbox" promoters: Differential effects of *katF* mutations and the role of σ^{70}, *J. Bacteriol.* **173**:4482–4492.

Bonnefoy, E., and Rouviere-Yaniv, J., 1991, HU and IHF, two homologous histone-like proteins of *Escherichia coli*, form different protein–DNA complexes with short DNA fragments, *EMBO J.* **10**:687–696.

Boos, W., Ehmann, U., Bremer, E., Middendorf, A., and Postma, P., 1987, Trehalase of *Escherichia coli*, *J. Biol. Chem.* **262**:13212–13218.

Boos, W., Ehmann, U., Forkl, H., Klein, W., Rimmele, M., and Postma, P., 1990, Trehalose transport and metabolism in *Escherichia coli*, *J. Bacteriol.* **172**:3450–3461.

Botsford, J. L., and Harman, J. G., 1992, Cyclic AMP in prokaryotes, *Microbiol. Rev.* **56**:100–122.

Buchanan, C. E., and Sowell, M. O., 1982, Synthesis of penicillin-binding protein 6 by stationary phase *Escherichia coli*, *J. Bacteriol.* **151**:491–494.

Buttner, M. J., 1989, RNA polymerase heterogeneity in *Streptomyces coelicolor* A3(2), *Mol. Microbiol.* **3**:1653–1659.

Calder, P. C., 1991, Glycogen structure and biogenesis, *Int. J. Biochem.* **23**:1335–1352.

Chater, K. F., Bruton, C. J. Plaskitt, K. A., Buttner, J. J., and Mendez, C., 1989, The developmental fate of *S. coelicolor* hyphae depends upon a gene product homologous with the motility σ factor of *B. subtilis*, *Cell* **59**:133–143.

Connell, N., Han, Z., Moreno, F., and Kolter, R., 1987, An *E. coli* promoter induced by the cessation of growth, *Mol. Microbiol.* **1**:195–201.

Crowe, J. H., Crowe, L. M., and Chapman, D., 1984, Preservation of membranes in anhydrobiotic organisms: The role of trehalose, *Science* **223**:701–703.

Crowe, J. H., Crowe, L. M., Carpenter, J. F., Rudolph, A. S., Aurell Wistrom, C., Spargo, B. J., and Anchordoguy, T. J., 1988, Interactions of sugars with membranes, *Biochim. Biophys. Acta Biomembr. Rev.* **947**:367–384.

Csonka, L. N., 1989, Physiological and genetic responses of bacteria to osmotic stress, *Microbiol. Rev.* **53**:121–147.

Dassa, J., Fsihi, H., Marck, C., Dion, M., Kieffer-Bontempts, M., and Boquet, P. L., 1992, A new oxygen-regulated operon in *Escherichia coli* comprises the genes for a putative third cytochrome oxidase and for pH 2.5 acid phosphatase (*appA*), *Mol. Gen. Genet.* **229**:342–352.

Davis, B. D., Luger, S. M., and Tai, P. C., 1986, Role of ribosome degradation in the death of starved *Escherichia coli* cells, *J. Bacteriol.* **166**:439–445.

Demple, B., and Amabile-Cuevas, C. F., 1991, Redox redux: The control of oxidative stress responses, *Cell* **67**:837–839.

Demple, B., Halbrook, J., and Linn, S., 1983, *Escherichia coli xth* mutants are hypersensitive to hydrogen peroxide, *J. Bacteriol.* **153**:1079–1082.

Dersch, P., Schmidt, K., and Bremer, E., 1993, Synthesis of the *Escherichia coli* K-12 nucleoid-associated DNA-binding protein H-NS is subjected to growth phase control and autoregulation, *Mol. Microbiol.*, in press.

Dewar, S. J., Kagan-Zur, V., Begg, K. J., and Donachie, W. D., 1989, Transcriptional regulation of cell division genes in *Escherichia coli*, *Mol. Microbiol.* **3**:1371–1377.

Diaz-Guerra, L., Moreno, F., and SanMillan, J. L., 1989, *appR* gene product activates transcription of microcin C7 plasmid genes, *J. Bacteriol.* **171**:2906–2908.

Eisenstark, A., 1989, Bacterial genes involved in response to near-ultraviolet radiation, *Adv. Genet.* **26**:99–147.

Erickson, J. W., and Gross, C. A., 1989, Identification of the σ^E subunit of *Escherichia coli* RNA polymerase: A second alternate σ factor involved in high-temperature gene expression, *Genes Dev.* **3**:1462–1471.

Fang, R. C., Libby, S. J., Buchmeier, N. A., Loewen, P. C., Switala, J., Harwood, J., and Guiney,

D. G., 1992, The alternative σ factor KatF (RpoS) regulates *Salmonella* virulence, *Proc. Natl. Acad. Sci. USA* **89:**11978–11982.

Farr, S. B., and Kogoma, T., 1991, Oxidative stress responses in *Escherichia coli* and *Salmonella typhimurium*, *Microbiol. Rev.* **55:**561–585.

Farr, S. B., Natvig, D. O., and Kogoma, T., 1985, Toxicity and mutagenicity of plumbagin and the induction of a possible new DNA repair pathway in *Escherichia coli*, *J. Bacteriol.* **164:**1309–1316.

Garcia-Bustos, J. F., Pezzi, N., and Mendez, E., 1985, Structure and mode of action of microcin C7, an antibacterial peptide produced by *Escherichia coli*, *Antimicrob. Agents Chemother.* **27:**791–797.

Gayda, R. C., Avni, H., Berg, P. E., and Markovitz, A., 1979, Outer membrane protein a and other polypeptides regulate capsular polysaccharide synthesis in *E. coli* K-12, *Mol. Gen. Genet.* **175:**325–332.

Giaever, H. M., Styrvold, O. B., Kaasen, I., and Strøm, A. R., 1988, Biochemical and genetic characterization of osmoregulatory trehalose synthesis in *Escherichia coli*, *J. Bacteriol.* **170:**2841–2849.

Groat, R. G., Schultz, J. E., Zychlinski, E., Bockman, A. T., and Matin, A., 1986, Starvation proteins in *Escherichia coli*: Kinetics of synthesis and role in starvation survival, *J. Bacteriol.* **168:**486–493.

Gross, C. A., Straus, D. B., Erickson, J. W., and Yura, T., 1990, The function and regulation of heat shock proteins in *Escherichia coli*, in: *Stress Proteins in Biology and Medicine* (R. I. Morimoto, A. Tissières, and C. Georgopoulos, eds.), Cold Spring Harbor Laboratory Press, Cold Spring Harbor, N.Y., pp. 167–189.

Gutierrez, C., Barondess, J., Manoil, C., and Beckwith, J., 1987, The use of transposon Tn*phoA* to detect genes for cell envelope proteins subject to a common regulatory stimulus, *J. Mol. Biol.* **195:**289–297.

Gutierrez, C., Ardourel, M., Bremer, E., Middendorf, A., Boos, W., and Ehmann, U., 1989, Analysis and DNA sequence of the osmoregulated *treA* gene encoding the periplasmic trehalase of *Escherichia coli* K12, *Mol. Gen. Genet.* **217:**347–354.

Heimberger, A., and Eisenstark, A., 1988, Compartmentalization of catalases in *Escherichia coli*, *Biochem. Biophys. Res. Commun.* **154:**392–397.

Helmann, J. D., and Chamberlin, M. J., 1988, Structure and function of bacterial sigma factors, *Annu. Rev. Biochem.* **57:**839–872.

Hengge-Aronis, R., 1993, Survival or hunger and stress: the role of *rpoS* in stationary phase gene regulation in *Escherichia coli*, *Cell* **72:**165–168.

Hengge-Aronis, R., and Fischer, D., 1992, Identification and molecular analysis of *glgS*, a novel growth phase-regulated and *rpoS*-dependent gene involved in glycogen synthesis in *Escherichia coli*, *Mol. Microbiol.* **6:**1877–1886.

Hengge-Aronis, R., Klein, W., Lange, R., Rimmele, M., and Boos, W., 1991, Trehalose synthesis genes are controlled by the putative sigma factor encoded by *rpoS* and are involved in stationary phase thermotolerance in *Escherichia coli*, *J. Bacteriol.* **173:**7918–7924.

Hengge-Aronis, R., Lange, R., Henneberg, N., and Fischer, D., 1993, Osmotic regulation of *rpoS*-dependent genes in *Escherichia coli*, *J. Bacteriol.* **175:**259–265.

Hernandez-Chico, C., SanMillan, J. L., Kolter, R., and Moreno, F., 1986, Growth phase and OmpR regulation of transcription of microcin B17 genes, *J. Bacteriol.* **167:**1058–1065.

Hurst, A., and Hughes, A., 1978, Stability of Staphylococcus aureus S6 sublethally heated in different buffers, *J. Bacteriol.* **133:**564–468.

Ivanov, A. Y., and Fomchenkov, V. M., 1989, Dependence of surfactant damage to *Escherichia coli* cells on culture growth phase, *Microbiology (USSR)* **58:**785–791.

Jenkins, D. E., Schultz, J. E., and Matin, A., 1988, Starvation-induced cross-protection against heat or H_2O_2 challenge in *Escherichia coli*, *J. Bacteriol.* **170:**3910–3914.

Jenkins, D. E., Chaisson, S. A., and Matin, A., 1990, Starvation-induced cross-protection against osmotic challenge in *Escherichia coli*, *J. Bacteriol.* **172:**2779–2781.

Jenkins, D. E., Auger, E. A., and Matin, A., 1991, Role of RpoH, a heat shock regulator protein, in *Escherichia coli* carbon starvation protein synthesis and survival, *J. Bacteriol.* **173:**1992–1996.

Jung, J. U., Gutierrez, C., and Villarejo, M. R., 1989, Sequence of an osmotically inducible lipoprotein gene, *J. Bacteriol.* **171:**511–520.

Jung, J. U., Gutierrez, C., Martin, F., Ardourel, M., and Villarejo, M., 1990, Transcription of *osmB*, a gene encoding an *Escherichia coli* lipoprotein, is regulated by dual signals, *J. Biol. Chem.* **265:**10574–10581.

Kaasen, I., Falkenberg, P., Styrvold, O. B., and Strøm, A. R., 1992, Molecular cloning and physical mapping of the *otsBA* genes, which encode the osmoregulatory trehalose pathway of *Escherichia coli*: Evidence that transcription is activated by KatF(AppR), *J. Bacteriol.* **174:**889–898.

Klein, W., Ehmann, U., and Boos, W., 1991, The repression of trehalose transport and metabolism in *Escherichia coli* by high osmolarity is mediated by trehalose-6-phosphate phosphatase, *Res. Microbiol.* **142:**359–371.

Kohara, Y., Akiyama, K., and Isono, K., 1987, The physical map of the whole *Escherichia coli* chromosome: Application of a new strategy for rapid analysis and sorting of a large genomic library, *Cell* **50:**495–508.

Kubitschek, H. E., 1990, Cell volume increase in *Escherichia coli* after shifts to richer media, *J. Bacteriol.* **172:**94–101.

Kustu, S., Santero, E., Keener, J., Popham, D., and Weiss, D., 1989, Expression of sigma 54 (*ntrA*)-dependent genes is probably united by a common mechanism, *Microbiol. Rev.* **53:** 367–376.

Lange, R., and Hengge-Aronis, R., 1991a, Growth phase-regulated expression of *bolA* and morphology of stationary phase *Escherichia coli* cells is controlled by the novel sigma factor σ^S (*rpoS*), *J. Bacteriol.* **173:**4474–4481.

Lange, R., and Hengge-Aronis, R., 1991b, Identification of a central regulator of stationary-phase gene expression in *Escherichia coli*, *Mol. Microbiol.* **5:**49–59.

Lee, C. W. B., Waugh, J. S., and Griffin, R. G., 1986, Solid-state NMR study of trehalose/1,2-dipalmitoyl-*sn*-phosphatidylcholine interactions, *Biochemistry* **25:**3737–3742.

Lindahl, T., Sedgwick, B., Sekiguchi, M., and Nakabeppu, Y., 1988, Regulation and expression of the adaptive response to alkylating agents, *Annu. Rev. Biochem.* **57:**133–157.

Loewen, P. C., and Triggs, B. L., 1984, Genetic mapping of *katF*, a locus that with *katE* affects the synthesis of a second catalase species in *Escherichia coli*, *J. Bacteriol.* **160:**688–675.

Loewen, P. C., Switala, J., and Triggs-Raine, B. L., 1985, Catalase HPI and HPII in *Escherichia coli* are induced independently, *Arch. Biochem. Biophys.* **243:**144–149.

Lonetto, M., Gribskov, M., and Gross, C. A., 1992, The σ^{70} family: Sequence conservation and evolutionary relationships, *J. Bacteriol.* **174:**3843–3849.

Ma, M., and Eaton, J. W., 1992, Multicellular oxidant defense in unicellular organisms, *Proc. Natl. Acad. Sci. USA* **89:**7924–7928.

Maaloe, O., and Kjeldgaard, N. O., 1966, *Control of Macromolecular Synthesis*, Benjamin, New York.

McCann, M. P., Kidwell, J. P., and Matin, A., 1991, The putative σ factor KatF has a central role in development of starvation-mediated general resistance in *Escherichia coli*, *J. Bacteriol.* **173:**4188–4194.

Magasanik, B., and Neidhardt, F. C., 1987, Regulation of carbon and nitrogen utilization, in: *Escherichia coli and Salmonella typhimurium* (F. C. Neidhardt, ed.), American Society for Microbiology, Washington, D.C., pp. 1318–1325.

Matin, A., 1991, The molecular basis of carbon-starvation-induced general resistance in *Escherichia coli*, *Mol. Microbiol.* **5:**3–10.

May, G., Dersch, P., Haardt, M., Middendorf, A., and Bremer, E., 1990, The *osmZ* (*bglY*) gene encodes the DNA-binding protein H-NS (H1a), a component of the *Escherichia coli* K12 nucleoid, *Mol. Gen. Genet.* **224:**81–90.

Moreno, F., San Millán, J. L., del Castillo, I., Gómez, J. M., Rodriguez-Sáinz, M. C., González-Pastor, J. E., and Diaz-Guerra, L., 1992, *Escherichia coli* genes regulating the production of microcins MccB17 and MccC7, in: *Bacteriocins, Microcins and Lantibiotics* (R. James, C. Lazdunski, and F. Pattus, eds.), Springer-Verlag, Berlin Heidelberg, pp. 3–13.

Morita, R. Y., 1988, Bioavailability of energy and its relationship to growth and starvation survival in nature, *Can. J. Microbiol.* **34:**436–441.

Mulvey, M. R., and Loewen, P. C., 1989, Nucleotide sequence of *katF* of *Escherichia coli* suggest KatF protein is a novel σ transcription factor, *Nucleic Acids Res.* **17:**9979–9991.

Mulvey, M. R., Switala, J., Borys, A., and Loewen, P. C., 1990, Regulation of transcription of *katE* and *katF* in *Escherichia coli*, *J. Bacteriol.* **172:**6713–6720.

Norel, F., Robbe-Saule, V., Popoff, M. Y., and Coynault, C., 1992, The putative sigma factor KatF (RpoS) is required for the transcription of the *Salmonella typhimurium* virulence gene *spvB* in *Escherichia coli*, *FEMS Microbiol. Lett.* **99:**271–276.

Novoa, M. A., Diaz-Guerra, L., SanMillan, J. L., and Moreno, F., 1986, Cloning and mapping of the genetic determinants for microcin C7 production and immunity, *J. Bacteriol.* **168:**1384–1391.

Okita, T. W., Rodriguez, R. L., and Preiss, J., 1981, Biosynthesis of bacterial glycogen: Cloning of the glycogen enzyme structural genes of *Escherichia coli*, *J. Biol. Chem.* **256:**6944–6952.

Olsén, A., Arnqvist, A., Sukupolvi, S., and Normark, S., 1993, The RpoS sigma factor relieves H-NS-mediated transcriptional repression of *csgA*, the subunit gene of fibronectin binding curli in *Escherichia coli*, *Mol. Microbiol.* **7:**523–536.

Olsén, A., Jonsson, A., and Normark, S., 1989, Fibronectin binding mediated by a novel class of surface organelles on *Escherichia coli*, *Nature* **338:**652–655.

Preiss, J., and Romeo, T., 1989, Physiology, biochemistry and genetics of bacterial glycogen synthesis, *Adv. Microb. Physiol.* **30:**183–238.

Reeve, C. A., Bockman, A. T., and Matin, A., 1984, Role of protein degradation in the survival of carbon-starved *Escherichia coli* and *Salmonella typhimurium*, *J. Bacteriol.* **157:**758–763.

Repoila, F., and Gutierrez, C., 1991, Osmotic induction of the periplasmic trehalase in *Escherichia coli* K12: Characterization of the *treA* promoter, *Mol. Microbiol.* **5:**747–755.

Romeo, T., and Preiss, J., 1989, Genetic regulation of glycogen biosynthesis in *Escherichia coli*: In vitro effects of cyclic AMP and guanosine 5′-diphosphate 3′-diphosphate and analysis of in vivo transcripts, *J. Bacteriol.* **171:**2773–2782.

Roszak, D. B., and Colwell, R. R., 1987, Survival strategies of bacteria in natural environments, *Microbiol. Rev.* **51:**365–379.

Sak, B. D., Eisenstark, A, and Touati, D., 1989, Exonuclease III and the catalase hydroperoxidase II in *Escherichia coli* are both regulated by the *katF* product, *Proc. Natl. Acad. Sci. USA* **86:**3271–3275.

Sammartano, L. J., Tuveson, R. W., and Davenport, R., 1986, Control of sensitivity to inactivation by H_2O_2 and broad-spectrum near-UV radiation by the *Escherichia coli katF* locus, *J. Bacteriol.* **168:**13–21.

Saporito, S. M., Smith-White, B. J., and Cunningham, R. P., 1988, Nucleotide sequence of the *xthA* gene of *Escherichia coli K12*, *J. Bacteriol.* **170:**4542–4547.

Schellhorn, H. E., and Hassan, H. M., 1988, Transcriptional regulation of *katE* in Escherichia coli K-12, *J. Bacteriol.* **170:**4286–4292.

Schellhorn, H. E., and Stones, V. L., 1992, Regulation of *katF* and *katE* in *Escherichia coli* K-12 by weak acids, *J. Bacteriol.* **174:**4769–4776.

Siegele, D. A., Hu, J. C., Walter, W. A., and Gross, D. A., 1989, Altered promoter recognition by mutant forms of the σ^{70} subunit of *Escherichia coli* RNA polymerase, *J. Mol. Biol.* **206:**591–603.

Smythe, C., and Cohen, P., 1991, The discovery of glycogenin and the priming mechanism for glycogen biogenesis, *Eur. J. Biochem.* **200:**625–631.

Souzu, H., 1982, *Escherichia coli* B membrane stability related to cell growth phase: Measurement of temperature dependent physical state change of the membrane over a wide range, *Biochim. Biophys. Acta* **691:**161–170.

Spassky, A., Rimsky, S., Garreau, H., and Buc, H., 1984, H1a, an *E. coli* DNA-binding protein which accumulates in stationary phase, strongly compacts DNA *in vitro*, *Nucleic Acids Res.* **12:**5321–5340.

Storz, G., Tartaglia, L. A., and Ames, B. N., 1990a, Transcriptional regulator of oxidative stress-inducible genes: Direct activation by oxidation, *Science* **248:**189–194.

Storz, G., Tartaglia, L. A., Farr, S. B., and Ames, B. N., 1990b, Bacterial defenses against oxidative stress, *Trends Genet.* **6:**363–368.

Stragier, P., 1991, Dances with sigmas, *EMBO J.* **10:**3559–3566.

Stragier, P., and Losick, R., 1990, Cascades of sigma factors revisited, *Mol. Microbiol.* **4:**1801–1806.

Stragier, P., Kunkel, B., Kroos, L., and Losick, R., 1989, Chromosomal rearrangement generating a composite gene for a developmental transcription factor, *Science* **243:**507–512.

Strøm, A. R., and Kaasen, I., 1993, Trehalose metabolism in *Escherichia coli*: Stress protection and stress regulation of gene expression, *Mol. Microbiol.* **8:**205–210.

Tanaka, I., Appelt, K., Kijk, J., White, S. W., and Wilson, K. S., 1984, 3-Å resolution structure of a protein with histone-like properties in procaryotes, *Nature* **310:**376–381.

Tanaka, K., Shiina, T., and Takahashi, H., 1988, Multiple principal sigma factor homologs in eubacteria: Identification of the "*rpoD* box," *Science* **242:**1040–1042.

Tanaka, K., Takayanagi, Y., Fujita, N., Ishihama, A., and Takahashi, H., 1993, Heterogeneity of the principal sigma factor in *Escherichia coli*: the *rpoS* gene product, σ^{38}, is a second principal sigma factor of RNA polymerase in stationary phase *Escherichia coli*, *Proc. Natl. Acad. Sci. USA* **90:**3511–3515.

Tartaglia, L. A., Storz, G., and Ames, B. N., 1989, Identification and molecular analysis of *oxyR*-regulated promoters important for the bacterial adaptation to oxidative stress, *J. Mol. Biol.* **210:**709–719.

Torriani-Gorini, A., Rothman, F. G., Silver, S., Wright, A., and Yagil, E., 1987, *Phosphate Metabolism and Cellular Regulation in Microorganisms*, American Society for Microbiology, Washington, D.C.

Touati, E., Dassa, E., and Boquet, P. L., 1986, Pleiotropic mutations in *appR* reduce pH 2.5 acid phosphatase expression and restore succinate utilization in CRP-deficient strains of *Escherichia coli*, *Mol. Gen. Genet.* **202:**257–264.

Touati, E., Dassa, E., Dassa, J., and Boquet, P. L., 1987, Acid phosphatase (pH 2.5) of *Escherichia coli*: Regulatory characteristics, in: *Phosphate Metabolism and Cellular Regulation in Microorganisms* (A. Torriani-Gorini, F. G. Rothman, S. Silver, A. Wright, and E. Yagil, eds.), American Society for Microbiology, Washington, D.C., pp. 31–40.

Touati, E., Dassa, E., Dassa, J., Boquet, P. L., and Touati, D., 1991, Are *appR* and *katF* the same *Escherichia coli* gene encoding a new sigma transcription initiation factor? *Res. Microbiol.* **142:**29–36.

Tsuchido, T., Katsui, N., Takeuchi, A., Takano, M., and Shibasaki, I., 1985, Destruction of the outer membrane permeability barrier of *Escherichia coli* by heat treatment, *Appl. Environ. Microbiol.* **50:**298–303.

Tsuchido, T., Aoki, I., and Takano, M., 1989, Interaction of the fluorescent dye 1-N-phenylnaphthylamine with *Escherichia coli* cells during heat stress and recovery from heat stress, *J. Gen. Microbiol.* **135:**1941–1947.

Tuveson, R. W., 1981, The interaction of a gene (*nur*) controlling near-UV sensitivity and the *polA1* gene in strains of *E. coli* K12, *Photochem. Photobiol.* **33:**919–923.

Ullman, A., and Danchin, A., 1983, Role of cyclic AMP in bacteria, *Adv. Cyclic Nucleotide Res.* **15:**1–51.

vanBogelen, R. A., Acton, M. A., and Neidhardt, F. C., 1987, Induction of the heat shock regulon does not produce thermotolerance in *Escherichia coli*, *Genes Dev.* **1:**525–531.

van Laere, A., 1989, Trehalose, reserve and/or stress metabolite? *FEMS Microbiol. Rev.* **63:** 201–210.

Vicente, M., Kushner, S. R., Garrido, T., and Aldea, M., 1991, The role of the "gearbox" in the transcription of essential genes, *Mol. Microbiol.* **5:**2085–2091.

von Ossowski, I., Mulvey, M. R., Leco, P. A., Borys, A., and Loewen, P. C., 1991, Nucleotide sequence of *Escherichia coli katE*, which encodes catalase HPII, *J. Bacteriol.* **173:**514–520.

Waldburger, C., Gardella, T., Wong, R., and Susskind, M. M., 1990, Changes in conserved region 2 of *Escherichia coli* σ^{70} affecting promoter recognition, *J. Mol. Biol.* **215:**267–276.

Walkup, L. K. B., and Kogoma, T., 1989, *Escherichia coli* proteins inducible by oxidative stress mediated by the superoxide radical, *J. Bacteriol.* **171:**1476–1484.

Weichart, D., Lange, R., Henneberg, N., and Hengge-Aronis, R., 1993, Identification and characterization of stationary phase-inducible genes in *Escherichia coli*, *Mol. Microbiol.*, submitted for publication.

Womersley, C., Uster, P. S., Rudolph, A. S., and Crowe, J. H., 1986, Inhibition of dehydration-induced fusion between liposomal membranes by carbohydrates as measured by fluorescence energy transfer, *Cryobiology* **23:**245–255.

Wu, J., and Weiss, B., 1991, Two divergently transcribed genes, *soxR* and *soxS*, control a superoxide response regulon of *Escherichia coli*, *J. Bacteriol.* **173:**2869–2871.

Wu, J., and Weiss, B., 1992, Two-stage induction of the *soxRS* (superoxide response) regulon of *Escherichia coli*, *J. Bacteriol.* **174:**3915–3920.

Yamamori, T., and Yura, T., 1982, Genetic control of heat-shock protein synthesis and its bearing on growth and thermal resistance in *Escherichia coli* K-12, *Proc. Natl. Acad. Sci. USA* **79:** 860–864.

Yim, H. H., and Villarejo, J., 1992, *osmY*, a new hyperosmotically inducible gene, encodes a periplasmic protein in *Escherichia coli*, *J. Bacteriol.* **174:**3637–3644.

9

Starvation-Stress Response (SSR) of Salmonella typhimurium

Gene Expression and Survival during Nutrient Starvation

Michael P. Spector and John W. Foster

1. INTRODUCTION

The environments that many pathogenic bacteria encounter as they cycle from their host organism(s) to the external aquatic or terrestrial environments commonly share one general characteristic: they are frequently limiting for bacterial growth (Koch, 1971). A wide variety of environmental factors can play an important role in restricting bacterial growth, e.g., pH, osmolarity, and the availabilities of oxygen as well as essential nutrients (Tempest *et al.*, 1983; Roszak and Colwell, 1987). The availability of essential nutrients such as phosphate (P), carbon (C), or nitrogen (N) in these environments is of particular importance in limiting bacterial growth (Harder and Dijkhuizen, 1983). As a consequence, non-spore-forming bacteria, e.g., *Salmonella*, frequently undergo drastic metabolic readjustments in an effort to survive until more favorable conditions are encountered.

Michael P. Spector • Department of Biomedical Sciences, College of Allied Health, and Department of Microbiology and Immunology, College of Medicine, University of South Alabama, Mobile, Alabama 36688.   *John W. Foster* • Department of Microbiology and Immunology, College of Medicine, University of South Alabama, Mobile, Alabama 36688.

Starvation in Bacteria, edited by Staffan Kjelleberg. Plenum Press, New York, 1993.

Salmonella are common agents of gastrointestinal-based diseases in humans, e.g., gastroenteritis and typhoid fever. As many as 40,000 cases are reported each year resulting in some 500 deaths and huge financial costs; and even so this is likely to represent only a small percentage of actual salmonelloses each year. In humans, the organism is most commonly acquired following ingestion of contaminated food or water, e.g., poultry products. *Salmonella typhimurium* causes a normally self-limiting "food poisoning" or gastroenteritis in humans, remaining localized within the small intestine. However, in immunosuppressed patients, such as individuals with AIDS, the organism can invade deeper tissue, enter the reticuloendothelial system, and cause a serious systemic disease (Finlay and Falkow, 1988, 1989). The enormous medical and financial burden of salmonelloses has stimulated interest in not only determining how these organisms cause disease, but also how they survive outside their hosts while maintaining virulence potential.

Salmonella, as a consequence of its life-style, endures extended periods of nutrient deprivation in natural aquatic and terrestrial environments while retaining its pathogenic potential. Even within the host, many of the microenvironments this organism encounters may be nutrient limiting, e.g., the human gastrointestinal tract (Koch, 1971). Therefore, the ability of *Salmonella*, as well as other enteric pathogens, to survive lengthy periods of nutrient deprivation can have a significant influence on virulence, and epidemiology of salmonellosis. In fact, recent evidence from several laboratories implicates a number of environmental factors, including nutrient deprivation, osmolarity, and anaerobiosis, in modulating the expression of *Salmonella* virulence factors (Miller *et al.*, 1989; Fang *et al.*, 1991; Galán and Curtiss, 1990; Ernst *et al.*, 1990; Lee and Falkow, 1990). This implies an empirical relationship between survival in nature and survival in the host organism.

Relatively little is known about how *S. typhimurium*, or other enteric pathogens, survives prolonged nutrient starvation. Over the last several years we have addressed this problem by identifying and characterizing genes/proteins required for survival during nutrient starvation in order to define the complex genetic and physiological events involved in this process.

2. ALTERATIONS OF CELLULAR CONSTITUENTS DURING NUTRIENT STARVATION IN SALMONELLA

Druilhet and Sobek (1984) reported that a number of cell components are degraded during starvation of *Salmonella enteritidis*. Chief among these is the degradation of cellular RNA and proteins. This is in agreement with previous and subsequent studies advocating a role for peptidases and proteases in starvation-survival (reviewed in Miller, 1975, 1987; Reeve *et al.*, 1984; Lazdunski, 1989). The physiological roles and importance of these enzymes have been presented in great detail and will not be discussed here.

3. TWO-DIMENSIONAL GEL ELECTROPHORETIC ANALYSIS OF THE STARVATION-STRESS RESPONSE OF SALMONELLA

Most suboptimal environments elicit elegant molecular responses from the stressed microorganism. These responses include macromolecular reorientations involving the altered expression of specific, and overlapping, sets of stress-response genes/proteins. A well-characterized example of this phenomenon is the heat shock response where many proteins become induced following a transition from 25°C to 42°C (reviewed in Neidhardt and VanBogelen, 1987). The ultimate purpose of this and other macromolecular realignments is the survival of the organism until conditions improve. Consequently, one can gain useful insight into an organism's molecular response to an adverse environment through the comparative examinations of pre- and poststress polypeptides. We have employed two-dimensional polyacrylamide gel electrophoresis (2-D-PAGE) to study the response of *S. typhimurium* to 11 different stress conditions, including nutrient starvation. We previously published a standard 2-D-PAGE map of *S. typhimurium* polypeptides (Spector *et al.*, 1986). Figure 1 summarizes the results of many studies and provides coordinates, based on the standard map, of proteins induced during the variety of stresses tested. Only those proteins induced during the indicated stress condition are included; the many proteins whose expression decreased during these conditions are not listed although they too may prove to be important. Of the 188 total stress proteins identified, 97 were induced by some form of nutrient deprivation (phosphate, carbon, nitrogen, NAD, iron). This figure becomes extremely useful when comparing the overlap between different stress modulons. Our cumulative data reveal that 38 of the 97 starvation-inducible proteins respond to two or more stress conditions, and 16 of these respond to three or more. Furthermore, there are at least 6 starvation-inducible proteins which increase in response to three or more nutrient starvation conditions. Members of this core set of starvation-regulated proteins are of particular interest for both their regulatory and functional significance. One would predict that proteins induced in response to several starvation conditions may prove of integral importance to survival. The StiA protein (54×77; SIN-8, Spector *et al.*, 1986) supports this prediction. This protein is induced in response to multiple nutrient starvations and its absence leads to a significant decrease in survival during prolonged nutrient deprivation (Spector and Cubitt, 1992).

The results also demonstrate an additional point of interest, that no one modulon is entirely distinct from the others (Fig. 1), since there is some degree

	pH 5-7 Gel	STARVATION^a				ANI^b		Oxi^c	HEAT^d	Iron^e	ASP^f	ATR^f
	Coordinates	NA	PO₄	NH₄	G	Glu	Gly/NO₃					
IRO-1	00X52									■		
ANI-29	03X19					■						
NSI-1	03X58			■								
OXI-1	05X40							■				
SIN	05X66				■							
PSI-1	06X82		■									
IRO-2	07X67									■		
GSI-1/ANI-28	09X29				■	■						
HSP-1	09X57								■			
ATR-1	10X64											■
PSI-2	10X77		■									
IRO-23	10X95									■		
OXI-2/IRO-14	11X86							■		■		
OXI-3	12X56							■				
IRO-3	16X78									■		
IRO-5	18X67									■		
ANI-1	18X74					■						
SIN-1	19X89	■			■					■		
HSP-2	19X67								■			
HSP-3	20X67								■			
ANI	20X97						■					
SIN-27	20X108					■						
IRO-4/ATR-2	21X94									■		■
ANI-2	22X61					■						
PSI-3	22X83		■									
ANI-3/SIN	22X85					■				■		
ANI-30	23X57						■					
IRO-6	23X67									■		
NIC-1	23X78	■										
NSI-2	22X90			■								
ANI-31	24X59						■					
IRO-7	25X50									■		
NSI-3	25X92			■								
ANI-4	26X75					■						
PSI-4	26X62		■									
PSI-5	27X68		■									
GSI-2	27X101				■							
ANI-5	28X39						■					
NSI-4	29X58			■								
IRO-5	30X63									■		
ANI-6	31X96					■						
OXI-3	32X72						■	■				
ANI-32	33X19						■					
OXI-4/ATR-3	33X67						■	■				
ANI-33	34X27						■					
GSI-3	34X88				■							
SIN-2	34X27				■					■		

Figure 1. Two-dimensional gel electrophoretic analysis of the separate and overlapping expression of nutrient starvation-inducible and other stress-inducible polypeptides of *Salmonella typhimurium*. Labeling and two-dimensional-PAGE analysis of NAD-, phosphate-, carbon-, and nitrogen-starvation-inducible^a, anaerobiosis-inducible^b, aerobiosis-inducible^c, heat shock-

	pH 5-7 Gel	STARVATION[A]				AN[b]		Oxi[c]	HEAT[d]	Iron[e]	ASP[f]	ATR[f]
	Coordinates	NA	PO_4	NH_4	G	Glu	Gly/NO_3					
ANI-34	35X54						■					
ANI-7	35X79					■	■					
ANI-35	35X90					■						
ANI-8	35X96					■	■					
ANI-9	35X94					■						
ANI-10	35X105					■	■					
HSP-4	35X79								■			
HSP-5	36X71								■		■	
IRO-10	36X81									■		
IRO-11	36X88									■		
HSP-6	36X98								■			
HSP-7	39X26								■			
ATR-4	40X84											■
OXI-6	40X91							■				
NIC-2	41X64											
NIC-3	42X63											
IRO-15	42X80									■		
IRO-12	42X88									■		
SIN-3	42X100	■	■									
ANI-11	42X104					■						
PSI-6	43X67	■										
ANI-12/IRO	43X71					■				■		
OXI-7	44X57							■				
ANI-13	44X100					■	■					
ANI-14/ASP-11	44X84					■					■	
NSI-5	45X87			■								
ANI-15	45X91					■						
NSI-6	46X24			■								
OXI-8	46X63							■				
IRO-18	46X81									■		
OXI-9/ATR-5	46X85							■				■
NSI-7	46X86			■								
IRO-16/OXI10/ATR-6	46X98							■		■		■
NIC-4	47X26		■									
SIN-4/ASP-14	47X38	■									■	
IRO-20	47X62									■		
ATR-7	47X74											■
SIN-5	48X58	■	■									
PSI-7	48X64	■										
SIN-6	49X28											
SIN-7	49X45											
ANI-26	49X102					■						
GSI-4	49X103				■							
HSP-8	50X56								■			
OXI-11	50X72							■				
NSI-8	50X100			■								

inducible[d], iron-regulated[e], and acid-regulated[f] (ASP and ATR) proteins were performed as previously described (Spector *et al.*, 1986; Foster and Hall, 1990; Foster, 1991; Foster and Hall, 1992).

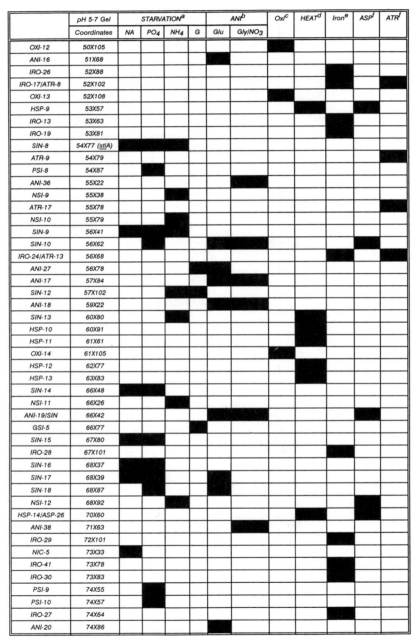

	pH 5-7 Gel	STARVATION[a]				ANI[b]		Oxi[c]	HEAT[d]	Iron[e]	ASP[f]	ATR[f]
	Coordinates	NA	PO₄	NH₄	G	Glu	Gly/NO₃					
OXI-12	50X105											
ANI-16	51X68											
IRO-26	52X88											
IRO-17/ATR-8	52X102											
OXI-13	52X108											
HSP-9	53X57											
IRO-13	53X63											
IRO-19	53X81											
SIN-8	54X77 (stiA)											
ATR-9	54X79											
PSI-8	54X87											
ANI-36	55X22											
NSI-9	55X38											
ATR-17	55X78											
NSI-10	55X79											
SIN-9	56X41											
SIN-10	56X62											
IRO-24/ATR-13	56X68											
ANI-27	56X78											
ANI-17	57X84											
SIN-12	57X102											
ANI-18	59X22											
SIN-13	60X80											
HSP-10	60X91											
HSP-11	61X61											
OXI-14	61X105											
HSP-12	62X77											
HSP-13	63X83											
SIN-14	66X48											
NSI-11	66X26											
ANI-19/SIN	66X42											
GSI-5	66X77											
SIN-15	67X80											
IRO-28	67X101											
SIN-16	68X37											
SIN-17	68X39											
SIN-18	68X87											
NSI-12	68X92											
HSP-14/ASP-26	70X60											
ANI-38	71X63											
IRO-29	72X101											
NIC-5	73X33											
IRO-41	73X78											
IRO-30	73X83											
PSI-9	74X55											
PSI-10	74X57											
IRO-27	74X64											
ANI-20	74X86											

Figure 1. (Continued)

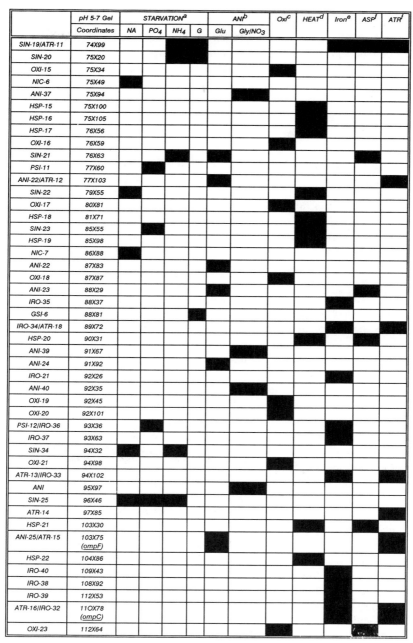

	pH 5-7 Gel	STARVATION^a				ANI^b		Oxi^c	HEAT^d	Iron^e	ASP^f	ATR^f
	Coordinates	NA	PO$_4$	NH$_4$	G	Glu	Gly/NO$_3$					
SIN-19/ATR-11	74X99											
SIN-20	75X20											
OXI-15	75X34											
NIC-6	75X49											
ANI-37	75X94											
HSP-15	75X100											
HSP-16	75X105											
HSP-17	76X56											
OXI-16	76X59											
SIN-21	76X63											
PSI-11	77X60											
ANI-22/ATR-12	77X103											
SIN-22	79X55											
OXI-17	80X81											
HSP-18	81X71											
SIN-23	85X55											
HSP-19	85X98											
NIC-7	86X88											
ANI-22	87X83											
OXI-18	87X87											
ANI-23	88X29											
IRO-35	88X37											
GSI-6	88X81											
IRO-34/ATR-18	89X72											
HSP-20	90X31											
ANI-39	91X67											
ANI-24	91X92											
IRO-21	92X26											
ANI-40	92X35											
OXI-19	92X45											
OXI-20	92X101											
PSI-12/IRO-36	93X36											
IRO-37	93X63											
SIN-34	94X32											
OXI-21	94X98											
ATR-13/IRO-33	94X102											
ANI	95X97											
SIN-25	96X46											
ATR-14	97X85											
HSP-21	103X30											
ANI-25/ATR-15	103X75 (ompF)											
HSP-22	104X86											
IRO-40	109X43											
IRO-38	108X92											
IRO-39	112X53											
ATR-16/IRO-32	110X78 (ompC)											
OXI-23	112X64											

Figure 1. (Continued)

of overlap between most of the conditions examined. This suggests a complex regulatory network of overlapping control. These 2-D-PAGE studies in concert with genetic and physiologic investigations will eventually reveal which genes/ proteins are required for *S. typhimurium* starvation-survival.

4. GENETIC ANALYSIS OF THE STARVATION-STRESS RESPONSE OF SALMONELLA

4.1. Identification of Starvation-Regulated Loci Using Mud-lac Gene Fusions

Metabolites of glucose, ammonium, and phosphate metabolism are all required for the biosynthesis of NAD(P), and the limitation of any of these nutrients has a direct effect on NAD biosynthesis in *S. typhimurium*. NAD(P) also plays a central role as a redox coenzyme in numerous catabolic and anabolic reactions of the cell. With this in mind, we reasoned that starving the cell for NAD might result in the production of numerous stress signals that would result in the induction of several genes associated with the maintenance, or restoration of balanced growth and viability. In addition, it was reasonable to assume that many of these genes would also respond to signals generated during P-, C-, or N-starvation. Using this approach, we identified seven distinct *lac* operon fusions to NAD-starvation-inducible genes. In addition to NAD starvation, these seven loci are induced by one or more of the following conditions: P-, C-, or N-starvation. These fusions were designated as *s*tarvation-*i*nducible (*sti*)−*lac* fusions to reflect the pleiotropic nature of their regulation (Foster, 1983; Spector *et al.*, 1988; Spector, 1990). Based on which starvation conditions induced them, these *sti* loci were divided into several classes (Spector *et al.*, 1988; Spector, 1990). Class I loci are NAD-, P-, C-, and N-starvation-inducible and include the *stiA* and *stiC* loci. Class II loci are NAD-, P-, and C-starvation-inducible and are represented by the *stiB* and *stiH* loci. Class III loci are NAD-, P-, and N-starvation-inducible and are represented by the *stiE* and *stiG* loci. Class IV loci are NAD- and P-starvation-inducible and include the *stiD* locus.

In addition to initially selecting for NAD-starvation-inducible Mu*d-lac* operon fusions, we also have used sulfate (S)-starvation, P-starvation, and C-starvation as our initial selection condition. Loci identified using these screening procedures make up Class V (*stiF*; also P- and N-starvation-inducible), Class VI (*psi* loci, induced by P-starvation only; Foster and Spector, 1986), and Class VII (*csi* loci, induced by C-starvation; Spector and Snell, unpublished results), respectively. One interesting aspect of these alternative selection procedures was the fact that, despite several attempts, we were unable to isolate any fusions initially identified as P-starvation-inducible that were also induced by other starvation conditions. However, all our sti fusions were P-starvation-inducible.

Several of these starvation-regulated loci have been mapped on the *S. typhimurium* linkage map (Sanderson and Roth, 1988). The *stiA* and *stiD* loci are genetically linked based on cotransduction and map at approximately 30 min. The *stiE* locus maps at 41 min between the *his* operon and the *flh-fli* (flagella synthesis) operon (F. G. del Portillo and B. B. Finlay, personal communication; Spector, unpublished results). The *stiG* and *stiH* loci map near 85 and 55 min, respectively. The *stiC* locus maps at approximately 75 min (Spector *et al.*, 1988). The *psiA* and *psiB* loci map at around 72 and 89 min, respectively. The *psiC* locus and its negative regulator *psiR* map at approximately 10 and 81 min, respectively.

4.2. Additional Aspects of the Physiological Regulation of sti Gene Expression

From the original seven *sti–lac* fusions, *stiA*, *stiB*, *stiC*, and *stiE* have been studied in greatest detail. Of these four loci, only the three N-starvation-inducible loci (*stiA*, *stiC*, and *stiE*) are also induced during Ile-Val, adenine (purines), or thiamine starvation; *stiB* is not (Spector *et al.*, 1988; Spector, unpublished results). This suggests that starvation for these nutrients induce signal-regulatory mechanisms common to N starvation but not C-starvation. However, it is also possible that this is a unique characteristic of *stiB* regulation. We will expand on these findings later in this chapter when we discuss the role of the stringent response in the regulation of these loci.

As will become apparent, the regulation of *stiB* is relatively unusual compared with other *sti* loci described here. One unusual aspect became apparent when we examined the effect of carbon source-energy downshifts. One would expect loci induced during C starvation (i.e., *stiA*, *stiB*, and *stiC*) to become induced during energy-source downshifts as well, since both of these conditions should produce similar metabolic effects (Magasanik and Neidhardt, 1987; Cashel and Rudd, 1987). Both *stiA* and *stiC* are induced on shifting from a preferential C-source, such as glucose, to a C-source that is more difficult to metabolize, i.e., acetate, succinate, and to a lesser degree glycerol. However, *stiB* is not induced following energy-source downshifts involving any of these C-sources, even though it is induced during C-starvation. This suggests that both common and specific signals are generated during C-starvation and energy-source downshifts. We will discuss possible common and specific signal-regulatory mechanisms later in this chapter.

Another interesting aspect of the regulation of these four loci is that none are induced by shifts to or growth during anaerobic or reduced oxygen growth conditions. These results controvert the idea that *sti* gene induction is simply a result of slow growth rates. Sti expression must be tied to more than just growth rate, since conditions like anaerobiosis or sulfate starvation as well as other conditions do not result in *sti* gene induction but do result in slow growth rates.

4.3. Temporal Expression of Starvation-Regulated lac Gene Fusions during Carbon Starvation

As mentioned previously, there is a dramatic molecular reorientation that occurs as *S. typhimurium*, and other enteric bacteria (Matin, 1991; Nyström *et al.*, 1990; Nyström, this volume; Östling *et al.*, this volume), move from logarithmic growth (nonlimiting conditions) into stationary phase (nutrient starvation conditions). Numerous proteins disappear or are reduced, while many others appear or show significant increase when whole cell extracts are examined by 2-D-PAGE. Preliminary work in *S. typhimurium* (Spector and Foster, unpublished results), and work in other organisms (Matin, 1991; Nyström *et al.*, 1990), using 2-D-PAGE analysis has indicated that increases in starvation-induced proteins occur in sequential phases following entry into starvation conditions. Therefore, we determined at what point following entry into starvation-invoked stationary phase the various starvation-induced *lac* fusions we have identified are "turned on." The induction of the more than 80 starvation-regulated *lac* fusions is found to occur in at least three phases during C-starvation (Fig. 2). Phase 1 genes/proteins are induced within the first hour of C-starvation. Some of the phase 1 genes continue to be expressed over at least the next 24 h and are under stringent control (see below). The *stiA*, *stiB*, and *stiC* loci represent this type of phase 1 gene (Spector and Cubitt, 1992). Phase 2 genes are induced between 1 and 2 h of starvation, and the phase 3 genes are represented, so far, by only a single *lac* fusion that is induced after about 4 to 5 h of starvation (Spector and Snell, unpublished results). The vast majority of starvation-inducible fusions we have identified to date are induced during phase 1. In addition, as discussed below, several of the phase 1 *sti* loci are necessary for starvation-survival.

4.4. Starvation-Survival in Salmonella typhimurium

4.4.1. Starvation-Survival in Salmonella Requires Protein Synthesis

As expected from the genetic fusion and 2-D gel electrophoretic analysis discussed above, the long-term survival of *Salmonella* is found to require protein synthesis. When chloramphenicol (CAM) is added to cultures at different times following entry into starvation-induced stationary phase, there is a dramatic effect on starvation-survival (Fig. 3). The earlier CAM is added

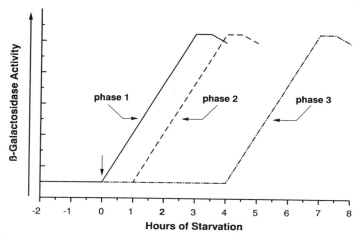

Figure 2. The sequential expression of starvation-regulated gene fusions during carbon starvation. Cells were starved and treated as described previously (Spector and Cubitt, 1992). β-Galactosidase activity is expressed in Miller units (Miller, 1972). ↓ indicates the point of nutrient exhaustion from the medium.

following the onset of starvation, the more striking is the effect on starvation-survival, e.g., addition after 0, 2, 4, and 6 h of starvation. This is in agreement with results obtained for *E. coli* and the marine *Vibrio* sp. strain CCUG 15956 (Nyström, this volume). Interestingly enough, if CAM is added to cultures which have been starving for as long as 24 h, there is still a significant effect on culture viability. Thus, the synthesis of new and existing proteins during the first 6 h is of critical importance to the long-term starvation-survival of this bacterium. Furthermore, the continual synthesis of at least some of these, and perhaps additional proteins, even after 24 h of starvation is also required for long-term survival under these conditions.

4.4.2. Starvation-Survival and Starvation-Inducible Loci

4.4.2a. stiA, stiB, and stiC Are Starvation-Survival Loci. The original premise of our studies was that by selecting for genes responsive to nutrient deprivation we could subsequently identify genetic loci necessary for *S. typhimurium* survival during prolonged nutrient starvation. The findings that (1) protein synthesis is required for long-term starvation-survival and (2) several of the *sti–lac* fusions are regulated by multiple starvation conditions made these loci likely candidates for being starvation-survival genes. Indeed, three of the four most studied *sti* loci are required for the long-term starvation-

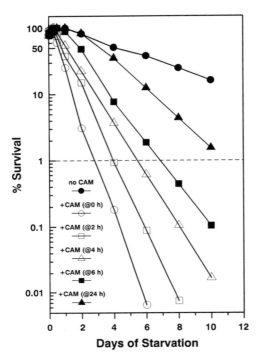

Figure 3. The effect of chloramphenicol (CAM) addition at various times during PCN starvation on starvation-survival. Cultures were grown and treated (no CAM) as previously described (Spector and Cubitt, 1992), with the exception that 100 µg CAM was added at the onset of starvation (@0 h), after 2 h (@2 h), 4 h (@4 h), 6 h (@6 h), or 24 h (@24 h) of PCN starvation.

survival of *S. typhimurium* (Spector and Cubitt, 1992). Strains lacking either *stiA*, *stiB*, or *stiC* (as a result of deletions or stable insertions) exhibited reduced survival during prolonged nutrient starvation (see below).

Typically, *S. typhimurium* LT-2 in transition between nonlimiting and starvation conditions reaches a maximal culture viability of ca. 3×10^8 colony-forming units (cfu) per milliliter. This level remains relatively constant over the next 48 h, or so, of starvation. Subsequently, culture viability declines to around 5–6% over the next 7–8 days of starvation (Fig. 4; Spector and Cubitt, 1992). Viability of the culture then remains relatively constant for up to 6 months (the maximum time period tested).

Figure 4A shows that strains deficient in *stiA*, *stiB*, or *stiC* exhibit significantly reduced survival over prolonged simultaneous phosphate, carbon, and nitrogen (PCN) starvation. Viability for the mutants continues to drop beyond the point where the viability of the parent culture levels off. After about 20 days of starvation, the survival of the *stiA* and *stiC* cultures is only about 1.4% that of the parent. In addition, the death phase kinetics, for the *stiA* and *stiC* mutants, are much more rapid than for wild type (Spector and Cubitt, 1992).

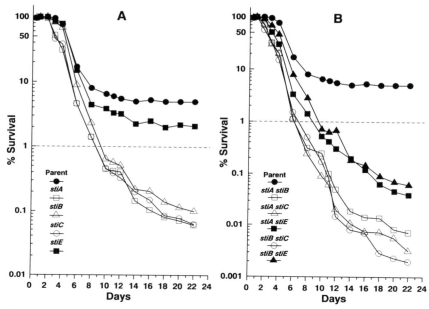

Figure 4. Survival of *Salmonella typhimurium* strains carrying individual (A) or combinations (B) of *sti* mutations during prolonged PCN starvation. Cells were grown and treated as previously described (Spector and Cubitt, 1992).

The effect of a *stiB* mutation on survival, however, is delayed compared with *stiA* or *stiC*. A *stiB* mutant routinely displays an initial death rate similar to wild type; however, between 6 and 7 days of starvation, viability of these cultures rapidly declines to levels equivalent to the *stiA* and *stiC* cultures. As with the *stiA* and *stiC* mutants, the viability of *stiB* cultures continues to drop, eventually reaching a level of survival of only about 2.2% of the parent culture. This would seem to indicate that *stiA* and *stiC* have a more immediate role in starvation-survival relative to *stiB*. A *stiE* mutation has no significant effect on starvation-survival (Spector and Cubitt, 1992). There is one other important finding suggested by these studies, namely, C-starvation appears to be the most critical factor in signaling the mechanisms important for the long-term survival of the bacterium. This is inferred from the results demonstrating that the three C-starvation-inducible loci, *stiA*, *stiB*, and *stiC*, are all required for starvation-survival, whereas *stiE*, which is not C-starvation-inducible, is not important for starvation-survival in the bacterium. This is in agreement with a recent report by Nyström *et al.* (1992) in marine vibrios, which demonstrated that C- or PCN-starved cells exhibit increasing resistance to subsequent exposures to a variety of different stresses, whereas cells starved for N or P separately did not develop significant resistance to other stresses and survive poorly relative to C-starved cells. A detailed analysis of these responses is given in Östling *et al.* (this volume).

4.4.2b. stiA, stiB, and stiC Exhibit an Epistatic Relationship in Starvation-Survival. It is clear that *stiA, B,* and *C* are important for survival during prolonged nutrient deprivation. Yet, individual mutations in these loci did not completely eliminate starvation-survival. However, when combinations of these mutations are introduced into the same genetic background, very dramatic reductions in culture survival occur. Figure 4B illustrates this point; combinations of *stiA stiB, stiA stiC,* or *stiB stiC* result in 500- to 2000-fold reductions in survival after 20 days of starvation. Furthermore, these *sti* double mutants die off even more quickly than strains carrying single *sti* mutations. An intriguing finding from these studies is that when two (or presumably more) *sti* mutations are combined, the effect on starvation-survival is much greater than the additive effects of each *sti* mutation alone. Thus, the *stiA, stiB,* and *stiC* loci exhibit an epistatic relationship where if one function, or pathway, is "missing" the presence of alternative survival pathways can partially compensate, allowing for some survival but at a reduced level. However, the loss of two, or more, survival pathways is too severe for the remaining functions to compensate. As a result, viability is reduced to levels considerably less than would be predicted for the loss of multiple components in a single pathway (Spector and Cubitt, 1992).

4.5. Negative Regulation of Starvation-Survival Loci by the cAMP Receptor Protein (CRP) in cAMP-Dependent and -Independent Manners

The role of the cAMP–CRP regulatory system in the bacterial response to altered or diminished carbon- or nitrogen-source availability is well known (Magasanik and Neidhardt, 1987). The fact that intracellular cAMP levels increase during C starvation (Botsford and Drexler, 1978) initially suggested that this regulatory system might play a role in *sti* gene induction. However, the result of introducing a *crp* (null) mutation into the various *sti–lac* fusion strains is the derepression of *sti* expression, rather than a failure to induce these loci (Fig. 5A–D; Spector and Cubitt, 1992). This indicates that CRP acts, directly or indirectly, as a negative regulator of *stiA, stiB, stiC,* and *stiE*.

As expected, a *cya* (null) mutation, which eliminates cAMP synthesis, has the same effect as a *crp* mutation (i.e., derepression) on *stiA, C,* and *E* (Fig. 5A, C, and D). In striking contrast, *cya* (i.e., cAMP) is not required for *stiB* regulation (Fig. 5B; Spector and Cubitt, 1992). Thus, CRP must act alone or with a signal molecule other than cAMP in the repression of *stiB*.

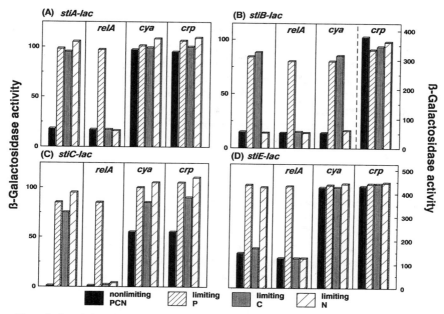

Figure 5. Regulation of *stiA–* (A), *stiB–* (B), *stiC–* (C), and *stiE–lac* (D) gene fusion expression by the *relA*, *cya*, and *crp* genes (Spector and Cubitt, 1992). In (B) the bars to the left of the dashed line correspond to the left-hand scale and those to the right correspond to the right-hand scale. β-Galactosidase activity is expressed in Miller units (Miller, 1972).

Moreover, the cAMP–CRP-dependent repression of *stiA*, *C*, and *E* is independent of fluctuations in intracellular cAMP levels. Growth on C-sources, such as mannitol, which results in increased cAMP levels does not perturb the typical N- or P-starvation regulation observed for these loci. Thus, simply increasing intracellular cAMP levels does not result in augmented repression of these *sti* loci. Logically this makes sense, since these loci are typically induced during conditions that result in increased cAMP inside the bacterial cell, e.g., C-starvation (Botsford and Drexler, 1978).

4.6. Stringent Control and Regulation of Starvation-Survival Loci

The role of stringent control, and ppGpp in particular, in the bacterial response to amino acid and carbon-energy source starvation has been studied for many years (Cashel and Rudd, 1987; Gallant *et al.*, 1976). The fact that the *stiA*, *B*, *C*, and *E* loci are induced during the period where ppGpp accumulates to relatively high levels inside the cell made this nucleotide a likely candidate as a signal molecule involved in *sti* induction. A *relA21*::Tn*10* will eliminate RelA-dependent ppGpp accumulation during the stringent response (Rudd *et al.*, 1985). Introduction of a *relA21*::Tn*10* into the various *sti–lac* fusion strains prevents both the C- and N-starvation-induction of *stiA*, and *stiC*, as well as the C-starvation-induction of *stiB* and the N-starvation-induction of *stiE* (Fig. 5A– D; Spector and Cubitt, 1992). This suggests a role for ppGpp as a signal molecule in the induction of these four *sti* loci during C- or N-starvation. These findings have several important implications.

Both C- and N-starvation result in ppGpp accumulation; however, accumulation of ppGpp was thought to be *relA*-independent during C-starvation (Cashel and Rudd, 1987). Our findings appear to contradict this model. In spite of this difference and because N-limitation is known to trigger ppGpp accumulation, one would expect a *relA*-dependent C-starvation-inducible locus to also be N-starvation-inducible, and vice versa. However, the fact that *stiB* exhibits *relA*-dependent C-starvation-induction but is not N-starvation-inducible, while *stiE* displays an opposite phenotype, indicates this is not always the case. These results suggest that ppGpp acts through separate pathways during C- and N-starvation, at least for *stiB* and *stiE* and possibly for *stiA* and *stiC* as well. The hypothesis of separate C- and N-starvation ppGpp induction pathways is further supported by the data showing that only the N-starvation-inducible loci *stiA*, *C*, and *E* are also induced by Ile-Val starvation, while *stiB* (not N-starvation-inducible) is not. The idea of separate C-starvation and N-starvation pathways is supported, in part, by recent studies in marine *Vibrio* sp. strain S14 (Nyström *et al.*, 1992).

However, such a model does pose an intriguing problem since ppGpp is thought to regulate gene transcription by interacting with a subunit(s) of σ^{70}:RNA polymerase holoenzyme (Travers *et al.*, 1982). How this occurs or which RNA polymerase subunit is involved is still a matter of debate (Nene and Glass, 1983; Baracchini *et al.*, 1984; Igarashi *et al.*, 1989; Gentry *et al.*, 1991). In any case, the direct interaction of ppGpp with σ^{70}:RNA polymerase holoenzyme could not in itself account for the observed differential *relA* dependence of *stiB* and *stiE* during C- and N-starvation, respectively. On the other hand, it is possible that ppGpp interacts with different forms of RNA polymerase holoenzyme, i.e., core RNA polymerase with alternative σ factors, that may function during C- and N-starvation, respectively.

Models to explain the differential *relA*-dependent C-starvation-induction of *stiB* and N-starvation-induction of *stiE* could involve one or more of several known alternative σ factors. One of these alternative σ factors may be σ^{54} (a.k.a. RpoN, GlnL, NtrA; Hunt and Magasanik, 1985; Reitzer and Magasanik, 1987) involved in ammonium assimilation and, therefore, N starvation. Another possible alterative σ factor may be the heat-shock σ factor, σ^{32} (a.k.a. HtpR, RpoH; Grossman *et al.*, 1984; Neidhardt and VanBogelen, 1987). This σ

factor was recently found to be C-starvation-inducible and required for the C-starvation-induction of three heat-shock proteins (DnaK, GroEL, and HtpG) of *E. coli* (Jenkins *et al.*, 1991). A third alternative σ factor that may be involved in *sti* gene induction is the KatF (a.k.a. RpoS) protein which has relatively recently been found to exhibit a high degree of homology to other σ factors (Mulvey and Loewen, 1989; Hengge-Aronis, this volume).

KatF and KatF-regulated proteins share many common regulatory features with *sti* genes. For example, the *katF* gene is negatively regulated by cAMP–CRP (Mulvey *et al.*, 1990; Lange and Hengge-Aronis, 1991). Although *katF* itself is not C-starvation-inducible, *katF*-regulated proteins were found to be C-starvation-inducible as well as being required for C-starvation-survival in *E. coli* (Lange and Hengge-Aronis, 1991; McCann *et al.*, 1991). This would suggest that KatF might interact with another signal (possibly ppGpp) to increase the transcription of *katF*-regulated C-starvation-inducible proteins since the level of KatF itself does not increase under this condition. This makes KatF a prime candidate for *sti* gene regulation in *S. typhimurium*. In fact, KatF has been shown to regulate several C-starvation-inducible (*csi*) and *sti* loci in *Salmonella typhimurium* (C. R. O'Neal, W. M. Gabriel, A. Turk, and M. P. Spector, 1993, Abstract H-115, Abstracts of General Meeting of the American Society for Microbiology).

A *Salmonella* homologue of *katF* has recently been identified (F. C. Fang *et al.*, 1992). The *Salmonella katF* appears to have an analogous function to the *E. coli katF* since disruption of the *katF* gene results in decreased survival during nutrient starvation and oxidative stress (F. Fang *et al.*, 1992). In addition, the *Salmonella katF* mapped at 59 min on the *S. typhimurium* chromosome, the same region as the *E. coli katF*. Interestingly, a functional KatF appears to be important for virulence of *S. typhimurium* since *katF* mutants are attentuated in their virulence in mice (F. Fang *et al.*, 1992). This appears to stem at least partially from the fact that genes present on the resident *Salmonella* virulence plasmid are expressed at only 15–20% of normal levels in a *katF* background (F. Fang *et al.*, 1992). One of these genes, *vsdC* (*spvB*), was previously reported to be C-starvation-inducible in a manner independent of *phoP*, *ompR*, or *cya/crp* (Fang *et al.*, 1992). The role of KatF as a regulator of virulence plasmid genes may only partially explain the significant effect of a *katF* mutation on the virulence of *S. typhimurium*. KatF is likely to influence the expression of additional chromosomal genes related to *Salmonella* virulence, although this has yet to be demonstrated (F. Fang and S. Libby, personal communication). Likewise, it is not known whether the *sti* loci described here play a role in this process.

Another model that may explain the specific *relA*-dependent C-starvation-induction of *stiB* and N-starvation-induction of *stiE* involves ppGpp acting as an inducer molecule interacting with separate C-starvation and N-starvation specific activator proteins to enhance the transcription from distinct sets of *sti* promoters during each condition. Both the alternative sigma factor and separate activator protein models are currently under investigation.

Both the *katF* regulon and the *sti* modulon are clearly needed for the long-term survival of this bacterium during nutrient starvation. In addition, mechanisms independent of KatF and the stringent response also seem to be important. T. Nyström and F. Neidhardt (1992, Abstract H-235, Abstracts of General Meeting of the American Society for Microbiology) have recently reported on the identification of a locus, *uspA* (*u*niversal *s*tress *p*rotein), in *E. coli* encoding a 13.5-kDa protein that is induced by 16 different starvation and stress conditions. The induction of this gene is independent of several known global regulatory loci including *relA*, *rpoH*, *katF*, *ompR*, *osmZ*, and *phoB*. Thus, if similar mechanisms exist in *Salmonella*, this would indicate that at least three separate regulatory networks are important for the starvation response of the organism: one KatF-dependent, one ppGpp-dependent, and one KatF- and ppGpp-independent.

4.7. P-Starvation-Induction of Starvation-Survival Loci Is Independent of relA and phoP

Figure 5A–D also illustrates that the P-starvation-induction of *stiA*, *B*, *C*, and *E* is independent of *relA*. This indicates that a separate, ppGpp-independent, pathway(s) is involved in the P-starvation-induction of *sti* loci. The signal(s) and mechanism(s) involved in P-starvation-specific induction are still unclear. One system that does not appear to be involved is the *phoP* regulatory system (Kier *et al.*, 1977, 1979; Groisman *et al.*, 1989; Miller *et al.*, 1989), although this regulatory system is involved in the P-starvation-induction of the *psiD* locus (Foster and Spector, 1986; Groisman *et al.*, 1989). A system similar to the *pho* regulon of *E. coli* (Wanner, 1987) has been proposed for *S. typhimurium* (Schlesinger and Olsen, 1968; Yagil and Hermoni, 1976; Foster and Spector, 1986). Unfortunately, except for the discovery of P-starvation-regulated genes (Foster and Spector, 1986; Spector *et al.*, 1988) and the *phoP* regulon, very little has been reported on the genetics of P-starvation regulation in *S. typhimurium*. Figure 6 schematically presents the various aspects of *sti* regulation currently under investigation.

5. COMPARISON OF S. TYPHIMURIUM STI PROTEINS WITH Pex PROTEINS OF E. COLI AND THE Sti PROTEINS OF MARINE VIBRIO

Abdul Matin and co-workers (reviewed in Matin, 1991) and Staffan Kjelleberg and his colleagues (Nyström *et al.*, 1988, 1990, 1992; Nyström and

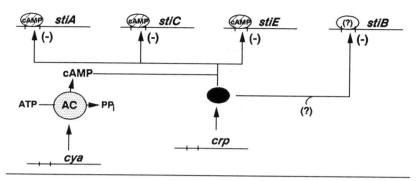

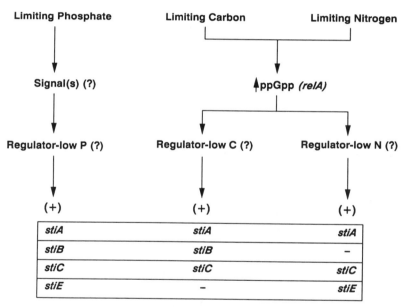

Figure 6. Schematic overview of the cAMP-dependent and -independent negative regulation by CRP and positive regulation by ppGpp of *sti* gene expression. *crp* gene encodes the cyclic 3', 5'-adenosine monophosphate (cAMP) receptor protein (CRP). *cya* gene encodes adenylate cyclase (AC), which catalyzes the conversion of ATP to cAMP and inorganic pyrophosphate (PP_i). The solid oval represents the conformation of CRP as it is synthesized from *crp* mRNA. The right-hatched ovals represent the conformation of CRP bound with cAMP which represses the synthesis of the *stiA*, *stiC*, and *stiE* loci. The open oval represents the conformation of CRP which, either alone or with a signal molecule other than cAMP, represses the synthesis of the *stiB* gene. (?) represents portions of the model currently under investigation. See text for further explanation and for alternative models.

Kjelleberg, 1989) have performed similar studies in *E. coli* and marine *Vibrio* species, respectively. Both groups utilized 2-D gel electrophoresis to demonstrate that each of these organisms sequentially expresses several temporal classes of proteins during the course of nutrient starvation (Nyström, this volume; Östling *et al.*, this volume). We have demonstrated a similar phenomenon in *S. typhimurium* using Mu*d-lac* fusions (Spector and Snell, unpublished results), as discussed earlier.

Furthermore, studies in both *E. coli* and marine *Vibrio* sp. strain CCUG 15956 indicate that proteins synthesized over the first 4 h of starvation are absolutely required for survival of the bacteria during prolonged nutrient starvations. Nyström *et al.* (1990) also reported that the earliest phase (phase 1) of the response of marine *Vibrio* sp. strain CCUG 15956 to multiple nutrient starvation resembles a stringent response. In addition, the proteins synthesized during phase 1 of this response partially disrupt the temporal synthesis of subsequent starvation proteins and prevent the synthesis of several late proteins. Most importantly, this early temporal class of proteins are also the most essential for the long-term starvation-survival of this bacterium (Nyström, this volume).

Studies in *E. coli* (Matin, 1991) have identified a core group of C-starvation-inducible proteins, referred to as Pex proteins. Pex proteins are thought to be required for the survival of *E. coli* during prolonged nutrient starvation as well as other stress conditions. Some of the Pex proteins are inducible by starvation and stress conditions in addition to C-starvation, e.g., DnaK, GroEL, and HtpG. These Pex proteins are distinguished from other C-starvation-inducible proteins (Cst proteins), in that they do not require cAMP for their induction, although several of these Pex proteins are negatively regulated by cAMP–CRP.

The *stiA*, *B*, *C*, and *E* loci of *S. typhimurium*, discussed here, exhibit many of the characteristics of the *E. coli* Pex proteins, and phase 1 Sti proteins of marine *Vibrio* sp. strain CCUG 15956. Most significantly, the *stiA*, *B*, and *C* loci are required for the long-term starvation-survival of this bacterium. The induction of these *sti* loci is independent of cAMP–CRP, but they are negatively regulated by CRP, similar to the Pex proteins of *E. coli*. Furthermore, the C- and/or N-starvation-induction of these *S. typhimurium sti* loci are under positive control by the *relA* gene, and occur during the transition between logarithmic growth and starvation-invoked stationary phase, analogous to the phase 1 Sti proteins described for marine *Vibrio*. Whether or not the *Salmonella* Sti proteins have functions analogous to the Pex proteins of *E. coli* or phase 1 Sti proteins of marine *Vibrio*, remains to be definitively established. It is clear that they all represent required functions for long-term starvation-survival of the respective organisms.

ACKNOWLEDGMENTS.   We thank Drs. Zarintaj Aliabadi, Simin Tirgari, and Yong K. Park, Ms. Holly K. Hall, Ms. Tania Gonzalez, Mr. Christopher Cubitt, Mr. Kevin Karem, Mr. Tom Penfound, Mr. Bob Morgan, and Mr. Kenneth Snell for helpful discussions and/or performance of various portions of the work discussed here.

We also thank Drs. Ferric Fang and Stephen Libby of the University of California at San Diego Medical Center for providing results prior to publication.

Various portions of this work were supported by grants 1 RO1 GM32595, 1 SO7 RR05870, and 1 RO1 GM34147 from the National Institutes of Health and DLB-89-04839 from the National Science Foundation awarded to J.W.F. Other portions of this work were supported by University of South Alabama Research Committee grants 3-61366, 3-61375, and 3-61407 and National Institutes of Health grant 1 R15 GM47628 awarded to M.P.S.

REFERENCES

Baracchini, E., Glass, R., and Bremer, H., 1984, Studies *in vivo* on *Escherichia coli* RNA polymerase mutants altered in the stringent response, *Mol. Gen. Genet.* **213**:379–387.

Botsford, J. L., and Drexler, M., 1978, The cyclic 3′,5′-adenosine monophosphate receptor protein and regulation of cyclic 3′,5′-adenosine monophosphate synthesis in *Escherichia coli*, *Mol. Gen. Genet.* **165**:47–56.

Cashel, M., and Rudd, K. E., 1987, The stringent response, in: *Escherichia coli and Salmonella typhimurium: Cellular and Molecular Biology*, Vol. 2 (F. C. Neidhardt, J. L. Ingraham, K. B. Low, B. Magasanik, M. Schaechter, and H. E. Umbarger, eds.), American Society for Microbiology, Washington, D. C., pp. 1410–1438.

Druilhet, R. E., and Sobek, J. M., 1984, Degradation of cell constituents during starvation of *Salmonella enteritidis*, *Microbios* **39**;73–82.

Ernst, R. K., Dombroski, D. M., and Merrick, J. M., 1990, Anaerobiosis, type 1 fimbriae, and growth phase are factors that affect invasion of HEp-2 cells by *Salmonella typhimurium*, *J. Bacteriol.* **58**:2014–2016.

Fang, F. C., Krause, M., Roudier, C., Fierer, J., and Guiney, D. G., 1991, Growth regulation of a *Salmonella* plasmid gene essential for virulence, *J. Bacteriol.* **173**:6783–6789.

Fang, F. C., Libby, S. J., Buchmeier, N. A., Loewen, P. C., Switala, J., Harwood, J., and Guiney, D. G., 1992, The alternative factor KatF (Rpos) regulates *Salmonella* virulence, *Proc. Natl. Acad. Sci. USA* **89**:11978–11982.

Finlay, B. B., and Falkow, S., 1988, Virulence factors associated with *Salmonella* species, *Microbiol. Sci.* **5**:324–328.

Finlay, B. B., and Falkow, S., 1989, *Salmonella* as an intracellular parasite, *Mol. Microbiol.* **3**: 1833–1841.

Foster, J. W., 1983, Identification and characterization of a *relA*-dependent starvation-inducible locus (*sin*) in *Salmonella typhimurium*, *J. Bacteriol.* **156**:424–428.

Foster, J. W., 1991, *Salmonella* acid shock proteins are required for the adaptive tolerance response, *J. Bacteriol.* **173**:6896–6902.

Foster, J. W., and Hall, H. K., 1990, Adaptive acidification tolerance response of *Salmonella typhimurium*, *J. Bacteriol.* **172**:771–778.

Foster, J. W., and Hall, H. K., 1992, Effect of *salmonella typhimurium* ferric uptake regulator (*fur*) mutations on iron- and pH-regulated protein synthesis, *J. Bacteriol.* **174**:4317–4323.

Foster, J. W., and Spector, M. P., 1986, Phosphate starvation regulon of *Salmonella typhimurium*, *J. Bacteriol.* **166**:666–669.

Galán, J. E., and Curtiss, R., III, 1990, Expression of *Salmonella typhimurium* genes required for invasion is regulated by changes in DNA supercoiling, *Infect. Immun.* **58**:1879–1885.

Gallant, J., Shell, L., and Bittner, R., 1976, A novel nucleotide implicated in the response of *E. coli* to energy source downshift, *Cell* **7**:75–84.

Gentry, D., Xiao, H., Burgess, R., and Cashel, M., 1991, The omega subunit of *Escherichia coli* K-12 RNA polymerase is not required for stringent RNA control in vivo, *J. Bacteriol.* **173**:3901–3903.

Groisman, E. A., Chiao, E., Lipps, C. J., and Heffron, F., 1989, *Salmonella typhimurium phoP* virulence gene is a transcriptional regulator, *Proc. Natl. Acad. Sci. USA* **86**:7077–7081.

Grossman, A. D., Erickson, J. W., and Gross, C. A., 1984, The *htpR* gene product of *E. coli* is a sigma factor for heat shock promoters, *Cell* **38**:383–390.

Harder, W., and Dijkhuizen, L., 1983, Physiological responses to nutrient limitation, *Annu. Rev. Microbiol.* **37**:1–23.

Hunt, T. P., and Magasanik, B., 1985, Transcription of *glnA* by purified *Escherichia coli* components: Core RNA polymerase and the products of *glnF*, *glnG*, and *glnL*, *Proc. Natl. Acad. Sci. USA* **82**:8453–8457.

Igarashi, K., Fujita, N., and Ishihama, A., 1989, Promoter selectivity of *Escherichia coli* RNA polymerase: Omega factor is responsible for the ppGpp sensitivity, *Nucleic Acids Res.* **17**:8755–8765.

Jenkins, D. E., Auger, E. A., and Matin, A., 1991, Role of RpoH, a heat shock regulator protein, in *Escherichia coli* carbon starvation protein synthesis and survival, *J. Bacteriol.* **173**:1992–1996.

Kier, L. D., Weppelman, R., and Ames, B. N., 1977, Regulation of two phosphatases and a cyclic phosphodiesterase of *Salmonella typhimurium*, *J. Bacteriol.* **130**:420–428.

Kier, L. D., Weppelman, R., and Ames, B. N., 1979, Regulation of nonspecific acid phosphatase in *Salmonella typhimurium*: *phoN* and *phoP* genes, *J. Bacteriol.* **138**:155–161.

Koch, A. L., 1971, The adaptive responses of *Escherichia coli* to a feast or famine existence, *Adv. Microb. Physiol.* **6**:147–217.

Lange, R., and Hengge-Aronis, R., 1991, Identification of a central regulator of stationary-phase gene expression in *Escherichia coli*, *Mol. Microbiol.* **5**:49–59.

Lazdunski, A. M., 1989, Peptidases and proteases of *Escherichia coli* and *Salmonella typhimurium*, *FEMS Microbiol. Rev.* **63**:265–276.

Lee, C. A., and Falkow, S., 1990, The ability of *Salmonella* to enter mammalian cells is affected by bacterial growth state, *Proc. Natl. Acad. Sci. USA* **87**:4304–4308.

Magasanik, B., and Neidhardt, F. C., 1987, Regulation of carbon and nitrogen utilization, in: *Escherichia coli and Salmonella typhimurium: Cellular and Molecular Biology* (F. C. Neidhardt, J. L. Ingraham, K. B. Low, B. Magasanik, M. Schaechter, and H. E. Umbarger, eds.), American Society for Microbiology, Washington, D.C., pp. 1318–1325.

Matin, A., 1991, The molecular basis of carbon-starvation-induced general resistance in *Escherichia coli*, *Mol. Microbiol.* **5**:3–10.

McCann, M. P., Kidwell, J. P., and Matin, A., 1991, The putative σ (sigma) factor KatF has a central role in the development of starvation-mediated general resistance in *Escherichia coli*, *J. Bacteriol.* **173**:4811–4194.

Miller, C. G., 1975, Peptidases and proteases of *Escherichia coli* and *Salmonella typhimurium*, *Annu. Rev. Microbiol.* **29**:485–504.

Miller, C. G., 1987, Protein degradation and proteolytic modification, in: *Escherichia coli and Salmonella typhimurium: Cellular and Molecular Biology* (F. C. Neidhardt, J. L. Ingraham, K. B. Low, B. Magasanik, M. Schaechter, and H. E. Umbarger, eds.), American Society for Microbiology, Washington, D. C., pp. 680–691.

Miller, J. H. (ed), 1972, *Experiments in Molecular Genetics*, Cold Spring Harbor Laboratory Press, Cold Spring Harbor, N.Y.

Miller, S. I., Kukral, A. M., and Mekalanos, J. J., 1989, A two component regulatory system (*phoP phoQ*) controls *Salmonella typhimurium* virulence, *Proc. Natl. Acad. Sci. USA* **86**:5054–5058.

Mulvey, M. R., and Loewen, P. C., 1989, Nucleotide sequence of *katF* of *Escherichia coli* suggests KatF protein is a novel σ transcription factor, *Nucleic Acids Res.* **17**:9979–9991.

Mulvey, M. R., Switala, J., Borys, A., and Loewen, P. C., 1990, Regulation of transcription of *katE* and *katF* of *Escherichia coli*, *J. Bacteriol.* **172**:6713–6720.

Neidhardt, F. C., and VanBogelen, R. A., 1987, Heat shock response, in: *Escherichia coli and Salmonella typhimurium: Cellular and Molecular Biology* (F. C. Neidhardt, J. L. Ingraham, K. B. Low, B. Magasanik, M. Schaechter, and H. E. Umbarger, eds.), American Society for Microbiology, Washington, D.C., pp. 1334–1345.

Nene, V., and Glass, R. E., 1983, Relaxed mutants of *Escherichia coli* RNA polymerase, *FEBS Lett.* **153**:307–310.

Nyström, T., and Kjelleberg, S., 1989, Role of protein synthesis in the cell division and starvation induced resistance to autolysis of a marine vibrio during the initial stage of starvation, *J. Gen. Microbiol.* **135**:1599–1606.

Nyström, T., Albertson, N., and Kjelleberg, S., 1988, Synthesis of membrane and periplasmic proteins during starvation of a marine *Vibrio* sp., *J. Gen. Microbiol.* **134**:1645–1651.

Nyström, T., Flärdh, K., and Kjelleberg, S., 1990, Responses to multiple-nutrient starvation in marine *Vibrio* sp. strain CCUG 15956, *J. Bacteriol.* **172**:7085–7097.

Nyström, T., Olsson, R. M., and Kjelleberg, S., 1992, Survival, stress resistance, and alteration in protein expression in the marine *Vibrio* sp. strain S14 during starvation for different individual nutrients, *Appl. Environ. Microbiol.* **58**:55–65.

Reeve, C. A., Bockman, A. T., and Matin, A., 1984, Role of protein degradation in the survival of carbon-starved *Escherichia coli* and *Salmonella typhimurium*, *J. Bacteriol.* **157**:758–763.

Reitzer, L. J., and Magasanik, B., 1987, Ammonia assimilation, in: *Escherichia coli and Salmonella typhimurium: Cellular and Molecular Biology* (F. C. Neidhardt, J. L. Ingraham, K. B. Low, B. Magasanik, M. Schaechter, and H. E. Umbarger, eds.), American Society for Microbiology, Washington, D.C., pp. 302–320.

Roszak, D. B., and Colwell, R. R., 1987, Survival strategies of bacteria in the natural environment, *Microbiol. Rev.* **51**:365–379.

Rudd, K. E., Bochner, B. R., Cashel, M., and Roth, J. R., 1985, Mutations in the *spoT* gene of *Salmonella typhimurium*: Effects on *his* operon expression, *J. Bacteriol.* **163**:534–542.

Sanderson, K. E., and Roth, J. R., 1988, Linkage map of *Salmonella typhimurium*, edition VII, *Microbiol. Rev.* **52**:485–532.

Schlesinger, M. J., and Olsen, R., 1968, Expression and localization of *Escherichia coli* alkaline phosphatase synthesized in *Salmonella typhimurium* cytoplasm, *J. Bacteriol.* **96**:1601–1605.

Spector, M. P., 1990, Gene expression in response to multiple nutrient-starvation conditions in *Salmonella typhimurium*, *FEMS Microbiol. Ecol.* **74**:175–184.

Spector, M. P., and Cubitt, C. L., 1992, Starvation-inducible loci of *Salmonella typhimurium*: Regulation and roles in starvation-survival, *Mol. Microbiol.* **6**:1467–1476.

Spector, M. P., Aliabadi, Z., Gonzalez, T., and Foster, J. W., 1986, Global control in *Salmonella typhimurium*: Two-dimensional gel electrophoretic analysis of starvation-, anaerobiosis- and heat shock-inducible proteins, *J. Bacteriol.* **168:**420–424.

Spector, M. P., Park, Y. K., Tirgari, S., Gonzalez, T., and Foster, J. W., 1988, Identification and characterization of starvation-regulated genetic loci in *Salmonella typhimurium* by using Mu*d*-directed *lacZ* operon fusions, *J. Bacteriol.* **170:**345–351.

Tempest, D. W., Neijssel, O. M., and Zevenboom, W., 1983, Properties and performance of microorganisms in laboratory culture: Their relevance to growth in natural ecosystems, in: *Microbes in Their Natural Environments* (J. H. Slater, R. Whittenbury, and J. W. T. Wimpenny, eds.), Cambridge University Press, London, pp. 119–152.

Travers, A. A., Lamond, A. I., and Mace, H. A., 1982, ppGpp regulates the binding of two RNA polymerase molecules to the *tyrT* promoter, *Nucleic Acids Res.* **10:**5043–5057.

Wanner, B. L., 1987, Phosphate regulation of gene expression in *Escherichia coli* in: *Escherichia coli and Salmonella typhimurium: Cellular and Molecular Biology* (F. C. Neidhardt, J. L. Ingraham, K. B. Low, B. Magasanik, M. Schaechter, and H. E. Umbarger, eds.), American Society for Microbiology, Washington, D. C., pp. 1326–1333.

Yagil, E., and Hermoni, E., 1976, Repression of alkaline phosphatase in *Salmonella typhimurium* carrying a *phoA$^+$ phoR$^-$* episome from *Escherichia coli*, *J. Bacteriol.* **128:**661–664.

10

The Impact of Nutritional State on the Microevolution of Ribosomes

C. G. Kurland and Riitta Mikkola

1. INTRODUCTION

Very little is known about the microevolution of the translation apparatus in bacteria, so little that there is almost no literature that we could review by way of introduction to this chapter. This situation is surprising since the translation apparatus is expressed from 3 different ribosomal RNA genes that are duplicated several times, as well as from 45 different tRNA genes, some of which are duplicated several times (Komine *et al.*, 1990), from 20 different aminoacyl-tRNA synthetase genes, some of which have more than one subunit, from more than 50 ribosomal protein genes, and at least 9 different translation factor genes. All told, this complex is coded by more than 150 different genes and is therefore a large target for mutations.

In what follows we will recount some experiments suggesting that the microevolution of ribosomes is strongly influenced by the nutritional state of bacterial populations. We will suggest that normal laboratory culture conditions favor a very special growth phenotype. Further, we will show that this phenotype is not particularly well suited to survival during certain starvation conditions, which favor a very different sort of growth phenotype. Both growth phenotypes are associated with a particular ribosome phenotype that is characterized in part by kinetic parameters that are readily measured in laboratory experiments. Furthermore, mutants that manifest the fast growth phenotype

C. G. Kurland and Riitta Mikkola • Department of Molecular Biology, S-751 24 Uppsala, Sweden.

Starvation in Bacteria, edited by Staffan Kjelleberg. Plenum Press, New York, 1993.

can be selected from the starvation-favored phenotype in chemostats. Finally, we attempt to relate the various phenotypes of natural isolates to the decreased selective differences under poor growth conditions and, consequently, the increased importance of genetic drift and to the differential selection pressures for cells under rich growth conditions and under starvation. The interactions of these three microevolutionary forces generate a spectrum of growth phenotypes for natural populations of *Escherichia coli*.

2. THE VARIABILITY OF NATURAL ISOLATES

We measured in glucose minimal medium the generation times of 65 natural isolates of *E. coli* (Mikkola and Kurland, 1991a). These natural isolates had been collected and originally described by others (Ochman and Selander, 1984; Wadström *et al.*, 1978; Söderlind *et al.*, 1988). The generation times of the strains can be arranged in a distribution that is pleasing to the eye because it is an almost normal distribution (Fig. 1), but this normality hides an interesting

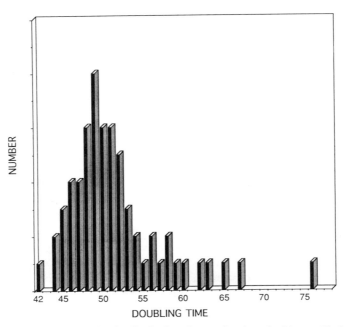

Figure 1. A histogram describing the distribution of generation times for 64 natural isolates. The bacterial strains are described in the text. Generation times were measured in glucose minimal medium. Data from Mikkola and Kurland (1991a).

discrepancy. We can compare the generation times of the natural isolates with wild-type laboratory strains that we use as reference strains: XAC, which grows with a 45-min doubling time, and 017, which grows with a 48-min doubling time. We notice that the generation times of the two laboratory strains are among the shortest within the distribution of generation times for natural isolates; indeed, only a few of the natural isolates are faster. Furthermore, one of the most interesting natural isolates is not included in Fig. 1 because its generation time is so long (124 min.). The immediate question is whether the apparent tendency for the natural isolates to grow slower than the laboratory strains is significant.

One way of addressing this question is to rephrase it: We can ask whether or not the relatively fast growth rates of the laboratory strains are a consequence of a selection that depends on the specific laboratory conditions that we employ for growing bacteria. For example, we normally cultivate bacteria only in the presence of a good carbon source, and we do not expose them to protracted starvation periods. In nature, good carbon sources and protracted periods without starvation are rare (Morita, 1988, this volume; Moriarty and Bell, this volume; van Elsas and van Oberbeek, this volume). Accordingly, we chose a group of seven natural isolates that more or less covered the range of growth rates for the whole collection. Then we cultivated them for 280 generations in glucose-limited chemostats at generation times close to 3 h. In this way we could ask whether continuous exposure to glucose would provide a condition that would select a faster growth rate from the natural isolates.

The answer to this question is seen in Fig. 2 and it is a very clear yes. We

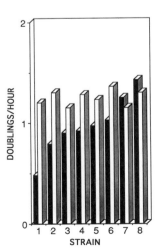

Figure 2. The growth rates of seven natural isolates and a laboratory wild type before (filled bars) and after (open bars) 280 generations of continuous culture in glucose-limited chemostats. Data from Mikkola and Kurland (1992).

find that the seven natural isolates that initially had generation times ranging between 42 and 124 min have generation times between 44 and 52 min after growth in the chemostats (Mikkola and Kurland, 1992). Indeed, all of the strains grown in the chemostats for 280 generations have growth rates that are virtually indistinguishable from the wild-type laboratory strains. Furthermore, if the growth is continued for an additional several hundred generations in the chemostats, there are no significant changes noted in the growth rates of the bacteria. It seems that a stable growth phenotype is selected after a few hundred generations in the continuous culture and this growth phenotype is remarkably similar to that of the laboratory wild-type strains.

Evidently, the growth phenotype of a bacterial population is dependent on its recent history. In particular, the uniformity of the results obtained in the chemostats suggests that the contrasting variation of the growth rates for the natural isolates is quite significant and that it reflects the variations in the recent histories of these isolates in nature. Likewise, the relative uniformity of the growth rates for standard wild type laboratory strains seems to reflect the relative uniformity of laboratory growth conditions. Finally, the evolution of growth rates and ribosome phenotypes in the chemostats is not only dramatic but also very rapid. Calculations based on the variation of growth rates with time in glucose-limited chemostats suggest that one to three mutations would be sufficient to generate the selected phenotypes (Mikkola and Kurland, 1992).

3. RIBOSOME MUTANTS

Earlier, we had established for mutant laboratory strains a very tight coupling between the growth rates of bacteria in batch cultures and the kinetic efficiency of their ribosomes (Bohman et al., 1984; Ruusala et al., 1984; Andersson et al., 1986). This correlation confirmed a view of bacterial "design" that we can refer to as the growth rate maximization (GRM) model (Ehrenberg and Kurland, 1984). According to this interpretation, every component of the optimum bacterium is to some extent rate limiting, and that extent is well defined by its relative contribution to the growth rate of the bacterium.

In the GRM model, departures from optimality of the sort associated with mutant phenotypes are weighted by two sorts of parameters (Ehrenberg and Kurland, 1984): One is the relative significance of a biosynthetic compartment in the growing system which is determined by the fraction of the total cellular metabolic activity that the compartment contributes under a given growth condition. The other is the kinetic efficiency of the compartment in a particular growth state; this is defined by its metabolic flow normalized to the mass investment in the compartment, which includes its substrates. Note that both of these weighting factors are explicitly dependent on the growth state of the bacterium. This is particularly important for assessing the performance of the translation system because as the Copenhagen School (Maaløe, 1979) has shown, the relative weight of the translation system compared with other compartments of the bacterium is quite dependent on growth conditions. Since the translation system is the dominant metabolic compartment of bacteria growing in normal laboratory batch cultures, the GRM model was very useful to account for the growth behavior of ribosome mutants.

The provocative observation that the GRM model could explain was that mutants with hyperaccurate ribosomes grow at rates that are slower than their less accurate, wild-type relatives. Thus, in the view of the GRM model, too much accuracy would lead to suboptimal ribosome kinetics, and these in turn would lower growth rates. The reason for this is that enhanced accuracy of aminoacyl-tRNA selection is obtained at the cost of the increased tendency of ribosomes to discard tRNA species from their binding sites. This tendency is expressed as a lower efficiency in the kinetics of the acquisition of tRNA species. As a consequence, there is a systematic correlation between the increased tendency to reject tRNA species, a decreased rate of polypeptide synthesis per ribosome, and a decreased growth rate for mutant bacteria (Andersson *et al.*, 1986).

It turns out that this was an incomplete explanation of why hyperaccurate mutants grow slower than wild-type strains. Not all translation errors involve missense substitutions. Some involve the abortive termination of the translation of an mRNA sequence, a so-called processivity error (reviewed in Kurland, 1992). Such errors occur when frameshift errors occur, when transcription is aborted, or when a false termination step is introduced. However, experiments suggest that these particular processivity errors are not quantitatively important (Jörgensen and Kurland, 1990; Jörgensen, Adamski, Tate, and Kurland, unpublished data; Dong and Kurland, unpublished data). Instead, the data suggest that so-called drop off events, in which the nascent polypeptidyl-tRNA is prematurely released from the ribosome (Menninger, 1976, 1977, 1978), are probably the dominant processivity error under normal laboratory growth conditions (Jörgensen and Kurland, 1990; Dong and Kurland, unpublished data; Kurland, 1992). Indeed, it seems that the reason that streptomycin-dependent bacteria cannot grow in the absence of antibiotic is that their processivity error rate is so great that they cannot complete the translation of large, essential proteins such as the β and β' subunits of RNA polymerase (Dong and Kurland, unpublished data; Kurland, 1992).

It seems that both unfavorable kinetics and destructive processivity errors can account for the suboptimal growth phenotypes of hyperaccurate mutants. In addition, there is a weaker growth rate inhibition following the accumulation of missense substitutions in proteins (Kurland, 1992). This effect together with ribosome assembly defects (Green and Kurland, 1971; Olsson *et al.*, 1974) and processivity errors (Dong and Kurland, unpublished data) will account for the slower growth rates of errorprone ribosome mutants. In summary, because of the large mass investment in the translation apparatus of bacteria growing under normal laboratory conditions, relatively subtle changes in the efficiency of translation are expressed in decreased growth rates for ribosome mutants.

4. RIBOSOMES OF NATURAL ISOLATES

The growth rate behavior of the natural isolate collection and their evolution in chemostats suggest that natural isolates are not growth rate maximized in nature. In addition, our experience with laboratory mutants predisposed us to guess that the variation of growth rates observed for the natural isolates would reflect a variation in their translation efficiencies. Accordingly, we purified the ribosomes from selected natural isolates and compared their performance characteristics with those of ribosomes from laboratory wild-type strains.

We found that there was, indeed, a marked variation in the efficiency with which aminoacyl-tRNA elongation factor GTP (ternary) complexes are processed by ribosomes of natural isolates (Mikkola and Kurland, 1991a). Furthermore, a similar relationship is found for the ribosomes from natural isolates and for those from mutants of laboratory strains: this consists of a positive correlation between the kinetic efficiency of ribosomes *in vitro* and the growth rates of the corresponding bacteria in a normal laboratory medium. In other words, the natural isolates produce variant ribosomes, and it is not necessary to invoke other mutational changes to explain their growth rate variability.

The ribosomes of the natural isolates do not by any means share a common translational phenotype. We may compare them to ribosomes from laboratory strains as well as from natural isolates that have been cultured in chemostats. These have kinetic properties as well as missense error rates that are virtually indistinguishable from each other. For example, Fig. 3 summarizes the variation of the maximum turnover rate (k_{cat}) in polypeptide synthesis for the ribosomes of natural isolates before and after the chemostat selection. Here the k_{cat} before the selection has a mean of 4.4/s with extremes between 3.5/s and 5.4/s and after the selection the mean k_{cat} is 7.9/s with extremes of 7.6/s and 8.3/s (Mikkola and Kurland, 1992). Note that the wild-type laboratory ribosomes have an initial k_{cat} close to 8.1/s and after the same chemostat selection it is 8.2/s (Mikkola and Kurland, 1992). Clearly, the mean selected ribosome has

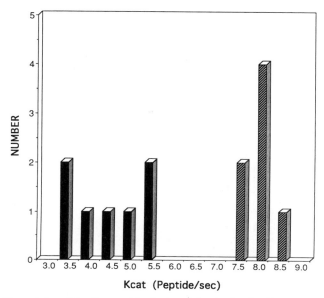

Kcat (Peptide/sec)

Figure 3. The variation of the ribosomal k_{cat} for translation *in vitro* for ribosomes extracted from natural isolates before (filled bars) and after (lined bars) 280 generations of cultivation in glucose-limited chemostats. Data from Mikkola and Kurland (1992).

more than doubled the initial k_{cat} and the distributions around the means are sharper for the chemostat-selected ribosomes.

A similar pattern is observed for the missense error rates. The ribosomes of seven natural isolates translate in a competition between tRNA^{Phe} and tRNA^{Leu4} in a poly(U)-primed translation system with a mean error rate close to 8.3×10^{-4}. After the chemostat selection the ribosomes of the seven strains translate with a mean error rate close to 3.8×10^{-4}. It is evident in Fig. 4 that the spread of error rates is initially much greater than that after the selection. Furthermore, ribosomes of the reference laboratory wild-type strain translate in the same system with error rates of 3.3×10^{-4} and 3.4×10^{-4} before and after the chemostat selection.

We are evidently justified in speaking about a laboratory wild-type phenotype for the ribosomes as well as for bacterial growth. Furthermore, it seems reasonable to identify both of these phenotypic characteristics with a growth rate-maximized bacterium of the sort described in the model of Ehrenberg and Kurland (1984). It seems equally clear that the natural isolate

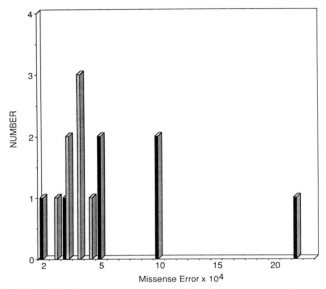

Figure 4. Variation of the missense error rates *in vitro* for ribosomes extracted from seven natural isolates and a laboratory wild type before (dark bars) and after (dotted bars) growth for 280 generations in glucose-limited chemostats. The error was measured as a competition between tRNA^{Phe} and tRNA^{Leu4} in an *in vitro* system primed by poly(U). Data from Mikkola and Kurland (1992).

populations are not properly described by this model. We next begin to explore the origins of departures from a growth rate-optimized arrangement.

5. DRIFT

Bacteria growing in poor media have a relatively small translation compartment that functions slowly compared with that of rapidly growing bacteria that utilize a maximum density of ribosomes (Maaløe, 1979). According to the GRM model the dominant role of the translation apparatus in determining the metabolic efficiency of the bacterium will decrease progressively as the quality of the medium decreases (Ehrenberg and Kurland, 1984). In genetic terms this means that the selective pressure on the efficiency of translation will decrease as the quality of the medium decreases. Accordingly, we expect there to be a significant tendency for bacteria that are spending most of their time under poor growth conditions to drift away from the optimum ribosome construction that is selected by competition at rapid growth rates, as for example under laboratory conditions. This may explain, at least in part, the distributions of the growth rates and ribosome efficiencies for the natural isolates.

In addition, there are ways for a bacterium with a suboptimal ribosome construction to compensate for the growth inhibitory effects of inefficient translation kinetics. For example, an overproduction of ribosomes might compensate for the slowness of individual ribosomes. However, we would expect this sort of compensation to be effective if the increase of the mass of the ribosomes is small compared with the overall mass of the cells (Ehrenberg and Kurland, 1984). Accordingly, this sort of strategy might be observed at slow growth rates where ribosome concentrations are low. However, we would not expect this to be an effective strategy at fast growth rates where the ribosomes are a significant fraction of the cellular mass.

In order to search for this sort of compensatory mechanism, we can compare the growth inhibitory influence of suboptimal ribosome mutants at different growth rates in different media. When this is done, it is found that the ratio of ribosome mutant growth rate to wild-type growth rate increases as the quality of the culture medium is decreased systematically (Mikkola and Kurland, 1988). For some mutants the differences between growth rates of mutant and wild-type bacteria nearly disappear in a minimal medium containing acetate as the sole carbon source (Mikkola and Kurland, 1988). Furthermore, when the ribosome concentrations in the mutant bacteria are compared with those of wild-type bacteria, a medium-dependent increase is seen in mutants (Mikkola and Kurland, 1991b). For example, streptomycin-resistant bacteria that grow significantly more slowly than wild type in amino acid-containing medium have the same ribosome concentrations as wild type in this medium. However, when they are growing in minimal medium containing acetate, they grow nearly as well as do wild type and they contain 40% higher ribosome concentrations than wild-type bacteria (Mikkola and Kurland, 1991b). Thus, the overproduction of ribosomes at low growth rates might allow bacteria in nature to escape from a strong selection for maximized translation kinetics.

Nevertheless, when we studied the production of ribosomes by natural isolates in different growth media, we were disappointed by the results. Several natural isolates were studied and the most extreme one of these could overproduce ribosomes to a level that is 11% higher than that of the laboratory wild-type bacteria in minimal medium containing acetate as the sole carbon source (Mikkola and Kurland, 1991b). Since this effect is much smaller than the sixfold lower kinetic efficiency of these particular ribosomes, it did not seem likely that biosynthetic compensation plays a very large role in the evolution of the ribosome phenotype of this particular natural isolate. We therefore searched for an alternative mechanism for natural isolates to escape the growth rate-maximizing selection for optimized ribosomes.

6. CARBON STARVATION

Starvation is a normal, recurrent state for bacteria in nature (Morita, 1988, this volume). Furthermore, gram-negative bacteria such as *E. coli* can enter well-defined physiological states that facilitate survival under starvation conditions (Matin, 1991; Spector *et al.*, 1988; Nyström and Kjelleberg, 1989; several chapters in this volume). Since *E. coli* is obviously capable of adapting to the rigors of starvation, we asked whether or not part of that adaptive strategy might be related to the tendency of some natural isolates to have a slow-growth-rate phenotype.

Bacteria were grown in glucose minimal medium and then the carbon source was rapidly removed by centrifuging the bacteria out of the medium. We found that the results of a rapid removal of carbon source are different from those of experiments in which the bacteria were allowed to exhaust the carbon source (Mikkola and Kurland, unpublished data). The difference is that when they exhaust the glucose by consuming it, the death rate during a continued incubation without a carbon source is very low. In contrast, rapid removal of the carbon source is associated with a significant death rate during the ensuing starvation (Mikkola and Kurland, unpublished data).

We find that laboratory wild-type strains die at a rate that is close to 13% per day, while natural isolates die at variable but slower rates (Mikkola and Kurland, unpublished data). Indeed, for the natural isolates that we studied, the death rates are correlated with the growth rates (correlation coefficient = 0.84) and with the kinetic efficiency of the ribosome–ternary complex interaction (correlation coefficient = 0.89) when the criterion of viability is viable counts (Mikkola and Kurland, unpublished data). For example, the most slowly growing natural isolate dies at less than half the rate at which the most rapidly growing isolate dies. In other words, the slower the growth rate and the slower the translation efficiency, the more resistant the bacteria are to the exigencies of starvation. This means that all other things being equal, the slower-growing natural isolates will selectively survive this sort of starvation regime.

We have also measured the attrition of the bacteria by using permeability to the fluorescent dye propidium iodide as a criterion of loss of viability. We had previously shown that viable counts and permeability to propidium iodide are well correlated under a variety of growth conditions for different laboratory strains (Jörgensen and Kurland, 1987). In contrast, our experiments with starving bacteria reveal a quantitative divergence in the results obtained with the two methods. Thus, when a major fraction of the natural isolates have lost the ability to yield viable cells, only a small fraction (10–15%) are unable to maintain a permeability barrier to the fluorescent dye (Mikkola and Kurland, unpublished data). It seems that rapid starvation leads to a situation in which cells remain alive in the metabolic sense that they can exclude the propidium iodide, but they are genetically dead in the sense that they cannot produce progeny. We suspect that this discrepancy will turn out to be important when we finally understand it.

After the natural isolates have been subjected to the chemostat selection, they lose their superior capacity to survive starvation. Indeed, they as well as standard ribosome mutants of laboratory strains die at the faster rates similar to those of laboratory wild-type strains (Mikkola and Kurland, unpublished data). These details of the starvation-induced death of laboratory strains are important for two reasons: First, they show that the augmented ribosome concentrations that some of the ribosomal mutants derived from laboratory strains exhibit do not help these mutants to survive the starvation regime. This suggests that models in which the enhanced survival of bacteria to this sort of starvation is dependent on extra reserves of accumulated ribosomes are not valid. Second, these data show that it is not slow growth or slow ribosome kinetics *per se* that enable the natural isolates to survive starvation. Rather the data suggest that there is a very specific ribosomal phenotype that is directly or indirectly associated with inefficient translation kinetics and that is responsible for the advantage these bacteria have during carbon starvation. We are persuaded that the identification of the relevant genetic differences between natural isolates and laboratory strains is likely to be useful in explaining the adaptations of natural isolates to starvation conditions.

7. CONCLUSIONS

Three factors, each associated with a particular nutritional state, seem to be relevant to the microevolution of natural isolates of *E. coli*. The one that we know the most about is the growth rate maximization selection that operates under conditions of fast growth, e.g., normal laboratory conditions. This sort of selection is extremely important for organisms in nature as indicated by a number of different observations. For example, although the ribosome phenotypes of natural isolates are distributed, they are a short mutational distance from the growth-maximized phenotype as shown by the chemostat experiments. Likewise, the biased codon patterns that are expressed in the major codon preference of *E. coli* are a clear indication of the impact on genomic structure of the growth rate maximization constraint (Emilsson and Kurland,

1990; Emilsson, Näslund, and Kurland, unpublished data; Kurland, 1991). It would seem that relatively short periods of intensive growth are sufficient for the growth rate maximization pattern to leave its imprint on natural isolates.

Second, we have presented evidence for the existence of a mechanism through which certain mutants can escape strong selection for optimal ribosome kinetics under slow growth conditions. Under these conditions, bacteria would be expected to drift away from the growth rate maximization design and we would expect mutations that are expressed as suboptimal ribosomes to appear in the populations with higher frequencies. So far we have not seen strong evidence for this sort of effect in the natural isolates.

Finally, we have presented evidence that slowly growing natural isolates have a greater capacity to survive the rapid withdrawal of a carbon source than do wild-type or mutant laboratory stains. In addition, during the conversion of natural isolates to the laboratory phenotype in chemostats a special adaptation to starvation is lost. It seems that the maximum growth rate optimization is antithetical to this adaptation to starvation conditions, and vice versa.

Our present interpretation is that the distributions of growth rates and ribosome phenotypes that we have observed in the natural isolates reflect the variation from population to population of the relative contributions of all three of these microevolutionary factors. Here, the fast growth representatives would have a dominant contribution from the growth rate maximization selection while the slowest ones would have a dominant contribution from the carbon starvation selection.

ACKNOWLEDGMENTS. We thank the Swedish Cancer Society and Natural Sciences Research Council for supporting our work.

REFERENCES

Andersson, D. I., Verseveld, H. W. v., Stouthamer, A. H., and Kurland, C. G., 1986, Suboptimal growth with hyper-accurate ribosomes, *Arch. Microbiol.* **144**:96–101.

Bohman, K. T., Ruusala, T., Jelenc, P. C., and Kurland, C. G., 1984, Kinetic impairment of streptomycin resistant ribosomes, *Mol. Gen. Genet.* **140**:91–100.

Ehrenberg, M., and Kurland, C. G., 1984, Costs of accuracy determined by a maximal growth rate constraint, *Q. Rev. Biophys.* **17**:45–82.

Emilsson, V., and Kurland, C. G., 1990, Growth rate dependence of transfer RNA abundance in Escherichia coli, *EMBO J.* **9**:4359–4366.

Green, M., and Kurland, C. G., 1971, Mutant ribosomal protein with defective RNA binding site, *Nature New Biol.* **234**:273–275.

Jörgensen, F., and Kurland, C. G., 1987, Death rates of bacterial mutants, *FEMS Microbiol. Lett.* **40**:43–46.

Jörgensen, F., and Kurland, C. G., 1990, Processivity errors of gene expression in *Escherichia coli*, *J. Mol. Biol.* **215**:511–521.

Komine, Y., Adachi, T., Inokuchi, H., and Ozeki, H., 1990, Genomic organization and physical mapping of the transfer RNA genes in *Escherichia coli* K12, *J. Mol. Biol.* **212:**579–598.

Kurland, C. G., 1991, Codon bias and gene expression, *FEBS Lett.* **285:**165–169.

Kurland, C. G., 1992, Translational accuracy and the fitness of bacteria, *Annu. Rev. Genet.* **26:** 29–50.

Maaløe, O., 1979, Regulation of the protein-synthesizing machinery ribosomes, tRNA, factors and so on, in: *Biological Regulation and Development* (R. F. Goldberger, ed.), Plenum Press, New York, pp. 487–542.

Matin, A., 1991, The molecular basis of carbon-starvation-induced general resistance in *Escherichia coli*, *Mol. Microbiol.* **5:**3–10.

Menninger, J. R., 1976, Peptidyl transfer RNA dissociates during protein synthesis from ribosomes of *Escherichia coli*, *J. Biol. Chem.* **251:**3392–3398.

Menninger, J. R., 1977, Ribosome editing and the error catastrophe hypothesis of cellular ageing, *Mech. Ageing Dev.* **6:**131–142.

Menninger, J. R., 1978, The accumulation as peptidyl-transfer RNA of isoaccepting transfer RNA families in *Escherichia coli* with temperature-sensitive peptidyl-transfer RNA hydrolase, *J. Biol. Chem.* **253:**6808–6813.

Mikkola, R., and Kurland, C. G., 1988, Media dependence of translational mutant phenotype, *FEMS Microbiol. Lett.* **56:**265–270.

Mikkola, R., and Kurland, C. G., 1991a, Is there a unique ribosome phenotype for naturally occurring *Escherichia coli*? *Biochimie* **73:**1061–1066.

Mikkola, R., and Kurland, C. G., 1991b, Evidence for demand-regulation of ribosome accumulation in *E. coli*, *Biochimie* **73:**1551–1556.

Mikkola, R., and Kurland, C. G., 1992, Selection of laboratory wild type phenotype from natural isolates of *E. coli* in chemostats, *Mol. Biol. Evol.* **9:**394–402.

Morita, R. Y., 1988, Bioavailability of energy and its relationship to growth and starvation survival in nature, *Can. J. Microbiol.* **34:**436–441.

Nyström, T., and Kjelleberg, S., 1989, Role of protein synthesis in the cell division and starvation induced resistance to autolysis of marine *Vibrio* during the initial phase of starvation, *J. Gen. Microbiol.* **135:**1599–1606.

Ochman, H., and Selander, R. K., 1984, Standard reference strains of *Escherichia coli* from natural populations, *J. Bacteriol.* **157:**690–693.

Olsson, M., Isaksson, L., and Kurland, C. G., 1974, Pleiotropic effects of ribosomal protein S4 studied in *Escherichia coli* mutants, *Mol. Gen. Genet.* **135:**191–202.

Ruusala, T., Andersson, D., Ehrenberg, M., and Kurland, C. G., 1984, Hyper-accurate ribosomes inhibit growth, *EMBO J.* **1:**741–745.

Söderlind, O., Thafvelin, B., and Möllby, R., 1988, Virulence factors in *Escherichia coli* strains isolated from Swedish piglets with diarrhoea, *J. Clin. Microbiol.* **26:**879–884.

Spector, M. P., Park, Y. K., Tirgari, S., Gonzalez, T., and Foster, J. W., 1988, Identification and characterization of starvation-regulated genetic loci in *Salmonella typhimurium* by using Mu d-directed *lacZ* operon fusions, *J. Bacteriol.* **170:**345–351.

Wadström, T., Smyth, C. J., Faris, A., Jonsson, P., and Freer, J. H., 1978, Hydrophobic adsorptive and hemagglutinating properties of enterotoxigenic *Escherichia coli* with different colonizing factors: K88, K99 and colonization factor antigens and adherence factor, *Proc. 2nd Int. Symp. Neonatal Diarrhoea*, pp. 30–55.

11

Formation of Viable but Nonculturable Cells

James D. Oliver

1. INTRODUCTION

It has long been realized that plate counts can dramatically underestimate the total number of bacteria (typically determined by direct microscopic examination) present in samples taken from natural environments. In the late 1970s, several easily performed noncultural methods (Zimmerman *et al.*, 1978; Kogure *et al.*, 1979) for determining cell viability allowed confirmation of earlier microautoradiographic studies (e.g., Hoppe, 1976) which demonstrated that many of these nonculturable cells are indeed viable and able to actively metabolize. Such studies led to the further realization that some bacteria, in response to certain environmental stresses, may lose the ability to grow on media on which they are routinely cultured, while remaining viable. This state is of considerable interest to our understanding of microbial ecology. It is of special concern when considering release into the environment of genetically engineered microorganisms, and for those indicator bacteria (e.g., coliforms) and human pathogens which enter this nonculturable state and are thus not detectable through routine bacteriological procedures.

A bacterium in this viable but nonculturable (VBNC) state is defined here as a cell which can be demonstrated to be metabolically active, while being incapable of undergoing the sustained cellular division required for growth in or on a medium normally supporting growth of that cell. It should also be pointed out that while VBNC or "nonculturable" or "nonplateable" have become the most commonly used phrases for describing these cells, all are rather poor

James D. Oliver • Department of Biology, University of North Carolina at Charlotte, Charlotte, North Carolina 28223.

Starvation in Bacteria, edited by Staffan Kjelleberg. Plenum Press, New York, 1993.

descriptions in that, under the proper conditions, it appears that these cells are able to "resuscitate" to the "normal" culturable state.

In this review, the discussion of cells entering the VBNC state is limited to those cells which respond to a natural environmental stress in such a manner. These stresses, e.g., a temperature downshift, are those which would normally be encountered by bacteria in their natural environment. Thus, this review does not include a discussion of the detrimental effects of such xenobiotic agents as antibiotics, chlorine, or other chemicals to which cells may be exposed and which may result in cell injury or death (e.g., see Singh and McFeters, 1987; Singh *et al.*, 1985, 1986). More subjective is my decision not to review here the many publications that exist on the so-called "injured" bacteria which may be found in a variety of samples, such as natural and drinking waters, or in foods which have been frozen and thawed. The sublethal injury incurred by these cells results in their inability to grow in/on the selective media designed for their culture, but not their ability to be cultured on nonselective media (e.g., see McFeters *et al.*, 1982).

The existence of the VBNC state may have been first realized by Dawe and Penrose (1978), who presented evidence that coliforms, when present in seawater, did not die off as had long been reported. Instead, they demonstrated through determination of ATP levels that seawater may in fact have a protective effect on coliforms, although the cells were thought to be "injured." At the same time, Stevenson (1978) speculated that the small bacterial cells routinely observed by direct microscopy of seawater may be in a "dormant" state. Such cells we now know can result from both nutrient starvation and entry of cells into the nonculturable state. The first publication to present clear experimental evidence of a VBNC state was that of Xu *et al.* (1982). These investigators used methods not requiring cell cultivation to demonstrate that both *Vibrio cholerae* and *Escherichia coli*, following incubation in artificial seawater (ASW) solutions, remained viable although they lost all ability to produce colonies on media routinely employed for their culture. These authors thus provided the first direct evidence for the VBNC state, and speculated that "the usefulness of the coliform and fecal coliform indices for evaluating water quality for public health purposes may be seriously compromised. . . ." Following the work of Xu *et al.* (1982), it was realized that the VBNC state exhibited by *V. cholerae* could explain the seasonality and distribution of this organism in regions of the world where cholera is endemic. Colwell and co-workers subsequently showed that the organism was present in many waters from which it could not be cultured (Brayton *et al.*, 1987; Xu *et al.*, 1984).

While the number of bacteria which are reported to enter into the VBNC state continues to grow, very few studies have examined the physiological, metabolic, or genetic basis for this phenomenon. To date, the best studied bacterium in this respect has been *Vibrio vulnificus*, and this review will concentrate on this organism. *V. vulnificus*, which occurs in estuarine waters, is found in high numbers in such filter-feeding shellfish as oysters and clams. The bacterium is of considerable interest owing to its ability to produce extremely rapid and often fatal infections, at rates of over 60%, in humans who have consumed raw or undercooked shellfish, especially oysters (for a review see Oliver, 1989). While *V. vulnificus* is readily isolated from seawater and shellfish during the warm months, researchers have been generally unsuccessful in culturing this, or other estuarine vibrios, during the winter months when seawater temperatures may drop below 5°C. We now realize that *V. vulnificus* enters the VBNC state in response to a temperature downshift.

2. GENERAL CHARACTERISTICS OF THE VBNC STATE

Bacteria which respond to an environmental stress by entrance into the VBNC state respond to various viability assays as demonstrated in Fig. 1. While cells lose plateability, a significant reduction in the total cell counts (○ in Fig. 1) does not occur. Probably most important are one or more of the direct viability

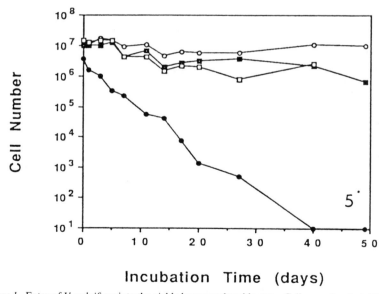

Incubation Time (days)

Figure 1. Entry of *V. vulnificus* into the viable but nonculturable state. Cells were incubated in an artificial seawater (ASW) microcosm at 5°C. Cells were enumerated by acridine orange direct counts (○), direct viable counts (□) by the method of Kogure *et al.* (1979), direct viable counts (■) by the INT method, and plate counts (●) on a marine agar. Taken from Wolf and Oliver (1992).

assays ($\square$ and $\blacksquare$, in Fig. 1) which reveal the number of viable cells present in the population. General characteristics of the VBNC response are described below.

2.1. Loss of Plateability

"Injured" cells demonstrate markedly reduced plateability on inhibitory media compared with nonselective media. In contrast, cells undergoing the VBNC response appear to have similar culturability on all media. For example, Byrd *et al.* (1991) reported little difference in the plateability of *Klebsiella pneumoniae* or *Enterobacter aerogenes* on trypticase soy agar versus Mac-Conkey agar as the cells became nonculturable. Similarly, Roszak *et al.* (1984) reported no significant difference in culturability of *Salmonella enteritidis* on veal infusion or XLD agar as these cells entered the VBNC state. Xu *et al.* (1982) found that as *V. cholerae* cells became nonculturable, they responded approximately equally on inoculation into alkaline peptone or tryptic soy broths, or to platings onto trypticase soy agar or the vibrio-specific medium TCBS. Linder and Oliver (1989) found the same to occur with *V. vulnificus* when plated onto heart infusion agar, a high-salt-containing marine agar, or TCBS agar.

2.2. Size Reduction and Ultrastructural Changes

Cells entering the VBNC state generally exhibit a reduction in size (Rollins and Colwell, 1986; Grimes and Colwell, 1986; Linder and Oliver, 1989; Oliver *et al.*, 1991; Nilsson *et al.*, 1991; Morgan *et al.*, 1991). In the case of *V. vulnificus*, the stationary-phase cells we have used to inoculate ASW microcosms have a size of ca. 1.5×0.7 μm, while the nonculturable cells which develop tend to be small cocci with a diameter of $0.8-1.0$ μm (Fig. 2). Despite this significant drop in biovolume, there is no evidence for the reductive division exhibited by cells entering the starvation state (Novitsky and Morita, 1976, 1977, 1978; Colwell *et al.*, 1985; Kjelleberg *et al.*, 1987). This size reduction may be a strategy to minimize cell maintenance requirements, as has been proposed for cells undergoing starvation. Indeed, Roszak and Colwell (1987a) have suggested that the size reduction seen may represent a sporelike stage for non-spore-forming bacteria.

In one of the only two studies reporting on ultrastructural changes accompanying entry of bacteria into the VBNC state, Linder and Oliver (1989) found VBNC cells of *V. vulnificus* to possess a significantly reduced density of ribosomal and nucleic acid material, but to have apparently maintained a normal cytoplasmic membrane (Fig. 3). An external acidic polysaccharide capsule was retained in the nonculturable cells. This study found all nonculturable cells to be rounded, with sizes only one-half those of culturable cells.

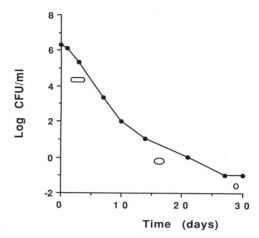

Figure 2. Reduction in cell morphology exhibited by *V. vulnificus* during entry into the viable but nonculturable state. Points indicate plate counts. Adapted from Nilsson *et al.* (1991).

Rollins and Colwell (1986) found an intact, but "asymmetric" membrane structure in transmission electron micrographs of nonculturable *C. jenuni* cells. A condensed cytosol was also observed, similar to that reported by Linder and Oliver (1989) for *V. vulnificus*.

2.3. Time Required to Enter the VBNC State

While cells often require many days, or even months, to become fully nonculturable, great variability exists between species, and even within the same species. Times to entry into the nonculturable state have ranged from 13 h reported for *E. coli* incubated in seawater (Grimes and Colwell, 1986) to as long as 40 days for *V. vulnificus* (Wolf and Oliver, 1992). Different strains of the same bacterial species have been found to enter the nonculturable state at different times, or not at all. Xu *et al.* (1982) and Linder and Oliver (1989) reported almost no drop in culturability of *E. coli* strain ATCC 25922 when incubated in ASW at 5 or 10°C, while Byrd and Colwell (1990) reported strain JM83 to require only 6 days to become nonculturable under similar conditions. Similarly, in their study of seven strains of *V. vulnificus*, Wolf and Oliver (1992) found six strains to become nonculturable at 5°C, while one clinical strain maintained a population of over 10^5 culturable cells during the same time period. Significantly different times required for entry of the same strain (C7184) of *V. vulnificus* into the nonculturable state have also been reported. Linder and Oliver (1989) and Wolf and Oliver (1992) reported entry in 23 and 40 days, respectively, in ASW at 5°C, whereas Oliver and Wanucha (1989) found only 5 days to be required when cells were incubated at 5°C in heart

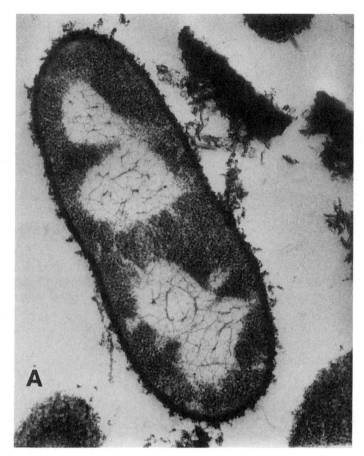

Figure 3. Electron micrographs of *V. vulnificus* at (A) T_0 in an ASW microcosm and (B) the point of nonculturability. Taken from Linder and Oliver (1989).

infusion broth. The time to nonculturability has been noted to vary, even when identical microcosm conditions are employed, as has been reported by Hoff (1989) for *V. salmonicida*. In one microcosm, this bacterium remained culturable for more than 5 months whereas cells were normally seen to become nonculturable within 4 weeks. Factors which have been reported to influence nonculturability are discussed later in this chapter.

While the time required for cells to become completely nonculturable may be quite long, it is also evident that, at least in the case of *V. vulnificus*, cells are able to detect and respond to a temperature downshift in an extremely rapid

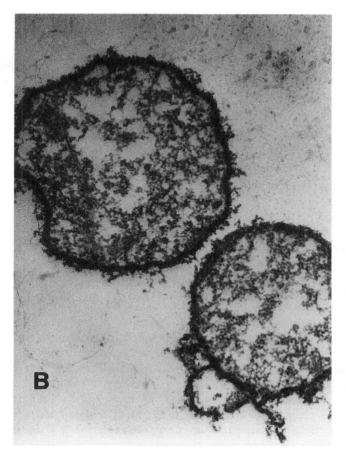

Figure 3. (Continued)

manner. Oliver *et al*. (1991) observed populations of *V. vulnificus* to exhibit a decrease in plate count within 15 min of a temperature downshift. During this time, with the cells placed into a 5°C incubator, the temperature of the culture only decreased from 25°C to 19°C.

2.4. Response to a Reversal of the Inducing Factor

Cells in the VBNC state differ from those in the starved state in that starved cells generally respond quickly to a reversal of the environmental factor (nutrient limitation) which induced the cells' response (Kjelleberg *et al*., 1987;

Nyström *et al.*, 1990). In contrast, the few studies that have examined exit ("resuscitation") from the nonculturable state suggest that the cells' response to a reversal of the inducing factor (e.g., a temperature upshift) is quite slow, requiring hours or days. For example, Nilsson *et al.* (1991) found cells of *V. vulnificus*, made nonculturable by incubation at 5°C, to become culturable only after 3 days of incubation at room temperature. Resuscitation from the nonculturable state is discussed in detail later in this chapter.

3. METHODS FOR DETERMINING THE VBNC STATE

A variety of methods may be employed to determine the percentage of cells of a population which are present in a VBNC state. All have advantages and problems. Most are compared to total direct counts, routinely determined through acridine orange staining and fluorescence microscopy (Daley and Hobbie, 1975; Hobbie *et al.*, 1977), in order to determine the percentage of cells in the population which are viable.

3.1. Direct Viable Count

Probably most commonly used to determine the nonculturable state is the "direct viable count" (DVC) originally reported by Kogure *et al.* (1979). In this procedure, small amounts of nutrient (typically yeast extract) and the DNA synthesis inhibitor, nalidixic acid, are added to the bacterial population under investigation which is then incubated for 6 or more hours. If the cells are able to respond to the nutrient addition (and, of course, many bacteria such as autotrophs would not be), they begin to elongate. Because DNA synthesis and septation are tightly coupled, however, the cells are unable to divide. The result is cells which are significantly elongated, and which are thus clearly viable. Recent studies comparing a variety of methods employed to determine viability have reported a good correlation between DVC counts, heterotrophic activity in natural samples (Kogure *et al.*, 1987), and metabolic activity as determined by microautoradiography (Roszak and Colwell, 1987b). Wolf and Oliver (1992) have also found an excellent correlation between DVC and INT counts (see below). Unfortunately, gram-positive bacteria are generally resistant to nalidixic acid, as are some gram-negative bacteria (e.g., *Aeromonas salmonicida* and *Legionella pneumophila*; Morgan *et al.* 991; Paszko-Kolva *et al.*, 1991). *Legionella* cells have even been reported to lyse when treated with nalidixic acid (Hussong *et al.*, 1987). In other cases, e.g., *Pseudomonas syringae*, the concentration of nalidixic acid must be adjusted in order not to inhibit cell elongation (Wilson and Lindow, 1992).

Another problem with the use of the direct viability assay is its application to natural environments. If the activity of a specific bacterium is to be investigated, then unless a specific fluorescently-tagged antibody is available (see below), it is generally not possible to determine which DVC-positive (elongated) cells are those of the bacterium under study. One solution to this problem was recently described by Wilson and Lindow (1992), who employed a rifampin-resistant strain of *P. syringae* in their study of epiphytic bacterial populations. The addition of rifampin to the nalidixic acid incubation mixture prevented multiplication of nalidixic acid-resistant contaminants, inhibited the elongation of nalidixic acid-sensitive contaminants, yet allowed the determination of metabolically active *P. syringae* cells.

3.2. Detection of Respiration

A second method for demonstrating cell viability involves the use of INT (*p*-iodonitrotetrazolium violet), an electron acceptor which diverts electrons from an active electron transport chain (Zimmerman *et al.*, 1978). Reduction of the soluble INT by metabolizing cells leads to the formation of insoluble INT-formazan, resulting in a visible precipitate in the cell membrane. This method has the distinct advantage over the DVC method in that no metabolizable nutrient is involved and all cells, whether heterotrophic or autotrophic, that possess an active electron transport chain can be detected. The method is also significantly faster (incubation times of less than an hour are routinely employed) than the DVC method. However, we have found (Oliver and Wanucha, 1989) that as cells become nonculturable and undergo a size reduction, the INT precipitates become extremely difficult to detect, and alternate methods, such as DVC, must be used. Fortunately, we have also seen a very high correlation between these two methods, at least in *V. vulnificus* (Fig. 1).

A newer method reported for direct viable counts employs the redox dye 5-cyano-2,3-ditolyl tetrazolium chloride (CTC). Like INT, oxidized CTC is colorless and nonfluorescent, whereas reduction of the compound through electron transport activity results in an CTC-formazan which fluoresces red and accumulates intracellularly. Counterstaining with DAPI (see 3.4 below) allows enumeration of both total cell number and the active subpopulation in the same preparation. Rodriquez *et al.* (1992) employed this method for the first time to examine bacterial cells, and reported 6% (seawater) to 88% (groundwater) of the populations observed to be CTC-positive. It is not yet known, however, whether all bacteria are capable of reducing CTC.

We have compared the activity of *V. vulnificus*, as indicated by both INT and CTC reduction and by the direct viable count method of Kogure *et al.* (1970), when incubated in artificial seawater at both 5°C and room temperature (unpublished observations). We found CTC to indicate lower numbers of viable cells at both incubation temperatures than either direct viable counts or INT reduction, which gave identical results. This was true whether or not exogenous nutrient was added (Rodriguez *et al.*, 1992), and at two CTC concentrations (2 and 5 mM). Nevertheless, CTC reduction allows determination of viability on the same filter preparation (not possible with nalidixic acid–yeast extract direct viable counting) and is easily visualized even in dwarfed cells.

3.3. Monoclonal Antibodies

A third method, which combines the DVC method with the use of fluorescently labeled monoclonal antibodies (MAbs), may prove to be one of the more powerful tools in studying nonculturable cells present in natural environments. The method requires the production of MAbs specific to the bacterium under investigation, which are then either directly or indirectly tagged with a fluorescent compound (typically fluorescein isothiocyanate). This technique allows the determination of viable cells (as a result of the DVC method), with identification of the bacterium under study through its specific reaction with the MAb. The method has been used extensively by Colwell and colleagues (Xu *et al.*, 1982, 1984; Grimes and Colwell, 1986; Roszak *et al.*, 1984; Brayton *et al.*, 1987; Paszko-Kolva *et al.*, 1991).

A potential problem with this method is that, to date, all MAbs that have been employed to detect VBNC cells have been produced against bacterial cells grown in the laboratory. Since it is likely that cells undergo surface changes as they enter the VBNC state, a loss of specificity to a VBNC target bacterium could occur. That such changes occur as cells encounter new environments is well documented in both gram-positive (Wicken and Knox, 1984) and gram-negative (Lugtenberg and Van Alphen, 1983) bacteria. Further, while some investigators (e.g., Colwell *et al.*, 1990) have gone to great lengths to demonstrate that no cross-reactions occur with their MAbs, no studies which have employed MAbs to detect nonculturable bacteria in the natural environment have demonstrated that the bacteria detected actually were the target species.

3.4. Cellular Integrity and Staining of Nucleic Acid with Acridine Orange or DAPI

While not generally considered a method of demonstrating viability, the simple examination of cellular integrity may be sufficient to indicate viability. Gónalez *et al.* (1992) suggest that a dead bacterium should be defined not as one which has lost culturability, but as one which has lost its morphological integrity (i.e., has undergone cellular lysis). As a consequence, they suggest that the only way to detect cell death is to observe decreases in direct counts over time. The determination of morphological integrity can routinely be performed through the acridine orange direct count (AODC) procedure.

The AODC procedure has been suggested as a means to identify living versus dead cells (Strugger, 1948; Korgaonkar and Ranade, 1966). Because acridine orange fluoresces orange when interacting with RNA, but green with DNA, it has been suggested that orange cells could be deemed viable due to the excess of RNA relative to DNA resulting from active metabolism. However, it is now realized that the color of the fluorescence is highly dependent on such factors as pH, incubation time, acridine orange concentration, growth medium, growth phase, and the method by which cells are fixed prior to staining (McFeters *et al.*, 1990; Roszak and Colwell, 1987b), and thus is not a valid indicator of viability. Further, the staining of stationary-phase cells, starved cells, and VBNC cells generally results in green fluorescence because of the lack of large amounts of RNA in cells in these states.

More appropriate, but to date less commonly employed, may be staining by DAPI (4′,6-diamidino-2-phenylindole). This fluorochrome is said to be specific for DNA (Porter and Feig, 1980; Hoff, 1988), and has been extensively used by Enger and colleagues in their studies on the survival of *V. salmonicida*, *V. anguillarum*, and *A. salmonicida* in aquatic environments (reviewed in Enger *et al.*, 1990). They have found DAPI staining to be especially valuable when used as a counterstain to confirm the presence of intact DNA in cells initially detected with fluorescent antibodies.

3.5. Loss of Radiolabel

Garcia-Lara *et al.* (1991) have suggested that loss of culturability relating to the VBNC state might be determined by prelabeling cells with [^{3}H]thymidine, and monitoring of label in the trichloroacetic acid-insoluble fraction in those water samples to which the labeled cells have been added. Using this method, the authors reported radiolabel loss to be an order of magnitude lower than the rate of culturability loss for fecal bacteria added to coastal seawater. This method, coupled with the INT assay, has also been employed to assess the dynamics of *E. coli* (Martinez *et al.*, 1989) and five freshly isolated marine bacterial populations (Penón *et al.*, 1991) in seawater.

Similarly, Paszko-Kolva *et al.* (1991) used the ^{3}H-thymidine-labeling method, compared with viable counts, direct fluorescent counts, and acridine orange counts, to assess the survival of *Legionella pneumophila* in aquatic environments. While the possibility of protozoan predation was not directly examined by these investigators, removal of protozoans through filtration still left a significant decrease in culturable counts, compared with the other viability assays. They attributed this to the VBNC state.

3.6. Intracellular ATP Levels

Dawe and Penrose (1978) were the first to show that cells of *E. coli*, exposed to sewage-contaminated seawater, maintained intracellular levels of ATP, although colony-forming units decreased rapidly with time. More recently, Roth *et al*. (1988) monitored ATP levels in *E. coli* cells exposed to 0.8 M NaCl. These cells underwent a 90% loss of culturability within 3 h of the osmotic stress, but exhibited no decrease in the total cell number as indicated by direct phase microscopy. Most interesting was the finding that ATP levels decreased transiently following the osmotic upshift and then rose, reaching a level 1.5-fold greater than the level in growing cells within 4 h after exposure to the 0.8 M Nacl. This finding provides strong evidence for the presence of an osmotic shock-induced VBNC state in *E. coli* cells.

Beumer *et al*. (1992), studying cells of *Campylobacter jejuni* incubated in physiological saline at 20°C, reported culturable cell counts to decrease from over 10^5 to nondetectable levels after only 4 days, whereas the levels of ATP remained fairly constant for 3 weeks. Under the same conditions, *Salmonella typhimurium* retained both ATP levels and culturability.

3.7 Flow Cytometry

Morgan *et al*. (1991) have employed flow cytometry to characterize cells of *Aeromonas hydrophila* during their nonculturable response in sterile river water. They observed the population to become less varied with time, and to form a tighter distribution, suggesting that the cells had become more uniform as they underwent a size reduction. Kaprelyants and Kell (1991), however, have shown using this technique that cells of *Micrococcus luteus* grown in a chemostat were extremely heterogeneous as regards their ability to accumulate the lipophilic cationic dye rhodamine 123. Based on their studies, they felt able to discern between "viable," "nonviable," and "nonviable but resuscitable" cells.

4. BACTERIA REPORTED TO ENTER THE VBNC STATE

The list of bacteria reported to enter the VBNC state has been growing rapidly. Table 1 lists those bacteria for which the VBNC state has been investigated as of the time of this review. (Note that those studies examining survival rates of various bacteria, but which did not specifically investigate the VBNC state, are not included here.)

Most of the bacteria that have been examined for their entry into the VBNC

Table 1
Bacteria for Which the Nonculturable State Has Been Investigated

Bacterium	References
Aeromonas salmonicida	Allen-Austin *et al.* (1984), Rose *et al.* (1990), Morgan *et al.* (1991, 1993)
Agrobacterium tumefaciens	Byrd *et al.* (1991)
Campylobacter jejuni	Rollins and Colwell (1986), Medema *et al.* (1992), Beumer *et al.* (1991)
Enterobacter aerogenes	Byrd *et al.* (1991)
Enterococcus faecalis	Barcina *et al.* (1990)
Escherichia coli	Dawe and Penrose (1978), Xu *et al.* (1982), Grimes and Colwell (1986), Flint (1987), Roth *et al.* (1988), Linder and Oliver (1989), Barcina *et al.* (1989, 1990), Byrd and Colwell (1990), Gonzales *et al.* (1992)
Klebsiella pneumoniae	Byrd *et al.* (1991)
Legionella pneumophila	Hussong *et al.* (1987), Paszko-Kolva *et al.* (1991)
Pseudomonas putida	Morgan *et al.* (1989)
Salmonella enteritidis	Roszak *et al.* (1984)
Shigella sonnei	Colwell *et al.* (1985)
S. flexneri	Colwell *et al.* (1985)
S. dysenteriae	Islam *et al.* (1993)
Vibrio anguillarum	Hoff (1989)
V. campbelli	Wolf and Oliver (1992)
V. cholerae	Xu *et al.* (1982), Colwell *et al.* (1985, 1990), Wolf and Oliver (1992), Xu and Colwell (1989)
V. mimicus	Wolf and Oliver (1992)
V natriegens	Wolf and Oliver (1992)
V. parahaemolyticus	Wolf and Oliver (1992)
V. proteolyticus	Wolf and Oliver (1992)
V. salmonicida	Hoff (1989)
V. vulnificus	Linder and Oliver (1989), Oliver and Wanucha (1989), Oliver *et al.* (1991), Nilsson *et al.* (1991), Wolf and Oliver (1992)

state are gram-negative. The reason for this may simply be that the nalidixic acid employed in the DVC assay does not generally inhibit gram-positive bacteria. In a study of nonculturability in a variety of bacteria, Byrd *et al.* (1991) reported *Micrococcus flavus* and *Streptococcus faecalis* to lose plateability within 7 days when incubated at 25°C in drinking water. Under the same conditions, *Bacillus subtilis* did not lose culturability over the 25-day study period, although the production of spores was not determined. It must be noted that in none of these gram-positives was viability determined by the DVC procedure, however, and thus whether these bacteria became nonculturable or simply died during the course of the study cannot be stated.

5. FACTORS INDUCING THE NONCULTURABLE RESPONSE

5.1. Temperature

Xu *et al.* (1982) were the first to demonstrate that temperature plays a significant role in the plateability of *V. cholerae*. When incubated in nutrient-free microcosms at 10 or 25°C, little decrease in culturability was seen. In contrast, when the cells were incubated at 4–6°C, a rapid and dramatic decline in culturability was observed, although the DVC method verified the continued viability of the population.

Oliver *et al.* (1991) showed that *V. vulnificus* enters the VBNC state when placed at 5°C, but not room temperature. This is one of the few studies that has demonstrated that a bacterium may have more than one survival strategy. In this case, when incubated at 25°C, *V. vulnificus* undergoes a classic starvation response, whereas at 5°C the cells become nonculturable. Wolf and Oliver (1992) have characterized in more detail the significance of temperature as the inducer of the nonculturable state for this bacterium (Fig. 4). From an initial inoculum of 10^7 cfu/ml, stationary-phase cells of *V. vulnificus* became noncul-turable within 40 days at 5°C, although both DVC and INT assays revealed a

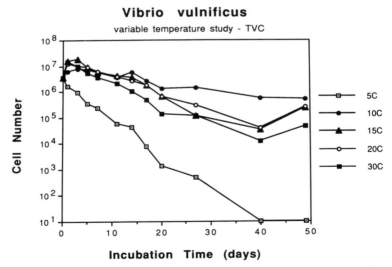

Figure 4. Effect of temperature on entry into the viable but nonculturable state by *V. vulnificus*. All data are from plate counts.

viable population exceeding 10^6 cells/ml. Cells at 10, 15, 20, and 30°C remained culturable (at least 10^4 cfu/ml) throughout the study.

Whereas low temperature is clearly the inducer of the nonculturable state in *V. vulnificus*, this is not always the case for other bacteria. Islam *et al.* (1993) reported *Shigella dysenteriae* to enter the VBNC state within 2–3 weeks at 25°C, depending primarily on the source of the environmental water employed. Rollins and Colwell (1986) reported that *C. jejuni* became rapidly (within 10 days) nonculturable in stream water only at 37°C; at 4°C the population retained a high level of culturability for over 120 days. A similar finding has been reported more recently by Medema *et al.* (1992) for this bacterium. Hussong *et al.* (1987) reported *L. pneumophila* to have a decimal rate of decline of 29 days at 4°C, but only 13 days at 37°C.

5.2. Physiological Age of the Culture

In a recent paper, Oliver *et al.* (1991) examined nine factors which might influence the nonculturable response of *V. vulnificus*: (1) microcosm incubation temperature, (2) inoculum incubation temperature (37 and 22°C), (3) the physiological age of the inoculum (logarithmic or stationary-phase cells), (4) the presence or absence of a nitrogen and phosphate source in the ASW microcosm, (5) whether the cells were washed or not prior to microcosm inoculation, (6) the initial cell density (between 6×10^5 and 4×10^7 cfu/ml), (7) the initial temperature of the plating medium (5 or 25°C), (8) the plate incubation temperature (15, 25, or 37°C), and (9) the salt content of the plating medium (0.5 to 3% NaCl). Other than microcosm incubation temperature, only the physiological age of the inoculum was found to have a significant effect on the time required for the cells to enter the nonculturable state. Cells taken from the stationary phase generally required about twice as many days to become nonculturable at 5°C than did logarithmic-phase cells. Such an effect is presumably related to the production of stationary phase-induced stress proteins, as described in Section 8.

5.3. Salt Levels

Many studies have indicated that salt, typically as present in natural or artificial seawater microcosms, results in an apparent loss of viability among enteric bacteria (e.g., Xu *et al.*, 1982; Roszak *et al.*, 1984; Grimes and Colwell, 1986; Roth *et al.*, 1988). For example, Roth *et al.* (1988) found that 80–90% of *E. coli* cells rapidly (2–3 h) lose plateability, but remain viable, on exposure to high salt (0.8 M NaCl). The study reported by Grimes and Colwell (1986) may be especially significant in that it examined the entry into the nonculturable state by *E. coli* in a natural environment. After suspending *E. coli*, contained in membrane chambers, into seawater off the Bahama coast, a total loss of culturability was observed within 13 h. Despite this decrease, no loss in the number of fluorescent antibody-labeled cells was seen, and direct viable counts revealed the continued presence of viable *E. coli* cells.

That high salt may represent a significant stress to enteric bacteria, and thus to induce a nonculturable state, is not surprising given the normal habitat of such bacteria. In contrast, salt does not appear to be a major inducer of the nonculturable state in the few marine and estuarine bacteria for which this factor has been studied. Xu *et al.* (1982), for example, found little effect on culturability when *V. cholerae* was incubated in microcosms of 5 or 25 ppt salinity. Similarly, Hoff (1989) found salinities of from 20 to 35 ppt to have little effect on the culturability of *V. salmonicida*, while 10 and 15 ppt resulted in rapid declines in plate counts. In all cases, however, total cell counts remained high throughout the 4-week study period. In the same study, *V. anguillarum* was found to undergo a rapid decrease in culturability only at salinities below 10 ppt, with total counts again showing no decreases in cell number.

5.4. Nutrient Levels

Most studies describing the VBNC state in bacteria have employed laboratory microcosms containing such relatively nutrient-depleted solutions as artificial seawater, river water, or tap water. Variations in nutrient concentrations as a factor affecting entrance into the nonculturable response have not been studied. That nutrient level is not a major factor in the nonculturable response of *V. vulnificus* is indicated by the observation that cells of *V. vulnificus*, incubated at 5°C, enter the nonculturable state even when incubated in heart infusion broth (Oliver and Wanucha, 1989).

5.5. Light

Hoff (1989) found no effects of incubation under light or dark conditions on the culturability of either *V. salmonicida* or *V. anguillarum*. In contrast, Barcina *et al.* (1989) reported dramatic effects of visible light on uptake and respiration of [^{14}C]glucose in *E. coli* cells suspended in freshwater microcosms. Cells in illuminated microcosms, but not nonilluminated microcosms, also exhibited a rapid decrease in culturability on both eosin methylene blue agar and trypticase soy agar. In all cases, however, INT determination of metabolic activity and AODC total counts showed no loss of viability in these cells.

A recent study suggests that the effect of visible light on culturability of *E. coli* is due to the photoproduct, hydrogen peroxide (Arana *et al.*, 1992).

5.6. Aeration

Rollins and Colwell (1986) found that aeration (produced by shaking) of microcosms of *C. jejuni* resulted in a more rapid entry (ca. 3 days) into the nonculturable state compared with microcosms held stationary (ca. 10 days).

5.7. Cell Washing

In contrast to our observations with *V. vulnificus* (Oliver *et al.*, 1991) and those of Xu *et al.* (1982) with *E. coli*, Hoff (1989) found washing to significantly affect survival of *V. salmonicida* cells. However, whether these cells were in the VBNC state or not is uncertain as DVC determinations were inconclusive in this study.

6. PHYSIOLOGICAL AND BIOCHEMICAL CHANGES IN VBNC CELLS

Although very few studies have examined in any detail the events occurring within cells undergoing entry into the VBNC state, some aspects of the metabolism and physiology of nonculturable cells have been studied and are summarized here.

6.1. Macromolecular Synthesis

Roth *et al.* (1988) examined protein synthesis in *E. coli* made nonculturable through osmotic shock. They found that the rate of [^{14}C]leucine incorporation into protein gradually decreased from that exhibited by exponentially growing cells (216 μmol/g protein per h) to a constant rate of 2.40 μmol/g protein per h in osmotically shocked, nonculturable, cells. While no net protein synthesis occurred during this period of osmotic upshock, protein synthesis was shown to be restored within 2 h after addition of betaine, an osmotic protectant.

We have examined macromolecular synthesis in cells of *V. vulnificus* grown at 22°C in glucose-minimal medium, then diluted into the nutrient-free minimal salts medium at 5°C (Morton *et al.*, 1992). At intervals pre- and posttransfer, the cells were pulse-labeled with either [^{3}H]thymidine, [^{3}H]uridine, or [^{3}H]leucine, to monitor DNA, RNA, and protein synthesis, respectively. We have observed a decrease of synthesis of all macromolecules immediately upon the temperature downshift (Fig. 5). These data are further evidence that rapid signal transduction occurs in *V. vulnificus* in response to a

Temperature vs. Time

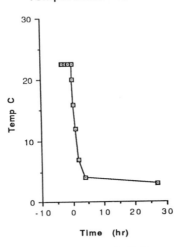

Macromolecular Incorporation of Tritium vs. Time

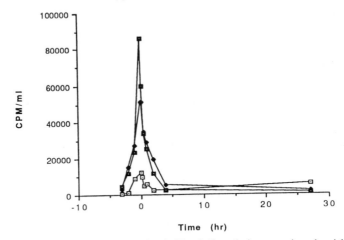

Figure 5. Macromolecular synthesis occurring in *V. vulnificus* during entry into the viable but nonculturable state at 5°C. Cells were assayed for protein (■), DNA (□), and RNA (♦) synthesis as described in the text. Top shows temperature decrease on transfer of culture to 5°C. Data from Morton and Oliver (unpublished).

temperature downshift; note in Fig. 5 that the cells had not reached 5°C until almost 3 h after the microcosm was placed at that temperature. Such a response may represent a protective mechanism designed to prevent the cells from continuing metabolism when nutrients are no longer available.

The dramatic decrease in macromolecular metabolism exhibited by VBNC cells is not unlike that observed in other marine bacteria undergoing a starvation response (Kjelleberg *et al.*, 1987). Further, whereas a temperature downshift from 37 to 5°C has been shown to quickly and significantly reduce the rate of protein synthesis in *E. coli* (Broeze *et al.*, 1978), a downshift from 37 to 10°C has also been shown to induce, or to greatly enhance, the synthesis of 13 proteins in this same organism (Goldstein *et al.*, 1990; Jones *et al.*, 1987). Thus, the nutrient-depletion and low-temperature conditions employed in our studies with *V. vulnificus* might not only induce starvation proteins, but cold shock proteins as well. Both classes of stress proteins are currently being studied in our laboratory.

6.2. Respiration

Morgan *et al.* (1991), in studying the respiration of $[^{14}C]$glucose to $^{14}CO_2$ by *A. hydrophila*, reported a rapid decline to undetectable levels ($< 10^2$ cpm) within 3 days of incubation in lake water at 10°C. INT reduction studies revealed respiration at day 2 of the study, but not after this period. The authors concluded that these nonculturable cells had lost their ability to take up glucose and release $^{14}CO_2$, although no uptake data were shown.

Oliver and Wanucha (1989) used ^{14}C-labeled mixed amino acids to study the uptake and respiration abilities of *V. vulnificus* which had been incubated for 4 h at temperatures from 2 to 13°C. Compared with rates at the highest temperature, we found the transport to be almost totally inhibited at 2°C (ca. 5% of the 13°C level after 60 min), but to be only slightly affected at 9°C. In contrast, respiration of the transported amino acids appeared to be relatively unaffected at even the lowest temperature.

Rollins and Colwell (1986) studied evolution of $^{14}CO_2$ from $[C^{14}]$glutamate in *C. jejuni* incubated in river water at 37°C (the temperature found to induce the VBNC state). Like the results reported by Oliver and Wanucha (1989), theses investigators found a high rate of respiration at the inducing temperature. They hypothesized that this rapid rate of glutamate respiration would be at the expense of other functions, leading to the rapid decline in culturability. No uptake/transport experiments were reported in their study.

Barcina *et al.* (1989), investigating the effects of visible light on the entrance of *E. coli* into the nonculturable state, reported that both the incorporation and respiration of $[^{14}C]$glucose decreased in illuminated cells compared with nonilluminated cells. Like the studies described above, these authors also observed an increase in the percentage of glucose respired to $^{14}CO_2$ as the cells became nonculturable.

6.3. Peptidoglycan

The dramatic change in cell morphology exhibited by *V. vulnificus*, as well as other bacteria entering the nonculturable state, suggests that significant cell envelope modifications must be occurring. Oliver and Kjelleberg (unpublished) found that addition of ampicillin (50 µg/ml final concentration) to cells of *V. vulnificus* incubated in ASW at 5°C resulted in an extremely rapid loss of culturability (Fig. 6). Control cells without ampicillin addition exhibited the expected gradual decline in plateability, with nonculturability reached in 26 days. These data suggest that the nonculturable response involve active peptidoglycan synthesis, at least in its initial stages. Continuous peptidoglycan

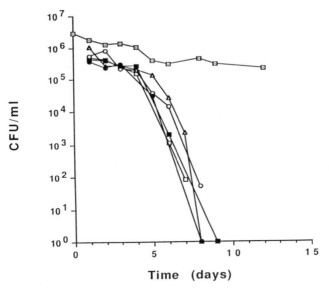

Figure 6. Effects of the peptidoglycan synthesis inhibitor, ampicillin (50 µg/ml), on entry of *V. vulnificus* into the viable but nonculturable state. Additions of ampicillin were made at 0 (□), 1 (○), 2 (Δ), 3 (■), and 4 (●) h after transfer of the cells to 5°C, and the incubation continued for 24 h before removal of the antibiotic. All points represent plate counts of the population as assayed at various intervals after the ampicillin treatment. Control cells (⊡) receiving no ampicillin addition are also shown. Data from Oliver and Kjelleberg (unpublished).

synthesis has been reported during the starvation response of *Vibrio* sp. S14, which presumably allows for the development of a resistant cell (Nyström and Kjelleberg, 1989).

6.4. Plasmids

van Overbeek *et al.* (1990) have shown that the plasmids present in *Pseudomonas* and *Klebsiella* species remained stable when the cells became nonculturable in agricultural drainage water. Similarly, Morgan *et al.* (1991) reported that probes made to three different plasmids hybridized with plasmid extracts prepared from cells of *A. salmonicida* which had become nonculturable. Grimes *et al.* (1986) reported that when *E. coli* H10407 cells were made nonculturable at 18°C, the plasmids harbored by this strain could no longer be detected on direct extraction and electrophoresis. However, when a portion of the nonculturable cells were grown in a nutrient broth, plasmid bands were present in the culturable cells. Similarly, Grimes and Colwell (1986) have reported that when the cells of the same strain of *E. coli*, which had been made nonculturable in natural seawater, were resuscitated in ligated rabbit ileal loops, the same pattern of plasmids as present in the original inocula were evident. Byrd and Colwell (1990), studying *E. coli* strains JM101 and JM83, also found the plasmids pUC8 and pBR322, respectively, to be maintained for at least 21 days when the cells were made nonculturable in ASW at 15°C. While not examined in their study, the possibility that such plasmids could remain transferable to autochthonous bacteria, or that resuscitation of the plasmid-bearing nonculturable population could occur, was discussed. We (Murphy and Oliver, unpublished) have observed that F^+ cells of *E. coli*, incubated at 5, 17, and 24°C, remained able to conjugate, resulting in antibiotic-resistant exconjugants, for up to 48 days. At this point, the donor cells were below the normal limits of detection by routine plating techniques. Such findings have significant and obvious importance for the potential of gene transfer in the environment, and along with potential virulence, are a major concern for the presence of nonculturable cells in the environment.

6.5. Chromosomal DNA

The ability to detect nonculturable cells, especially genetically engineered microorganisms that may be intentionally or unintentionally released to the natural environment, is of utmost concern. A possible solution to detecting nonculturable cells in the environment is the use of the polymerase chain reaction (PCR), a method to amplify DNA. Steffan and Atlas (1988) have shown in laboratory studies that the DNA from as few as 100 cells of a target bacterium (*Pseudomonas cepacia*) added to 100 g of river sediment could be amplified from a background of 10^{11} nontarget bacteria. Hassan *et al.* (1992) have recently employed this method to detect the cholera toxin gene in VBNC cells of *V. cholerae* present both in laboratory microcosms and in a diarrheal stool sample from a culture-negative patient with cholera symptoms.

Recently, Islam *et al.* (1993) employed PCR methodology, amplifying a portion of the unique invasion plasmid gene, *ipaH*, to detect cells of *Shigella dysenteriae* present in laboratory microcosms 3–4 weeks after their entry into the VBNC state. No discussion of any difficulties employing this method was described by these authors. In research reported from our laboratory, however, ca. 400 times more DNA extracted from cells of *V. vulnificus* in the VBNC state was found to be required for amplification by the PCR method (Brauns *et al.*, 1991). Recent reviews by both Steffan and Atlas (1991) and Bej and Mahbubani (1992) report extensive purification of nucleic acids to be necessary for successful PCR amplification of DNA from microorganisms obtained from environmental samples. It is likely that as cells enter the VBNC state, some change in the DNA may be occurring which prevents amplification by the PCR technique, and that the inability to amplify DNA taken from VBNC cells of *V. vulnificus* may be typical for microorganisms in general.

A number of studies have suggested modifications in DNA composition or structure may occur during starvation or nonculturability (e.g., see Baker *et al.*, 1983; Hood *et al.*, 1986; Hoff, 1989; Linder and Oliver, 1989; Moyer and Morita, 1989; Brauns *et al.*, 1991). Most indicate a condensation of the nuclear region, although no definitive studies on this point have been published.

6.6. Lipids

Only two studies have been published on changes in the membrane lipid composition of cells entering the VBNC state. Linder and Oliver (1989) found the major fatty acid species (C_{16}, $C_{16:1}$, C_{18}) of nonculturable cells of *V. vulnificus* to be decreased almost 60% compared with culturable cells. The percentage of fatty acids with chain lengths less than C_{16} increased as the cells became nonculturable, and long-chain acids (C_{19}, C_{20}, $C_{22:1}$) appeared which were not evident in culturable cells. A similar, but less dramatic, trend in fatty acid composition was also observed for *E. coli*.

Morgan *et al.* (1991) have reported similar observations for *A. hydrophila*. The incubation of these cells in river water at 10°C resulted in a reduction in the levels of $C_{16:1}$ and $C_{18:1}$, with the concomitant appearance of both shorter and longer chain acids.

Although no direct measurements were made in either of these studies, such changes presumably allow the cells to maintain membrane fluidity as they enter the nonculturable state.

6.7 Capsule

V. vulnificus exhibits two distinct (opaque and translucent) colony types, resulting from the presence or absence, respectively, of a surface polysaccharide. Of the two types, only the encapsulated form is virulent for laboratory animals (Simpson *et al.*, 1987). Wolf and Oliver (1992) observed no difference in the time required for the two morphotypes to enter the nonculturable state, suggesting no role for this polysaccharide in the VBNC state. Further, no loss of this capsule was observed when the cells entered the VBNC state (Linden and Oliver, 1989).

7. RESUSCITATION FROM THE VBNC STATE

Resuscitation is defined here as a reversal of the metabolic and physiologic processes that resulted in nonculturability, resulting in the ability of the cells to be culturable on those media normally supporting growth of the cell.

Resuscitation would appear to be essential to the VBNC state if this state is truly a survival strategy. Whereas the nonculturable state may in some manner protect the cell against one or more environmental stresses, resuscitation to a cell potentially capable of rapid metabolism would allow the cell to actively compete in the environment. We have theorized (Nilsson *et al.*, 1991) that cells of *V. vulnificus*, which become nonculturable when water temperatures are low, would resuscitate, even in low nutrient levels, when water temperatures become warmer. In the presence of nutrient, the cells would then likely initiate DNA replication and active growth.

7.1. In Vitro Studies

Unlike starved bacteria, cells in the VBNC state do not appear to respond quickly to a reversal of the factor which initially induced the nonculturable state. Roszak *et al.* (1984) were apparently the first to report resuscitation of cells which were present in the VBNC state. In their study, cells of *S. enteritidis*, made nonculturable (< 1 cfu/ml) in sterile river water, were reported to be resuscitated to nearly the original culturable cell density through the addition of nutrients (heart infusion broth) to 50-ml subsamples taken from the microcosm. A period of 25 h was required following nutrient addition before plateable cells appeared. These results were obtained when cells had been nonculturable 4 days; attempts to resuscitate cells after 21 days of nonculturability were unsuccessful. The authors suggested that longer periods of "dormancy" would require conditions other than simple nutrient addition to restore culturability.

Morgan *et al.* (1991) conducted resuscitation studies on cells of *A. salmonicida* made nonculturable in river water at 10°C. In their protocol, nutrient was added to cells contained in 1-, 10-, and 100-ml subsamples removed from the microcosms. After 5 days, samples were plated onto a nutrient medium to monitor growth. Resuscitation of *A. salmonicida* was found to occur only when culturable cells remained evident in the microcosms. Because the DVC method cannot be employed with this bacterium (it is resistant to nalidixic acid), the authors were unable to verify that the cells had indeed entered a VBNC state, and therefore whether resuscitation actually occurred. However, they noted that morphological integrity of the cells was maintained. Such a criterion has been suggested by Gónzalez *et al.* (1992) as a definition of viability. Morgan *et al.* (1991) point out that while Allen-Austin *et al.* (1984) reported *A. salmonicida* to enter a VBNC state, those authors were also only able to achieve "revival" of the "nonculturable" cells when culturable cells were present in the population. Morgan *et al.* (1991) suggested that growth of these few culturable cells resulted in a false resuscitation of nonculturable cells. Such a conclusion is in agreement with the findings of Rose *et al.* (1990), and emphasizes the point that differences should not exist in sample volumes between resuscitation and culture studies. Illustrating this concern is the finding of Morgan *et al.* (1991) that only when nutrient was added to 100-ml volumes taken from their microcosms, in which culturable cells could be demonstrated, but not the 1- or 10-ml volumes, could apparent resuscitation of nonculturable cells be obtained. The study of Roszak *et al.* (1984) involved culturable cells at a density of < 1 cfu/ml. Because 50-ml subsamples were removed for their resuscitation experiments, the possibility of culturable cells being present cannot be excluded.

We (Oliver *et al.*, 1991) have also pointed out the potential problem with resuscitation studies that rely on nutrient additions to nonculturable cells. Whether true resuscitation of the VBNC cells occurs, or whether a few culturable cells, which may have remained undetected following normal bacteriological culture methods, multiplied as a result of the added nutrient is difficult to prove. In an attempt to circumvent the nutrient problem, we (Nilsson *et al.*, 1991) reported for the first time the resuscitation of bacteria from ASW microcosms without nutrient addition. In these studies, cells of *V vulnificus* became nonculturable (< 0.1 cfu/ml) after 27 days at 5°C. The determination of such a low level of culturability was accomplished through filtration of 10-ml samples of the microcosm, with the filters being placed on a nonselective (LB agar) medium. When the nonculturable cells were subjected to a temperature upshift (to room temperature), the original bacterial numbers were detectable by plate counts after 3 days. No increase in total cell count was observed during the temperature-induced resuscitation, suggesting that the plate count increases were not the result of growth of a few culturable cells. In these studies, the subsamples removed for the resuscitation studies were of a 10-ml volume, which should have contained, on average, < 1 culturable cell of *V. vulnificus*. That resuscitation was not a result of a few culturable cells remaining present in the microcosm was suggested by microscopic examination which revealed no cocci present after 48 h (suggesting that the cocci present prior to the temperature upshift had resuscitated to rods). More recently, however, we have reported studies which indicate that, while no exogenous nutrients had been added to achieve resuscitation, the increases in cell numbers observed by Nilsson *et al.* (1991) could have been caused by regrowth of a few cells employing nutrients released by moribund or dead cells present in the population (Weichart *et al.*, 1992).

Roth *et al.* (1988) used a different approach to show that nonculturable cells of *E. coli* were capable of true resuscitation. When exposed to an osmotic stress (0.8 M NaCl), these cells lost culturability, but could be resuscitated through additions of the osmoprotectant betaine. Further, it was observed that betaine restored colony-forming ability (to 80% of the original level) even when protein synthesis was inhibited by the addition of chloramphenicol. Although nutrient was present in their system, the fact that the formation of new cells was prevented appears to rule out the possibility of cell proliferation as the cause of resuscitation.

As noted above, resuscitation, whether in the presence of exogenous nutrient or not, appears to require a considerable length of time. In the case of *V. vulnificus*, 2–4 days is typically required before plateable cells are observed following a temperature upshift. In examining the rates of macromolecular synthesis in cells of *V. vulnificus* which had been nonculturable for 2 days, Morton and Oliver (unpublished) observed a delay of 48 h before an increase in macromolecular synthesis could be detected following a temperature upshift. During this time, resuscitation is dramatically affected by additions of either ampicillin or chloramphenicol to inhibit peptidoglycan and protein synthesis, respectively (Nilsson *et al.*, 1991). Perhaps a more rapid recovery would be detrimental to the cell, placing it in a position where it is unable to survive a second round of stress if it comes quickly.

7.2. Do Nonculturable Cells of Pathogenic Bacteria Retain Virulence?

As with *in vitro* resuscitation studies, the complication of nutrient additions also exists in those studies which have employed animals. Despite this concern, any situation wherein cells of a bacterial pathogen are undetectable by standard methods, yet are capable of increasing in number to potentially hazardous levels, is of considerable public health concern. Thus, whether "true resuscitation" or division of a few "contaminating," but undetectable, culturable cells has occurred is unimportant.

A number of studies have employed animal models to monitor changes in virulence which might accompany entry into the VBNC state. As an example, we have reported (Birbari *et al.*, 1991) that VBNC cells of *V. vulnificus* could be resuscitated following passage through clams. The inability to detect such an important human pathogen in shellfish is obviously of considerable concern.

We (Simpson *et al.*, 1992) have also observed that when 0.5 ml of an ASW microcosm of cells of *V. vulnificus* (strain MO6), which had been nonculturable for over 8 months, was added to a murine macrophage culture, resuscitation to large numbers was observed within 6 h at 37°C. No resuscitation occurred in the RPMI cell culture medium alone, either with or without CO_2. Resuscitation was originally evident through the microscopic appearance of very long rods, which after ca. 24 h had divided to result in "normally appearing" cells of *V. vulnificus*.

Colwell *et al.* (1985, 1990) prepared nonculturable cells of an attenuated strain of *V. cholerae* O1 by incubation at 4°C. These were then fed to two human volunteers at a dosage of ca. 10^8 cells/ml (based on AODC; ca. 5×10^6 cells/ml based on DVC studies). Approximately 48 h after the challenge, one volunteer passed culturable *V. cholerae* O1 cells in his stool at a concentration of ca. 3×10^3 cells/g. After 5 days, the second volunteer also passed *V. cholerae* cells. At no time did incubation of the cells in a nutrient medium reveal the presence of culturable cells present among the nonculturable population, suggesting that resuscitation of the cells had occurred *in vivo*. Although cells which remained nonculturable for 1 month did not result in *in vivo* resuscitation, the authors believe their results support the hypothesis that nonculturable cells of *V. cholerae* can maintain pathogenic potential, and that human passage can provide a means of resuscitation for such cells.

Colwell *et al.* (1985) have also reported resuscitation and evidence of continued virulence for both *V. cholerae* and *E. coli* when these nonculturable cells were introduced into ligated rabbit ileal loops. Similarly, when cells of *E. coli*, following 4 days of nonculturability in natural seawater, were concentrated by centrifugation and introduced into ligated rabbit ileal loops, Grimes and Colwell (1986) reported successful reisolation of the cells after 36 h. Confirmation of the identity of the cells was accomplished through plasmid characterization. Although not directly tested, the authors considered the nonculturable cells to have retained virulence, as the plasmids harbored by this *E. coli* strain encoded colonizing factor antigen I, and both the heat-stable and heat-labile toxin subunits of the *E. coli* enterotoxin.

Hussong *et al.* (1987) injected chick embryos with nonculturable cells of *L. pneumophila* taken from a microcosm of sterile tap water. The lethality observed was considerably greater than they could account for by the number of culturable cells present, a result which the authors believed to be caused by virulence of the nonculturable cells. In contrast, Medema *et al.* (1992) were unable to demonstrate colonization of the intestines of 1-day-old chicks when 10^5 VBNC cells of *Campylobacter jejuni* were introduced orally. Recovery of cells following passage through the allantoic fluid of embryonated eggs was also unsuccessful. In that the infectious oral dose for cultured cells in this model was stated to be as low as 26 cfu, these authors questioned the significance of VBNC cells in the host–host transmission of campylobacters. Beumer *et al.* (1992) were also unsuccessful in demonstrating resuscitation of *C. jejuni* cells, made nonculturable in physiological saline, following transfer into simulated stomach, ileal, or colon environments. Nor were these authors able to demonstrate the presence of culturable cells in fecal samples taken for up to 14 days from laboratory animals or up to 30 days from human volunteers fed nonculturable *C. jejuni* cells.

Initial studies (Linder and Oliver, 1989) examining the effect of nonculturability on virulence of *V. vulnificus* in mice suggested that VBNC cells lost virulence. However, a relatively low inoculum of *V. vulnificus* (5×10^4 cells based on DVC data) was employed. In more recent studies (Manahan and Oliver, unpublished), we have found that injections into mice of 0.5 ml of a population of nonculturable (< 0.1 cfu/ml) *V. vulnificus* cells resulted in lethality. In this experimental protocol, less than 0.05 cfu of culturable *V. vulnificus* was injected, suggesting that *in vivo* resuscitation of the cells occurred.

Aeromonas salmonicida is the causative agent of furunculosis in fish. A number of studies have indicated this bacterium loses culturability in river and seawater (Allen-Austin *et al.*, 1984; Rose *et al.*, 1990), but whether a true VBNC state exists for this bacterium has been questioned (Rose *et al.*, 1990; Morgan *et al.*, 1991). Nevertheless, Michel and Dubois-Darnaudpeys (1980) have reported virulence to be maintained in this bacterium on prolonged storage in river sediments.

8. RELATIONSHIP BETWEEN THE VBNC AND STARVATION STATES

While starvation has not been systematically studied in as great a variety of bacteria as the nonculturable state, it is clear that at least some bacteria are able to enter both states, depending on the environmental conditions. In the case of *V. vulnificus*, low temperature (e.g., 5°C) induces the VBNC state, whereas at warmer temperatures (e.g., 20°C) cells enter into the starvation state (Fig. 7). It is difficult to interpret the relative importance of the two responses to the cell, but it seems highly likely that the starvation response plays a significant role in

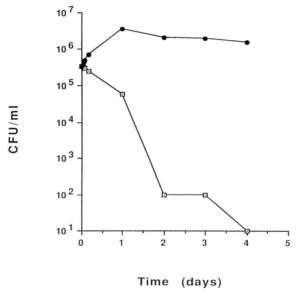

Time (days)

Figure 7. Comparison of the starvation (room temperature) and viable but nonculturable (5°C) responses of *V. vulnificus* cells incubated in ASW microcosms. Plate counts were made on cells grown to logarithmic phase at room temperature and incubated in ASW microcosms held at 5°C (□) or room temperature (●). Taken from Oliver *et al*. (1991).

the survival of bacteria in the natural environment. Starvation leads to "cross-protection," a response to a certain stress which generates protection not only against that particular stress but to other, unrelated stresses as well. This is most likely accomplished through the production of the so-called "heat shock," or stress, proteins. While the mode of action of stress proteins is only beginning to be understood, it is clear that their synthesis by a cell undergoing starvation results in protection against such potentially fatal factors as UV light, elevated temperature, and heavy metal toxicity (see chapters in this monograph for details). Cross-protection has not been demonstrated for bacteria entering the nonculturable state.

When cells of *V. vulnificus* were incubated under carbon starvation conditions at room temperature for as little as 4 h prior to incubation at 5°C, Oliver *et al*. (1991) observed a decrease in the rate at which cells became nonculturable. Cells prestarved for 24 h or more exhibited no decrease in culturability when transferred to 5°C. This raises interesting questions concerning the role of starvation-specific gene expression. The coordinated response to carbon starvation has been reported to involve the induction of stimulons and regulons, some of which overlap with other known stress responses such as the heat shock regulon (see Nyström, Spector, Hengge-Aronis, this volume). It appears that the starvation conditions employed in our study induced such responses in *V. vulnificus*, and that these repress the VBNC program displayed during low-temperature incubation. The interaction of these two stresses (nutrient starvation and temperature downshift) is likely to be complex and is currently being investigated in our laboratory.

9. CONCLUSIONS

The significance of the VBNC state is not yet fully known. This state may allow for survival during long periods in an adverse environment (e.g., low or high temperature, lack of nutrient, low salinity). Regardless of the mechanisms, it is clear that (1) many bacteria become nonculturable while remaining viable, (2) the environmental factors which induce this state vary considerably for different bacteria, (3) entrance into the VBNC state involves a variety of morphological, physiological, and biochemical changes in the cells, (4) cells in the VBNC state are capable of maintaining plasmids, many of which are potentially transferable, (5) the VBNC state appears to be under genetic control, and there exists some relationship with the starvation-survival response, and (6) while conclusive evidence does not yet exist for resuscitation of nonculturable cells in the laboratory, a number of pathogenic bacteria have been shown to become culturable, and to retain virulence, following *in vitro* and *in vivo* treatment.

In their classic study on the survival of starved bacteria, Postgate and Hunter (1962) reported on a variety of parameters which affect survival of cells undergoing nutrient starvation. In their study, the term "dead" was used to describe ". . . bacteria that failed to multiply in the arbitrary favourable environment provided. . . ." Such a definition has been widely held by microbiologists, who are primarily concerned with populations as opposed to single bacterial cells. However, as later realized by Postgate (1976), bacteria may lose their ability to form colonies, yet remain functional as individual cells. Indeed, Hoppe (1976, 1978) has suggested that nonculturable cells are primarily responsible for organic turnover in the open ocean. Whether VBNC cells play this role has yet to be conclusively demonstrated. However, the possibility of *in situ* resuscitation of such cells could provide a mechanism whereby cells, in a metabolically inactive state, would revert to a state of increased metabolism and the ability to become a functioning member of a microbial population.

REFERENCES

Allen-Austin, D., Austin, B., and Colwell, R. R., 1984, Survival of *Aeromonas salmonicida* in river water, *FEMS Microbiol. Lett.* **21:**143–146.

Arana, I., Muela, A., Iriberri, J., Egea, L., and Barcina, I., 1992, Role of hydrogen peroxide in loss of culturability mediated by visible light in *Escherichia coli* in a freshwater ecosystem, *Appl. Environ. Microbiol.* **58:**3903–3907.

Baker, R. M., Singleton, F. L., and Hood, M. A., 1983, Effects of nutrient deprivation on *Vibrio cholerae*, *Appl. Environ. Microbiol.* **46:**930–940.

Barcina, I., González, J. M., Iriberri, J., and Egea, L., 1989, Effect of visible light on progressive dormancy of *Escherichia coli* cells during the survival process in natural fresh water, *Appl. Environ. Microbiol.* **55:**246–251.

Barcina, I., González, J. M., Iriberri, J., and Egea, L., 1990, Survival strategy of *Escherichia coli* and *Enterococcus faecalis* in illuminated fresh and marine systems, *J. Appl. Bacteriol.* **68:**189–198.

Bej, A. K., and Mahbubani, M. H., 1992, Applications of the polymerase chain reaction in environmental microbiology, *PCR Methods Appl.* **1:**151–159.

Beumer, R. R., de Vries, J., and Rombouts, F. M., 1992, *Campylobacter jejuni* non-culturable coccoid cells, *Intern. J. Food Microbiol.* **15:**153–163.

Birbari, W., Rodrick, G. F., and Oliver, J. D., 1991, Uptake and resuscitation of viable but nonculturable *Vibrio vulnificus* by *Mercenaria campechiensis*, *Abstr. 16th Annu. Trop. Subtrop. Fish. Technol. Conf.*

Brauns, L. A., Hudson, M. C., and Oliver, J. D., 1991, Use of the polymerase chain reaction in detection of culturable and nonculturable *Vibrio vulnificus* cells, *Appl. Environ. Microbiol.* **57:**2651–2655.

Brayton, P., Tamplin, M., Huq, A., and Colwell, R., 1987, Enumeration of *Vibrio cholerae* O1 in Bagladesh waters by fluorescent-antibody direct viable count. *Appl. Environ. Microbiol.* **53:**2862–2865.

Broeze, R. J., Solomon, C. J., and Pope, D. H., 1978, Effects of low temperature on in vivo and in vitro protein synthesis in *Escherichia coli* and *Pseudomonas fluorescens*, *J. Bacteriol.* **134:**861–874.

Byrd, J. J., and Colwell, R. R., 1990, Maintenance of plasmids pBR322 and pUC8 in nonculturable *Escherichia coli* in the marine environment, *Appl. Environ. Microbiol.* **56:**2104–2107.

Byrd, J. J., Xu, H.-S., and Colwell, R. R., 1991, Viable but nonculturable bacteria in drinking water, *Appl. Environ. Microbiol.* **57:**875–878.

Colwell, R. R., Brayton, P. R., Grimes, D. J., Roszak, D. B., Huq, S. A., and Palmer, L. M., 1985, Viable but non-culturable *Vibrio cholerae* and related pathogens in the environment: Implications for the release of genetically engineered microorganisms, *Bio/Technology* **3:**817–820.

Colwell, R. R., Tamplin, M. L., Brayton, P. R., Gauzens, A. L., Tall, B. D., Herrington, D., Levine, M. M., Hall, S., Huq, A., and Sack, D. A., 1990, Environmental aspects of *Vibrio cholerae* in transmission of cholera, in: *Advances on Cholera and Related Diarrheas*, Vol. 7 (R. B. Sack and Y. Zinnaka, eds.), KTK Scientific, Tokyo, pp. 327–343.

Daley, R. J., and Hobbie, J. E., 1975, Direct counts of aquatic bacteria by a modified epifluorescence technique, *Limnol. Oceanogr.* **20:**875–882.

Dawe, L. L., and Penrose, W. R., 1978, "Bactericidal" property of seawater: Death or debilitation? *Appl. Environ. Microbiol.* **35:**829–833.

Enger, O., Hoff, K. A., Schei, G., and Dundas, I., 1990, Starvation survival of the fish pathogenic bacteria *Vibrio anguillarum* and *Vibrio salmonicida* in marine environments, *FEMS Microbiol. Ecol.* **74:**215–220.

Flint, K. P., 1987, Long term survival of *Escherichia coli* in river water, *J. Appl. Bacteriol.* **63:** 261–270.

Garcia-Lara, J., Menon, P., Servais, P., and Billen, G., 1991, Mortality of fecal bacteria in seawater, *Appl. Environ. Microbiol.* **57:**885–888.

Goldstein, J., Pollitt, N. S., and Inouye, M., 1990, Major cold shock protein of *Escherichia coli*, *Natl. Acad. Sci. USA* **87:**283–287.

Gónzalez, J. M., Iriberri, J., Egea, L., and Barcina, I., 1992, Characterization of culturability, protistan grazing, and death of enteric bacteria in aquatic ecosystems, *Appl. Environ. Microbiol.* **58:**998–1004.

Grimes, D. J., and Colwell, R. R., 1986, Viability and virulence of *Escherichia coli* suspended by membrane chamber in semitropical ocean water, *FEMS Microbiol. Lett.* **34:**161–165.

Grimes, D. J., Atwell, R. W., Brayton, P. R., Palmer, L. M., Rollins, D. M., Roszak, D. B., Singleton, F. L., Tamplin, M. L., and Colwell, R. R., 1986, The rate of enteric pathogenic bacteria in estuarine and marine environments, *Microbiol. Sci.* **3:**324–329.

Hassan, J. A. K., Shahabuddin, M., Huq, A., Loomis, L., and Colwell, R. R., 1992, Polymerase chain reaction for detection of cholera toxin genes in viable but nonculturable *Vibrio cholerae* O1, *Abstr. Annu. Meet. Am. Soc. Microbiol.* D138.

Hobbie, J. E., Daley, R. J., and Jasper, S., 1977, Use of Nuclepore filters for counting bacteria by fluorescence microscopy, *Appl. Environ. Microbiol.* **33:**1225–1228.

Hoff, K. A., 1988, Rapid and simple method for double staining of bacteria with 4′,6-diamidino-2-phenylindole and fluorescein isothiocyanate-labeled antibodies, *Appl. Environ. Microbiol.* **54:**2949–2952.

Hoff, K. A., 1989, Survival of *Vibrio anguillarum* and *Vibrio salmonicida* at different salinities, *Appl. Environ. Microbiol.* **55:**1775–1786.

Hood, M. A., Guckert, J. B., White, D. C., and Deck, F., 1986, Effect of nutrient deprivation on lipid, carbohydrate, DNA, RNA, and protein levels in *Vibrio cholerae*, *Appl. Environ. Microbiol.* **52:**788–793.

Hoppe, H. G., 1976, Determination and properties of actively metabolizing heterotrophic bacteria in the sea, investigated by means of microautoradiograpy, *Mar. Biol.* **36:**291–302.

Hoppe, H. G., 1978, Relations between active bacteria and heterotrophic potential in the sea, *Neth. J. Sea Res.* **12:**78–98.

Hussong, D., Colwell, R. R., O'Brien, M., Weiss, A. D., Pearson, A. D., Weiner, R. M., and Burge, W. D., 1987, Viable *L. pneumophila* not detectable by culture on agar media, *Bio/Technology* **5:**947–950.

Islam, M. S., Hasan, M. K., Miah, M. A., Sur, G. C., Felsenstein, A., Venkatesan, M., Sack, R. B., and Albert, M. J., 1993, Use of the polymerase chain reaction and fluorescent-antibody methods for detecting viable but nonculturable *Shigella dysenteriae* type 1 in laboratory microcosms, *Appl. Environ. Microbiol.* **59:**536–540.

Jones, P. G., vanBogelen, R. A., and Neidhardt, F. C., 1987, Induction of proteins in response to low temperature in *Escherichia coli*, *J. Bacteriol.* **169:**2092–2095.

Kaprelyants, A. S., and Kell, D. B., 1991, Rapid assessment of bacterial viability and vitality using rhodamine 123 and flow cytometry, *J. Appl. Bacteriol.* **72:**410–422.

Kjelleberg, S., Hermansson, M., Mårdén, P., and Jones, G. W., 1987, The transient phase between growth and non-growth of heterotrophic bacteria, with emphasis on the marine environment, *Annu. Rev. Microbiol.* **41:**25–49.

Kogure, K., Simidu, U., and Taga, N., 1979, A tentative direct microscopic method for counting living marine bacteria, *Can. J. Microbiol.* **25:**415–420.

Kogure, K., Simidu, U., Taga, N., and Colwell, R. R., 1987, Correlation of direct viable counts with heterotrophic activity for marine bacteria, *Appl. Environ. Microbiol.* **53:**2332–2337.

Korgaonkar, K. S., and Ranade, S. S., 1966, Evaluation of acridine orange fluorescence test in viability studies of *Escherichia coli*, *Can J. Microbiol.* 12:185–190.

Linder, K., and Oliver, J. D., 1989, Membrane fatty acid and virulence changes in the viable but nonculturable state of *Vibrio vulnificus*, *Appl. Environ. Microbiol.* 55:2837–2842.

Lugtenberg, B., and Van Alphen, L., 1983, Molecular architecture and functioning of the outer membrane of *Escherichia coli* and other gram-negative bacteria, *Biochim. Biophys. Acta* 737:51–115.

Martinez, J., Garcia-Lara, J., and Vebes-Rego, J., 1989, Estimation of *Escherichia coli* mortality in seawater by the decrease in ^{3}H-label and electron transport system activity, *Microb. Ecol.* 17:219–225.

McFeters, G. A., Cameron, S. C., and LeChavalier, M. W., 1982, Influence of diluents, media, and membrane filters on detection of injured waterborne coliform bacteria, *Appl. Environ Microbiol.* 43:97–103.

McFeters, G. A., Singh, A., Williams, S., Byun, S., and Callis, P. R., 1990, Acridine orange staining as an index of physiological activity in *E. coli*, *Abstr. Annu. Meet. Am. Soc. Microbiol.* Q133, p. 310.

Medema, G. J., Schets, F. M., van de Giessen, A. W., and Havelaar, A. H., 1992, Lack of colonization of 1 day old chicks by viable, non-culturable *Campylobacter jeuni*, *J. Appl. Bacteriol.* 72:512–516.

Michel, C., and Dubois-Darnaudpeys, A., 1980, Persistence of the virulence of *Aeromonas salmonicida* strains kept in river sediments, *Ann. Rech. Vet.* 11:375–380.

Morgan, J. A. W., Cranwell, P. A., and Pickup, R. W., 1991, Survival of *Aeromonas salmonicida* in lake water, *Appl. Environ. Microbiol.* 57:1777–1782.

Morgan, J. A. W., Rhodes, G., and Pickup, R. W., 1993, Survival of nonculturable *Aeromonas salmonicida* in lake water, *Appl. Environ. Microbiol.* 59:874–880.

Morgan, J. A. W., Winstanley, C., Pickup, R. W., Jones, J. G., and Saunders, J. R., 1989, Direct phenotypic and genotypic detection of a recombinant pseudomonad population released into lake water, *Appl. Environ. Microbiol.* 55:2537–2544.

Morton, D., El-Janne, M., and Oliver, J. D., 1992, Macromolecular synthesis in *Vibrio vulnificus* during starvation and entry into the viable but nonculturable state, *Abstr. Annu. Meet. Am. Soc. Microbiol.*

Moyer, C. L., and Morita, R. Y., 1989, Effect of growth rate and starvation-survival on the viability and stability of a psychrophilic marine bacterium, *Appl. Environ. Microbiol.* 55:1122–1127.

Nilsson, L., Oliver, J. D., and Kjelleberg, S., 1991, Resuscitation of *Vibrio vulnificus* from the viable but nonculturable state, *J. Bacteriol.* 173:5054–5059.

Novitsky, J. A., and Morita, R. Y., 1976, Morphological characterization of small cells resulting from nutrient starvation of a psychrophilic marine vibrio, *Appl. Environ. Microbiol.* 32:617–662.

Novitsky, J. A., and Morita, R. Y., 1977, Survival of a psychrotrophic marine vibrio under long-term nutrient starvation, *Appl. Environ. Microbiol.* 33:635–641.

Novitsky, J. A., and Morita, R. Y., 1978, Possible strategy for the survival of marine bacteria under starvation conditions, *Mar. Biol.* 48:289–295.

Nyström, T., and Kjelleberg, S., 1989, Role of protein synthesis in the cell division and starvation induced resistance to autolysis of a marine *Vibrio* during the initial phases of starvation, *J. Gen. Microbiol.* 135:1599–1606.

Nyström, T., Albertson, N. H., Flärdh, K., and Kjelleberg, S., 1990, Physiological adaptation to starvation and recovery from starvation by the marine *Vibrio* S14, *FEMS Microbiol. Ecol.* 74:129–140.

Oliver, J. D., 1989, *Vibrio vulnificus*, in: *Foodborne Bacterial Pathogens* (M. P. Doyle, ed.), Dekker, New York, pp. 569–600.

Oliver, J. D., and Wanucha, D., 1989, Survival of *Vibrio vulnificus* at reduced temperatures and elevated nutrient, *J. Food Saf.* **10**:79–86.

Oliver, J. D., Nilsson, L., and Kjelleberg, S., 1991, Formation of nonculturable *Vibrio vulnificus* cells and its relationship to the starvation state, *Appl. Environ. Microbiol.* **57**:2640–2644.

Paszko-Kolva, C., Shahamat, M., Yamamoto, H., Sawyer, T., Vives-Rego, J., and Colwell, R. R., 1991, Survival of *Legionella pneumophila* in the aquatic environment, *Microbiol. Ecol.* **22**:75–83.

Penón, F. J., Martinez, J., Vives-Rego, J., and Garcia-Lara, J., 1991, Mortality of marine bacterial strains in seawater, *Antonie van Leeuwenhoek* **59**:207–213.

Porter, K. G., and Feig, Y. S., 1980, The use of DAPI for identifying and counting aquatic microflora, *Limnol. Oceanogr.* **25**:943–948.

Postgate, J. R., 1976, Death in microbes and macrobes, in: *The Survival of Vegetative Microbes*, Vol. 26 (T. R. G. Gray and J. R. Postgate, eds.), Cambridge University Press, London, pp. 1–19.

Postgate, J. R., and Hunter, J. R., 1962, The survival of starved bacteria, *J. Gen. Microbiol.* **26**:1–18.

Rodriguez, G. G., Phipps, D., Ishiguro, K., and Ridgway, H. F., 1992, Use of a fluorescent redox probe for direct visualization of actively respiring bacteria, *Appl. Environ. Microbiol.* **58**:1801–1808.

Rollins, D. M., and Colwell, R. R., 1986, Viable but nonculturable stage of *Campylobacter jejuni* and its role in survival in the natural aquatic environment, *Appl. Environ. Microbiol.* **52**:531–538.

Rose, A. S., Ellis, A. E., and Munro, A. L. S., 1990, Evidence against dormancy in the bacterial fish pathogen *Aeromonas salmonicida* subsp. salmonicida, *FEMS Microbiol. Lett.* **68**:105–108.

Roszak, D. B., and Colwell, R. R., 1987a, Survival strategies of bacteria in the natural environment, *Microbiol. Rev.* **51**:365–379.

Roszak, D. B., and Colwell, R. R., 1987b, Metabolic activity of bacterial cells enumerated by direct viable count, *Appl. Environ. Microbiol.* **53**:2889–2983.

Roszak, D. B., Grimes, D. J., and Colwell, R. R., 1984, Viable but nonrecoverable stage of *Salmonella enteritidis* in aquatic systems, *Can. J. Microbiol.* **30**:334–338.

Roth, W. G., Leckie, M. P., and Dietzler, D. N., 1988, Restoration of colony-forming activity in osmotically stressed *Escherichia coli* by betaine, *Appl. Environ. Microbiol.* **54**:3142–3146.

Simpson, L. M., White, V. K., Zane, S. F., and Oliver, J. D., 1987, Correlation between virulence and colony morphology in *Vibrio vulnificus*, *Infect. Immun.* **55**:269–272.

Simpson, L. M., Travis, J. C., and Oliver, J. D., 1992, Interaction of opaque and translucent variants of *Vibrio vulnificus* with peritoneal cells, *Abstr. Annu. Meet. Am. Soc. Microbiol.* B342.

Singh, A., and McFeters, G. A., 1987, Survival and virulence of copper- and chlorine-stressed *Yersinia enterocolitica* in experimentally infected mice, *Appl. Environ. Microbiol.* **53**:1768–1774.

Singh, A., LeChavallier, M. W., and McFeters, G. A., 1985, Reduced virulence of *Yersinia enterocolitica* by copper-induced injury, *Appl. Environ. Microbiol.* **50**:406–411.

Singh, A., Yeager, R., and McFeters, G. A., 1986, Assessment of in vivo revival, growth, and pathogenicity of *Escherichia coli* strains after copper- and chlorine-induced injury, *Appl. Environ. Microbiol.* **52**:832–837.

Steffan, R. J., and Atlas, R. M., 1988, DNA amplification to enhance the detection of genetically engineered bacteria in environmental samples, *Appl. Environ. Microbiol.* **54**:2185–2191.

Steffan, R. J., and Atlas, R. M., 1991, Polymerase chain reaction: Applications in environmental microbiology, *Annu. Rev. Microbiol.* **45**:137–161.

Stevenson, L. H., 1978, A case for bacterial dormancy in aquatic systems, *Microb. Ecol.* **4:** 127–133.

Strugger, S., 1948, Fluorescence microscope examination of bacteria in soil, *Can. J. Res.* **26:**188.

van Overbeek, L. S., van Elsas, J. D., Trevors, J. T., and Starodub, M. E., 1990, Long-term survival of and plasmid stability in *Pseudomonas* and *Klebsiella* species and appearance of nonculturable cells in agricultural drainage water, *Microb. Ecol.* **19:**239–249.

Weichart, D., Oliver, J. D., and Kjelleberg, S., 1992, Low temperature induced nonculturability and killing of *Vibrio vulnificus*, *FEMS Microbiol. Ecol.* **100:**205–210.

Wicken, A. J., and Knox, K. W., 1984, Variable nature of the bacterial cell surface, *Aust. J. Biol. Sci.* **37:**315–322.

Wilson, M., and Lindow, S. E., 1992, Relationship of total viable and culturable cells in epiphytic populations of *Pseudomonas syringae*, *Appl. Environ. Microbiol.* **58:**3908–3913.

Wolf, P. W., and Oliver, J. D., 1992, Temperature effects on the viable but nonculturable state of *Vibrio vulnificus*, *FEMS Microbiol. Ecol.* **101:**33–39.

Xu, H., and Colwell, R. R., 1989, Overwintering of *Vibrio cholerae*—viable but non-culturable state and its determination, *J. Ocean Univ. Qingdao* (in Chinese) **19:**77–83.

Xu, H.-S., Roberts, N., Singleton, F. L., Atwell, R. W., Grimes, D. J., and Colwell, R. R., 1982, Survival and viability of nonculturable *Escherichia coli* and *Vibrio cholerae* in the estuarine and marine environment, *Microb. Ecol.* **8:**313–323.

Xu, H.-S., Roberts, N. C., Adams, L. B., West, P. A., Siebeling, R. J., Huq, A., Huq, M. I., Rahman, R., and Colwell, R. R., 1984, An indirect fluorescent antibody staining procedure for detection of *Vibrio cholerae* serovar O1 cells in aquatic environmental samples, *J. Microbiol. Methods* **2:**221–231.

Zimmerman, R., Iturriaga, R., and Becker-Birck, J., 1978, Simultaneous determination of the total number of aquatic bacteria and the number thereof involved in respiration, *Appl. Environ. Microbiol.* **36:**926–935.

Index